DFG

The MAK-Collection for Occupational Health and Safety

Part I: MAK Value Documentations

The MAK-Collection

Part I: MAK Value Documentations
ISSN 1860-496X

Part II: BAT Value Documentations
ISSN 1860-4978

Part III: Air Monitoring Methods
ISSN 1860-4986

Part IV: Biomonitoring Methods
ISSN 1860-4994

The MAK-Collection online
www.mak-collection.com

DFG Deutsche Forschungsgemeinschaft

The MAK-Collection for Occupational Health and Safety

Part I: MAK Value Documentations

Volume 26

Edited by Andrea Hartwig

Commission for the Investigation of Health Hazards of Chemical Compounds in the Work Area

WILEY-VCH

WILEY-VCH Verlag GmbH & Co. KGaA

Prof. Dr. Andrea Hartwig
Commission
for the Investigation of Health Hazards
of Chemical Compounds in the Work Area
Kaiserstr. 12
76131 Karlsruhe
Germany

Translators:
O. Bomhard
P. Brozena
J. Handwerker-Sharman
M. Ruff-König

Library of Congress Card No.: applied for
British Library Cataloging-in-Publication Data: A catalogue record for this book is available from the British Library.

Bibliographic information published by the Deutsche Nationalbibliothek
The Deutsche Nationalbibliothek lists this publication in the Deutsche Nationalbibliografie; detailed bibliographic data are available in the Internet at http://dnb.d-nb.de.

ISSN: 1860-496X
ISBN: 978-3-527-32306-7

Printed on acid-free paper

Layout: K. Schöpe
Printing: betz-druck GmbH, Darmstadt
Binding: Litges & Dopf Buchbinderei GmbH, Heppenheim

Printed in the Federal Republic of Germany

Note

In the present volume reference is frequently made to the various sections of the *List of MAK and BAT Values** (MAK List). The MAK List was introduced and its sections described in detail in the first volume of this series.

* Available from the publisher, WILEY-VCH, D-69451 Weinheim, both in the original German edition and as an English translation

The MAK-Collection Part I: MAK Value Documentations, Vol. 26. DFG, Deutsche Forschungsgemeinschaft

ISBN: 978-3-527-32306-7

Contents of Volume 26

The MAK-Collection Part I: MAK Value Documentations, Vol. 26. DFG, Deutsche Forschungsgemeinschaft

ISBN: 978-3-527-32306-7

Acetic acid

MAK value	**not yet established, see Section IIb of the *List of MAK and BAT Values***
Peak limitation	–
Absorption through the skin	–
Sensitization	–
Carcinogenicity	–
Prenatal toxicity	–
Germ cell mutagenicity	–
BAT value	–
Synonyms	Acidum aceticum (98–100%) ethanoic acid ethylic acid methanecarboxylic acid
Chemical name (CAS)	acetic acid
CAS number	64-19-7
Molecular formula	$C_2H_4O_2$
Structural formula	CH_3–COOH
Molecular weight	60.05
Melting point	16.6°C
Boiling point	118.5°C
Density at 15°C	1.0492 g/cm^3
Vapour pressure at 20°C	1.47 hPa
1 ml/m^3 (ppm) ≙ 2.5 mg/m^3	**1 mg/m^3 ≙ 0.4 ml/m^3 (ppm)**

The MAK-Collection Part I: MAK Value Documentations, Vol. 26. DFG, Deutsche Forschungsgemeinschaft

ISBN: 978-3-527-32306-7

1 Toxic Effects and Mode of Action

In vapour and aerosol form, acetic acid is highly irritating to the skin and mucous membranes. Aqueous solutions of 1% and above cause corrosion of the mucous membranes and solutions of 10% to 20% corrosion of the skin. The irritative effects on mucous membranes are manifest in itching and stinging in the nose, burning and stinging in the eyes, blepharospasms, burning in the throat, and coughing. After repeated and long-term exposure, considerable habituation to these irritative effects can, however, occur. The consequences of corrosion to the mucous membranes of the upper respiratory tract are laryngitis, tracheitis, bronchitis and bronchiolitis; forced inhalation of higher concentrations can cause obstructive pulmonary oedema. Corrosion of the eyes leads to permanent opacity of the cornea with or without anaesthesia; compared with other organic acids, acetic acid is neither more strongly penetrating nor causes stronger protein denaturation.

Long-term inhalation of low concentrations of acetic acid can lead to damage to the surface of the teeth. Erosion of the front surfaces of the teeth is known to be a typical occupational disease.

Acetic acid can be absorbed via both the lungs and the gastrointestinal tract. The resulting effects are not ascribed to the acetate ion, but to the undissociated acid. They consist (for example, after swallowing concentrated aqueous solutions) of intravascular haemolysis, haemorrhagic diathesis in the case of strong acidosis, and cardiovascular and kidney failure.

2 Mechanism of Action

There are no data available for the mechanism of action of acetic acid.

3 Toxicokinetics and Metabolism

The metabolism of absorbed acetic acid in the organism corresponds to that of the acetate ion occurring in intermediary metabolism, which on the one hand is used to produce numerous endogenous substances, and on the other can be degraded, forming CO_2.

4 Effects in Humans

4.1 Single exposures

Concentrated acetic acid doses of 20 to 50 g were reported to be lethal in humans (Wirth *et al.* 1971). In the case of survival, oesophageal constriction is the main effect treated. The odour threshold was found to be 1 ml/m^3 (Leonardos *et al.* 1969). Concentrations of 25 ml/m^3 cause nose and eye irritation, 50 ml/m^3 is intolerable for persons not habituated to the substance. Persons who are habituated to the substance can tolerate concentrations up to 30 ml/m^3 (Hygienic Guide Series 1972). Concentrations of 200 to 500 ml/m^3 are highly irritating, but have allegedly been tolerated by different persons for short periods of time (Hygienic Guide Series 1957). After an accident with acetic acid (glacial acetic acid), reversible airway obstruction and steroid-responsive interstitial pneumonitis were diagnosed in the exposed persons (Rajan and Davies 1989).

4.2 Repeated exposures

Despite its frequent use, there are only few reports of disease after occupational exposure to acetic acid.

The following three publications are based on the investigations of a laboratory which carried out numerous analyses of the concentrations of substances in workplace air in various branches of industry. The health of the workers employed there was monitored at the same time. It was assumed that the TLV value of 10 ml/m^3 (ACGIH 1948) represents the maximum tolerable acetic acid concentration.

The following symptoms were observed in 5 workers who were exposed by inhalation to acetic acid concentrations of 26 to 260 ml/m^3 for 7 to 12 years: black discoloration of the skin of the hands; hyperkeratosis with fissures; conjunctivitis with hyperaemia; oesophageal oedema and secretion; chronic bronchitis, asthma-like in 3 cases; discoloration and erosion of the teeth; gastritis (Baldi 1953; Parmeggiani and Sassi 1954). As early as 1953 it was noted that the clinico-toxicological studies necessary to be able to evaluate the importance of possible damage are lacking (Baldi 1953). The following conclusions were drawn on the basis of the available data: For decades people have been exposed during the production of vinegar from wine to concentrations that are most probably higher than 10 ml/m^3, yet there are hardly any known cases of poisoning. We assume that 20 and even 30 ml/m^3 can be tolerated without danger[*] (Vigliani and Zurlo 1955).

A 10% acetic acid solution caused only weak skin irritation in a 4-hour patch test (Griffiths *et al.* 1997; Nixon *et al.* 1975).

[*] translation of original German text

4.3 Local effects on skin and mucous membranes

Corrosion, necrosis and ulceration of the skin were described after occlusive self-treatment for 2 hours with a mixture of 50% flour : 50% rice vinegar (containing 4.5% acetic acid) (Kuniyuki and Oonishi 1997).

4.4 Allergenic effects

A 68-year-old woman developed type 1 sensitization reactions after ingesting alcoholic beverages, medicines containing alcohol and salad dressing containing acetic acid. On the basis of the anamnesis and the allergological tests, the authors concluded that acetic acid caused these reactions (Boehncke and Gall 1996). As a result of the exposure to a mixture of substances, however, these findings cannot be evaluated and a relationship between sensitization reactions and acetic acid is, therefore, not proven.

5 Animal Experiments and *in vitro* Studies

5.1 Acute toxicity

5.1.1 Inhalation

Groups of 7 to 12 guinea pigs were exposed by inhalation for one hour to acetic acid concentrations of 0, 5, 39, 119 or 568 ml/m^3. A significant increase in airway resistance with a simultaneous significant decrease in compliance was determined at all concentrations. The product of airway resistance and compliance increased in a concentration-dependent manner. There was a statistically significant decrease in the respiration rate and respiratory minute volume only at the two high concentrations. The animals exposed to concentrations of 119 ml/m^3 recovered one hour after the end of the experiment. The effects were not reversible, however, after exposure to 568 ml/m^3. The bronchoconstriction caused by acetic acid was more pronounced than that caused by formic acid or formaldehyde (Amrud 1961).

Concentrations above 1000 ml/m^3 caused irritation of the airways and conjunctiva in mice and guinea pigs. The LC_{50} for mice was 5620 ml/m^3 (Ghiringhelli and Di Fabio 1957).

Concentrations of 16000 ml/m^3, inhaled by rats for 4 hours, were lethal for 1 of 6 animals within 14 days (no other details) (Smyth *et al.* 1951).

5.1.2 Ingestion

The oral LD_{50} for mice was found to be 4960 mg/kg body weight and for rats 3200 to 3880 mg/kg body weight (Smyth *et al.* 1951).

5.1.3 Dermal absorption

For guinea pigs, the dermal LD_{50} of a 28% acetic acid solution was above 3.36 mg/kg body weight, that of a 5% solution was above 21 mg/kg body weight. Only moderate irritation was observed after application of the 28% solution (Hygienic Guide Series 1972).

5.2 Subacute, subchronic and chronic toxicity

5.2.1 Inhalation

There are no data available from inhalation studies in animals.

5.2.2 Ingestion

A daily dose of 1800 mg/kg body weight for 2 weeks is regarded as the maximum tolerable dose for rats; doses of 2400 mg/kg body weight are lethal within 3 to 5 days, while doses of 4800 mg/kg body weight and day, administered as sodium acetate, are, however, tolerable (von Oettingen 1960).

5.2.3 Dermal absorption

There are no data available for the effects of acetic acid after dermal absorption in animals.

5.3 Local effects on skin and mucous membranes

5.3.1 Skin

The skin of guinea pigs was severely damaged after the application of an 80% acetic acid solution, moderately to severely damaged after 50% to 80% solutions, and moderately damaged after a 50% solution. Acetic acid solutions of 5% to 10% had no effect (Katz and Guest 1994).

0.5 ml acetic acid was applied occlusively to the shaved dorsal skin and flanks of rabbits for 4 hours. Skin corrosion was not observed (Vernot *et al.* 1977).

A 10% acetic acid solution applied to the intact and abraded skin of rabbits and guinea pigs caused weak irritation (Nixon *et al.* 1975).

5.3.2 Eyes

Solutions with 10% acetic acid caused severe permanent eye damage in rabbits, 5% solutions caused damage that healed within 14 days (Hygienic Guide Series 1972).

5.4 Allergenic effects

There are no data available for the allergenic effects of acetic acid.

5.5 Reproductive toxicity

There are no data available for the reproductive toxicity of acetic acid.

5.6 Genotoxicity

In vitro

Acetic acid in concentrations up to 6666 µg/plate was not found to be mutagenic in the *Salmonella typhimurium* strains TA97, TA98, TA100 and TA1535 either with or without an added metabolic activation system (Zeiger *et al.* 1992). Nor was the substance found to have mutagenic effects in *Saccharomyces cerevisivae* (Katz and Guest 1994). Acetic acid concentrations of 10 to 14 mM (pH of about 6.0 or lower in the medium) induced chromosomal aberrations in hamster ovarian K1 cells both with and without an added metabolic activation system. Concentrations of 16 mM were cytotoxic. If the medium was buffered to a pH value of around 7, no clastogenic effects occurred. The authors concluded that acetic acid itself is not clastogenic, and the chromosomal aberrations are the result of the change in pH (Morita *et al.* 1990).

In vivo

Acetic acid did not cause a significant increase in somatic mutations and recombinations in *Drosophila melanogaster* after feeding; the authors do not completely exclude that it has an effect, however (Mollet 1976).

5.7 Carcinogenicity

There are no 2-year carcinogenicity studies available.

Acetic acid concentrations of 250 to 1500 mg/ml did not induce transformations in C3H/10T1/2 cells (Abernethy *et al.* 1982).

The application of up to 833 mmol acetic acid to the skin of female CD-1 mice after initiation with 12,13-ditetradecanoate-phorbol led to the weak stimulation of epidermal hyperplasia (Slaga *et al.* 1975).

In female Sencar mice (mouse skin model), acetic acid in concentrations of 667 mM was found to be a weak tumour promotor after initiation with dimethylbenzo[*a*]-anthracene. The mechanism probably responsible is thought to be a selective cytotoxicity of papilloma cells combined with compensatory cell growth (Rotstein and Slaga 1988).

6 Manifesto (MAK value/classification)

No epidemiological studies in humans have been published. Irritation of the eyes, skin and respiratory organs was observed in persons exposed at the workplace to high concentrations of more than 60 ml/m^3. Systemic effects did not occur. The MAK value of 10 ml/m^3 valid until now was established on the basis of a workplace study from 1953. This study reported that contact for decades with concentrations in the range of 20 ml/m^3 did not lead to any signs of intoxication (no other details). The authors concluded that concentrations of 20 and 30 ml/m^3 were tolerable with no risk.

There are no data available from which a NOAEL (no observed adverse effect level) can be deduced, and there are no suitable inhalation studies with animals; the previously valid MAK value of 10 ml/m^3 has therefore been withdrawn, and acetic acid has been classified in Section IIb of the *List of MAK and BAT Values*. There are no data for the reproductive toxicity and carcinogenicity of the substance. The few data for genotoxicity do not permit a conclusive evaluation. It cannot be decided whether designation with "S" or "H" (for substances which cause sensitization or can be absorbed by the skin) is necessary because there are no data.

7 References

Abernethy DJ, Frazelle JH, Boreiko CJ (1982) Effects of ethanol, acetaldehyde and acetic acid in the C3H/10T/1/2Cl 8 cell transformation system. *Environ Mutagen 4*: 331

ACGIH (American Conference of Governmental Industrial Hygienists) (1948) Acetic Acid. in: *Documentation of TLVs and BEIs*, ACGIH, Cincinnati, OH, USA

Amrud O (1961) The respiratory response of guinea pigs to the inhalation of acetic acid vapor. *Am Ind Hyg Assoc J 22*: 1–5

Baldi G (1953) Patologia professionale da acetone e derivati alogenati, acido acetico, anidride acetica, cloruro di acetile, acetil acetone (Occupational poisoning by acetone and halogenate derivatives, acetic acid, acetic anhydride, acetyl chloride and acetone acetyl) (Italian). *Med Lav 44*: 403–415

Boehncke WH, Gall H (1996) Ethanol metabolite acetic acid as causative agent for type-1 hypersensitivity-like reactions to alcoholic beverages. *Clin Exp Allergy 26*: 1089–1091

Ghiringhelli L, Di Fabio A (1957) Patologia da acido acetico: osservazioni negli animali da esperimento e nell'uomo (Pathology due to acetic acid: observations in experimental animals and man) (Italian). *Med Lav 48*: 559–565

Griffiths HA, Wilhelm K-P, Robinson MK, Wang XM, McFadden J, York M, Basetter DA (1997) Interlaboratory evaluation of a human patch test for the identification of skin irritation potential/hazard. *Food Chem Toxicol 35*: 255–260

Hygienic Guide Series (1957) Acetic acid. *Am Ind Hyg Assoc Q 18*: 80–81

Hygienic Guide Series (1972) Acetic acid. *Am Ind Hyg Assoc J 33*: 624–627

Katz G V, Guest D (1994) Aliphatic carboxylic acids. in: Clayton GD, Clayton FE (Eds) *Patty's industrial hygiene and toxicology*. 4th Edition, Vol. II, Part E, John Wiley and Sons, New York, 3532–3536

Kuniyuki S, Oonishi H (1997) Chemical burn from acetic acid with deep ulceration. *Contact Dermatitis 36*: 169–170

Leonardos G, Kendall D, Barnard N (1969) Odor threshold determinations of 53 odorant chemicals. *J Air Pollut Control Assoc 19*: 91–95

Mollet P (1976) Lack of proof of induction of somatic recombination and mutation in *Drosophila* by methyl-2-benzimid-azole carbamate, dimethyl sulfoxide and acetic acid. *Mutat Res 40*: 383–388

Morita T, Takeda K, Okumura K (1990) Evaluation of clastogenicity of formic acid, acetic acid and lactic acid on cultured mammalian cells. *Mutat Res 240*: 195–202

Nixon GA, Tyson CA, Wertz WC (1975) Interspecies comparisons of skin irritancy. *Toxicol Appl Pharmacol 31*: 481–490

von Oettingen WF (1960) The aliphatic acids and their esters: toxicity and potential dangers. *Arch Ind Health 21*: 28–32

Parmeggiani L, Sassi C (1954) Sui danni per la salute provocati dalla' acido acetico nella produzione degli acetati di cellulosa (Injuries to health caused by acetic acid in production of cellulose acetates) (Italian). *Med Lav 45*: 319–323

Rajan KG, Davies BH (1989) Reversible airways obstruction and interstitial pneumonitis due to acetic acid. *Br J Ind Med 46*: 67–68

Rotstein JB, Slaga TJ (1988) Acetic acid, a potent agent of tumor progression in the multistage mouse skin model for chemical carcinogenesis. *Cancer Lett 42*: 87–90

Slaga TJ, Bowden GT, Boutwell RK (1975) Acetic acid, a potent stimulator of mouse epidermal macromolecular synthesis and hyperplasia but with weak tumor-promoting ability. *J Nat Cancer Inst 55*: 983–987

Smyth HF, Carpenter CP, Weil CS (1951) Range-finding toxicity data: list IV. *Arch Ind Hyg 4*: 119–122

Vernot EH, MacEwen JD, Haun CC, Kinkead ER (1977) Acute toxicity and skin corrosion data for some organic and inorganic compounds and aqueous solutions. *Toxicol Appl Pharmacol 42*: 417–423

Vigliani EC, Zurlo N (1955) Erfahrungen der Clinica del Lavoro mit einigen maximalen Arbeitsplatzkonzentrationen (MAK) von Industriegiften (Experience of the Clinica del Lavoro with some maximum concentrations at the workplace (MAK values) for toxic industrial substances) (German). *Arch Gewerbepathol Gewerbehyg 21*: 528–534

Wirth W, Hecht G, Gloxhuber C (1971) *Toxikologie-Fibel* (Toxicology Lexicon) (German). 2nd ed., Thieme, Stuttgart, 221–223

Zeiger E, Anderson B, Haworth S (1992) Salmonella mutagenicity test V. Results from the testing of 311 chemicals. *Environ Mol Mutagen 19, Suppl 21*: 2–141

completed 29.11.2001

Acetic acid

Supplement 2008

MAK value (2007)	**10 ml/m³ ≙ 25 mg/m³**
Peak limitation	**Category I, excursion factor 2**
Absorption through the skin	**–**
Sensitization	**–**
Carcinogenicity	**–**
Prenatal toxicity (2007)	**Pregnancy risk group C**
Germ cell mutagenicity	**–**
BAT value	**–**

Since the documentation from 2002 was published, further studies of the acute effects in humans have been carried out, which allow a re-evaluation of the substance.

Effects in Humans

Single exposures

In one study the odour and irritation thresholds for acetic acid were determined using standardized psychophysical analyses (van Thriel *et al.* 2006). The irritation threshold was determined as a so-called lateralization threshold, based on a clear stimulation of intranasal nerve endings of the *N. trigeminus*. Stimulation of these nerve endings can—unlike in the case of olfactory receptors—be assigned to one side of the nose, in other words, lateralized (e.g. Kobal *et al.* 1989; Hummel 2000). In the age and gender-stratified random sample (n = 72), the odour threshold was found to be 0.6 ml/m³ and the irritation threshold 40 ml/m³. This study confirmed the findings of Cometto-Muniz (2001). As chemosensory threshold values are based on very short exposure times (in the range of seconds), no conclusions can be drawn, however, about possible effects on health after exposure for longer periods.

Changes in the nasal airway resistance were investigated in eight female and eight male persons who inhaled acetic acid concentrations of 15 ml/m³ for 15 minutes through the nose (Shusterman *et al.* 2005). Eight of the 16 persons suffered from seasonal

The MAK-Collection Part I: MAK Value Documentations, Vol. 26. DFG, Deutsche Forschungsgemeinschaft

ISBN: 978-3-527-32306-7

allergic rhinitis, which was correctly diagnosed by means of the prick test. The rhinitic persons were investigated when the relevant pollen was not in the air. Compared with under the control conditions (pure air), the nasal airway resistance of the rhinitic persons was significantly increased immediately after the exposure. This increase in the nasal airway resistance was still evident 15 minutes after inhalation of the substance. In the healthy control persons, the nasal airway resistance was not significantly increased at any time determined. As there are similar findings for chlorine gas (Shusterman *et al.* 2003), the authors conclude that the nasal hyperreactivity to allergens also intensifies the reaction to the local irritant acetic acid. These findings cannot, therefore, be used for establishing a MAK value. The two groups did not, however, differ in their subjective estimations of irritation and odour intensity. On the visual evaluation scale with six categories for symptom intensity (none, slight, moderate, strong, very strong, overpowering), the estimations for irritation were considered overall to be "weak", while odour intensity was evaluated as being "moderate". The estimations for irritation increased slightly during the 15-minute exposure phase, while the estimations for odour intensity decreased slightly over the same period. The symptoms of nasal congestion, rhinorrhea and postnasal drip were not significantly increased compared with under the control conditions. Overall, after concentrations of 15 ml/m^3 there was a significant increase in nasal irritation that was described as "slight" by the test persons. The NOAEC (no observed adverse effect concentration) of the study based on the nasal airway resistance in healthy persons and the subjective estimations of irritation is, therefore, 15 ml/m^3.

In another laboratory study, six male and six female persons were exposed for two hours to acetic acid concentrations of 0, 5 or 10 ml/m^3 (Ernstgård *et al.* 2006). Irritative effects were determined in this study on various levels using various methods (details of symptoms, psychophysiological analyses). Irritative and general symptoms were recorded by means of a questionnaire. The individual symptoms were: "discomfort in the eyes: burning, irritated, or running eyes"; "discomfort in the nose: burning, irritated, or runny nose"; "discomfort in the throat or airways"; "breathing difficulty"; "solvent smell"; "headache"; "fatigue"; "nausea"; "dizziness"; "feeling of intoxication". The standardized evaluation of the severity of the symptoms was carried out by means of a visual analog scale from 0 to 100 mm with seven categories ("not at all", "hardly at all", "somewhat", "rather", "quite", "very", "almost unbearable"). For effects on lung function, the parameters which were determined included the following: forced expiratory volume in 1 second (FEV1) and FEV1/vital capacity. Possible nasal swelling during or after the exposure was determined by means of acoustic rhinometry. The blinking frequency was determined electromyographically and evaluated in six blocks of 20 minutes. C-reactive protein and interleukin-6 were determined in the blood of the test persons as indicators of inflammation. Evaluation of the results of these investigations showed that none of the physiological parameters were significantly influenced by the level of exposure to acetic acid. There were some significant changes in the parameters for the lung function test, nasal swelling and the indicators of inflammation in the blood on comparison of the values for before and after exposure, but these changes were not significantly covariant with the level of exposure. The blinking frequency was slightly higher after concentrations of 10 ml/m^3 than after 0 and 5 ml/m^3, but this difference was not statistically significant. "Discomfort in the nose" was significantly increased compared with under the control conditions only after acetic acid concentrations of 10 ml/m^3. The symptom intensity, however, was estimated to be "hardly at all" at 10 ml/m^3. The estimations for the category "solvent smell" were significantly increased after acetic acid concentrations of both 5 and 10 ml/m^3. The estimations of the symptoms were to a certain extent concentration-dependent, but the statistical significance was not tested. The symptom intensity after acetic acid concentrations of 5 ml/m^3 was estimated to be "somewhat intense", while after 10 ml/m^3 it was between "somewhat" and "rather intense". The estimations for "solvent smell" decreased considerably over the 120-minute exposure phase. The study shows that slight irritation of the nose can occur after exposure to concentrations of 10 ml/m^3. It must be noted, however, that the symptom intensity was estimated to be "hardly at all" at 10 ml/m^3. A further limitation is that this value is based merely on a subjective symptom of irritation which was possibly influenced by the odour, as comparable estimations for "irritated eyes" did not increase with the concentration. None of the physiological parameters were significantly changed at this concentration, and the NOAEC for a 2-hour exposure to acetic acid is, therefore, 10 ml/m^3.

In another study, 11 female and 13 male test persons were exposed for four hours to acetic acid concentrations of 0.6, 5 or 10 ml/m^3 (HVBG 2007). The concentration of 0.6 ml/m^3 was based on the odour threshold (van Thriel *et al.* 2006) and served as the control, as trigeminal effects can be excluded at this level of exposure. The exposure to 5 ml/m^3 was carried out with four peak exposures of 10 ml/m^3. Under these exposure conditions the fluctuating concentrations of acetic acid were between 0.32 and 9.72 ml/m^3, so that the result was on average a value of 5 ml/m^3. The exposure to 10 ml/m^3 was constant. Acute symptoms and chemosensory sensations were recorded using computer-aided evaluation scales several times during the exposure. The following variables were investigated as physiological parameters: nasal airway resistance (anterior rhinometry), blinking frequency (electromyographic determination), and substance P in the nasal lavage as evidence of neurogenic inflammation. The estimations for odour symptoms, such as a feeling that the air was poor, for odour intensity and for annoyance were significantly increased after concentrations of 5 and 10 ml/m^3 compared with after exposure under the control conditions of 0.6 ml/m^3. These olfactory sensations were generally estimated, however, to be only "moderate" to "weak". The estimations for odour symptoms decreased over the 4-hour exposure phase like in the study of Ernstgård *et al.* (2006). Symptoms that indicate trigeminally mediated sensory irritation, such as eye irritation, did not differ from one another significantly at the three concentrations investigated. The different acetic acid concentrations had no influence on the investigated physiological parameters blinking frequency, nasal airway resistance and nasal lavage mediators. Physiological signs of trigeminal effects of acetic acid were, therefore, not evident in the investigated concentration range up to 10 ml/m^3 (HVBG 2007).

Reproductive toxicity

In a case–control study of risk factors for neural tube defects, such defects were identified from medical reports in 653 of 708129 births in California between June 1989 and May 1991. Children from the same hospital were chosen randomly as controls. 538 mothers of children with neural tube defects and 539 mothers of healthy children were interviewed. The mothers answered questions on 74 chemicals and exposure to them (probable, possible, or none). Twelve mothers of children with neural tube defects and 13 mothers of healthy children were exposed to acetic acid. The odds ratio was 0.92 (95% confidence interval 0.42–2.0) and, therefore, not increased (Shaw *et al.* 1999).

Animal Experiments and *in vitro* Studies

Reproductive toxicity

In vivo

There are no relevant studies of the toxic effects on development of acetic acid.

In vitro

Acetic acid was compared with haloacetic acids in whole embryo culture. Nine-day-old embryos of CD-1 mice that had been cultivated for up to 26 hours were used. The control medium had a pH of 8.37. The lowest acetic acid concentration used of 3 mM (pH 8.01) did not cause any neural tube defects. After concentrations of 4 to 10 mM (pH 7.89–7.4), the percentage of embryos with neural tube defects increased with the concentration, from 25% after 4 mM to 100% after 10 mM. After concentrations of 5 mM and above, defects were observed also in the pharyngeal arch and heart, and the number of somites was reduced. Investigations with hydrochloric acid showed that the malformations were not the result of lowering the pH; lowering the pH to 7.5 did not lead to malformations. For acetic acid, a concentration of 1.9 mM was calculated to be the benchmark concentration for increasing the incidence of neural tube defects by 5%. Compared with haloacetic acids, acetic acid (benchmark concentration > 1 mM) was in the range of trichloroacetic acid (1.3 mM), tribromoacetic acid (1.4 mM), dichloroacetic acid (2.5 mM), trifluoroacetic acid (3.7 mM) and difluoroacetic acid (5.9 mM). More effective were monofluoroacetic acid (0.7 mM), dibromoacetic acid (0.16 mM), monochloroacetic acid (0.09 mM), monobromoacetic acid (0.002 mM) and monoiodoacetic acid (0.0007 mM) (Hunter *et al.* 1996; Richard and Hunter 1996). It is apparent from the study that acetic acid has teratogenic potential *in vitro* in concentrations of 4 mM and above, but not in lower concentrations up to 3 mM.

Sodium acetate was not found to be teratogenic in two different micromass teratogen assays—in mouse blastocysts stimulated for differentiation (Newall and Beedles 1994, 1996) and in brain cells and limb buds from 13-day-old rat embryos (Uphill *et al.* 1990) in concentrations of 500 µg/ml and above (no other details).

Studies with other acetates

In a developmental toxicity study with inhalation of *n*-butyl acetate, maternal feed consumption and body weight gains were reduced after concentrations of 2000 ml/m^3 and above. Reduced foetal body weights were observed after concentrations of 3000 ml/m^3. The NOAEC for maternal toxicity was 1000 ml/m^3, the NOAEC for developmental toxicity 2000 ml/m^3 (Saillenfait *et al.* 2007; see Table 1).

Table 1. Studies of developmental toxicity with alkyl acetates

Species, strain, number of animals per group	Exposure	Results	References
***n*-Butyl acetate**			
rat, SD, groups of 19–21 ♀	**GD 6–20** 0, 500, 1000, 2000, 3000 ml/m^3, 6 hours/day examination GD 20	**1000 ml/m^3** dams: NOAEC **2000 ml/m^3 and above** dams: feed consumption decreased, body weight gains decreased; foetuses: NOAEC **3000 ml/m^3** foetuses: body weights decreased	Saillenfait *et al.* 2007
Octyl acetate			
rat, SD, groups of 20–22 ♀	**GD 6–15** 0, 100, 500, 1000 mg/kg body weight and day examination GD 20	**100 mg/kg body weight** dams: NOAEL **500 mg/kg body weight and above** dams: feed consumption decreased, body weight gains decreased; foetuses: NOAEL **1000 mg/kg body weight** foetuses: total external, visceral and skeletal malformations increased	Daughtrey *et al.* 1989

GD: gestation day; NOAEL: no observed adverse effect level

In a developmental toxicity study with oral administration of octyl acetate, maternal feed consumption and body weight gains were reduced after concentrations of 500 ml/m^3 and above. In the foetuses, the total number of external, visceral and skeletal malformations was increased after doses of 1000 mg/kg body weight and day. The NOAEL for maternal toxicity was 100 mg/kg body weight and day, the NOAEL for developmental toxicity 500 ml/m^3 (Daughtrey *et al.* 1989; see Table 1). The NOAEL for developmental toxicity corresponds for humans (70 kg body weight; 10 m^3 air inhaled in 8 hours) to a concentration of about 3500 mg/m^3 or 1400 ml/m^3.

Manifesto (MAK value, classification)

The irritation threshold of 40 ml/m^3 clearly shows that acetic acid is an irritant. The estimations of subjective irritation symptoms were significantly increased in two relevant studies (Ernstgård *et al.* 2006; Shusterman *et al.* 2005) at 10 and 15 ml/m^3, but amounted only to "hardly at all" to "somewhat" and were possibly influenced by the odour. In a third study (HVBG 2007), a NOAEC of 10 ml/m^3 was obtained for subjective effects of irritation in the nose and eyes. The physiological indicators of trigeminally mediated irritation, such as nasal airway resistance, blinking frequency or inflammation markers, did not reveal any significant changes in either study (Ernstgård *et al.* 2006; HVBG 2007) up to the highest exposure of 10 ml/m^3. The NOAEC for subjective sensory effects and objective local effects is, therefore, 10 ml/m^3. All three inhalation studies showed there to be a concentration-dependent increase in odour-mediated effects (odour perception, odour nuisance). It must be remembered, however, that these subjective estimations of the degree of intensity were in the range "somewhat to rather" and "slight to moderate" on the evaluation scales used. In addition, the two workplace-relevant studies with exposure for two and four hours revealed a marked decrease in odour perception over the investigation period. These adaptation processes confirm that these are purely olfactory effects, as these adaptation effects were not observed with trigeminally mediated effects (van Thriel *et al.* 2005; Wise *et al.* 2004). As a result of the adaptation processes and the quantitatively low level of odour nuisance, the described effects do not represent an unreasonable annoyance in the sense of the MAK value definition. A MAK value of 10 ml/m^3 has therefore been established for acetic acid.

Acetic acid has been classified in Peak limitation category I, as it is an irritant. Since the nasal airway resistance in healthy test persons was not increased after exposure for 15 minutes to concentrations of 15 ml/m^3 and the local symptoms in the other studies were generally regarded as weak after 10 ml/m^3, an excursion factor of 2 is justifiable.

There are no relevant studies available for evaluation of the prenatal toxicity of acetic acid. After inhalation exposure to acetic acid, prenatal toxicity is not to be expected, however, if the MAK value of 10 mg/m^3 (25 ml/m^3) is observed. A decrease in the pH or absorption of acetate must be considered as well. With acetic acid concentrations of 10 ml/m^3 (25 mg/m^3) and a respiratory volume of about 1.25 m^3 per hour, assuming 100 % absorption, 250 mg acetic acid (4.16 mmol) is absorbed per 8 hours or 31.25 mg (0.52 mmol) per hour. Based on the amount of acetic acid absorbed per hour (31.25 mg; 0.52 mmol) and assuming a blood volume of 4.5 litres in humans, an acetic acid concentration of 6.9 mg/l (0.116 mmol/l) is obtained. Assuming 100 % dissociation, 0.116 mmol H^+ ions per litre blood causes a decrease in the bicarbonate concentration from 24 to 23.884 mmol/l blood. According to the Henderson-Hasselbalch equation (concentration of CO_2 in blood = 1.2 mmol/l, concentration of HCO_3^- = 24 mmol/l, pKs = 6.1) the physiological pH of 7.4 is changed to 7.399, which is not significant. As there are other buffer systems (phosphate, protein) in the organism, which are not included in the calculation, and the elimination of acetic acid was not taken into consideration, the actual change is certainly below that calculated. As it can be assumed that the buffer capacities in the maternal and foetal bloodstream are comparable, this change is still within the physiological range of the blood pH of 7.35 to 7.45 (Jungermann and Möhler 1984). The esters of acetic acid, methyl acetate (MAK value 100 ml/m^3) and ethyl acetate (MAK value 400 ml/m^3), from which acetate is rapidly formed in the organism, are classified in Pregnancy risk group C in analogy to ethanol (MAK value 500 ml/m^3), which is metabolized to acetate. For *n*-butyl acetate, from which acetate is also formed, the NOAEC for maternal toxicity was 1000 ml/m^3 and the NOAEC for developmental toxicity 2000 ml/m^3 (Saillenfait *et al.* 2007). The data for octyl acetate present a similar picture (Daughtrey *et al.* 1989). In analogy to the acetic acid esters, ethanol and *n*-butyl acetate, acetic acid can be classified in Pregnancy risk group C. There are still no data for the carcinogenicity of acetic acid; the few data for its genotoxicity still cannot be evaluated conclusively (see the documentation from 2001). Therefore, classification in these categories is not possible.

It cannot be decided whether designation with "Sa", "Sh" or "H" is necessary because there are still no data (see the documentation from 2001).

References

Cometto-Muniz JE (2001) Physicochemical basis for odor and irritation potency of VOCs. in: Spengler JD, Samet JM, McCarthy JF (Eds) *Indoor air quality handbook*, McGraw-Hill, New York, 20.1–20.21

Daughtrey WC, Wier PJ, Traul KA, Biles RW, Egan GF (1989) Evaluation of the teratogenic potential of octyl acetate in rats. *Fundam Appl Toxicol 13*: 303–309

Ernstgård L, Iregren A, Sjögren B, Johanson G (2006) Acute effects of exposure to vapours of acetic acid in humans. *Toxicol Lett 165*: 22–30

Hummel T (2000) Assessment of intranasal trigeminal function. *Int J Psychophysiol 36*: 147–155

Hunter ESI, Rogers EH, Schmid JE, Richard A (1996) Comparative effects of haloacetic acids in whole embryo culture. *Teratology 54*: 57–64

HVBG (Hauptverband der Gewerblichen Berufsgenossenschaften) (2007) *Endbericht zum Verbundprojekt "Abgrenzung und Differenzierung irritativer und belästigender Effekte von Gefahrstoffen" (FF228)* (Final report of the project "Limitation and differentiation of irritative and annoying effects of hazardous substances" (FF228)) (German), IfADo, Institut für Arbeitsphysiologie an der Universität Dortmund

Jungermann K, Möhler H (1984) *Biochemie* (Biochemistry) (German) Springer Verlag, Berlin

Kobal G, van Toller S, Hummel T (1989) Is there directional smelling? *Experientia 45*: 130–132

Newall DR, Beedles KE (1994) The stem-cell test – a novel *in vitro* assay for teratogenic potential. *Toxicol In Vitro 8*: 697–701

Newall DR, Beedles KE (1996) The stem-cell test: an *in vitro* assay for teratogenic potential. Results of a blind trial with 25 compounds. *Toxicol In Vitro 10*: 229–240

Richard AM, Hunter ES (1996) Quantitative structure-activity relationships for the developmental toxicity of haloacetic acids in mammalian whole embryo culture. *Teratology 53*: 352–360

Saillenfait AM, Gallissot F, Sabate JP, Bourges-Abella N, Muller S (2007) Developmental toxic effects of ethylbenzene or toluene alone and in combination with butyl acetate in rats after inhalation exposure. *J Appl Toxicol 27*: 32–42

Shaw GM, Velie EM, Katz EA, Morland KB, Schaffer DM, Nelson V (1999) Maternal occupational and hobby chemical exposures as risk factors for neural tube defects. *Epidemiology 10*: 124–129

Shusterman D, Balmes J, Avila PC, Murphy MA, Matovinovic E (2003) Chlorine inhalation produces nasal congestion in allergic rhinitics without mast cell degranulation. Eur *Respir J 21*: 652–657

Shusterman D, Tarun A, Murphy MA, Morris J (2005) Seasonal allergic rhinitic and normal subjects respond differentially to nasal provocation with acetic acid vapor. *Inhal Toxicol 17*: 147–152

van Thriel C, Kiesswetter E, Schaper M, Blaszkewicz M, Golka K, Seeber A (2005) An integrative approach considering acute symptoms and intensity ratings of chemosensory sensations during experimental exposures. *Environ Toxicol Pharmacol 19*: 589–598

van Thriel C, Schaper M, Kiesswetter E, Kleinbeck S, Juran S, Blaszkewicz M, Fricke HH, Altmann L, Berresheim H, Brüning T (2006) From chemosensory thresholds to whole body exposures – experimental approaches evaluating chemosensory effects of chemicals. *Int Arch Occup Environ Health 79*: 308–321

Uphill PF, Wilkins SR, Allen JA (1990) *In vitro* micromass teratogen test: results from a blind trial of 25 compounds. *Toxicol In Vitro 4*: 623–626

Wise PM, Radil T, Wysocki CJ (2004) Temporal integration in nasal lateralization and nasal detection of carbon dioxide. *Chem Senses 29*: 137–142

completed 15.02.2007

Acrylic acid

MAK value (2005)	**10 ml/m³ (ppm) ≙ 30 mg/m³**
Peak limitation (2005)	**Category I, excursion factor 1**
Absorption through the skin	–
Sensitization	–
Carcinogenicity	–
Prenatal toxicity (2005)	**Pregnancy Risk Group C**
Germ cell mutagenicity	–
BAT value	–
Synonyms	acroleic acid ethylenecarboxylic acid vinylformic acid
Chemical name (CAS)	2-propenoic acid
CAS number	79-10-7
Structural formula	HOOC—CH=CH$_2$
Molecular formula	$C_3H_4O_2$
Molecular weight	72.06
Melting point	14°C (EU 2002)
Boiling point at 1013 hPa	141°C (EU 2002)
Density at 20°C	1.06 g/cm³ (EU 2002)
Vapour pressure at 20°C	3.8 hPa (EU 2002)
log P_{ow}*	0.46 (EU 2002)
pK_a#	4.25 (EU 2002)
1 ml/m³ (ppm) ≙ 2.990 mg/m³	**1 mg/m³ ≙ 0.334 ml/m³ (ppm)**

* *n*-octanol/water partition coefficient
\# dissociation constant

The MAK-Collection Part I: MAK Value Documentations, Vol. 26. DFG, Deutsche Forschungsgemeinschaft

ISBN: 978-3-527-32306-7

The present documentation is based on the EU Risk Assessment Report (EU 2002), in which many of the earlier studies are described and assessed; these will not be referred to here.

1 Toxic Effects and Mode of Action

Acrylic acid is corrosive to the skin and eyes and irritating to the respiratory tract. Irritation is also the main effect after repeated exposure. In a 13-week study, histopathologically evident irritation of the olfactory epithelium occurred in mice after concentrations of 5 ml/m^3 and above, but in rats not until a concentration of 75 ml/m^3. Comparative medium-term inhalation studies show that irritation of the nasal epithelium is considerably more pronounced in mice than in rats. These quantitative differences between rats and mice result from different dosimetry. On the basis of estimates it can be assumed that the concentrations to which the olfactory epithelium is exposed are not higher in humans than in rats.

Pure acrylic acid has no contact sensitizing potential in animal studies.

The substance was not genotoxic in *Salmonella* mutagenicity tests, in the HPRT (hypoxanthine guanine phosphoribosyl transferase) test in CHO (Chinese hamster ovary) cells or in the UDS (unsheduled DNA synthesis) assay. Positive results were obtained in mouse lymphoma tests and chromosome aberration tests *in vitro*. The results obtained in one of the mouse lymphoma tests suggest primarily clastogenic effects. Acrylic acid was not genotoxic in *in vivo* chromosome aberration tests in rats, a dominant lethal test in mice or an SLRL test (test for recessive lethal mutations on the X chromosome) in *Drosophila melanogaster*. The clastogenic effects observed *in vitro* thus do not manifest themselves systemically *in vivo*.

Acrylic acid was not found to be carcinogenic in a long-term study with rats after administration in the drinking water or in two well documented studies with mice after dermal application. A long-term study with mice with dermal application of acrylic acid, which produced local tumours, was not carried out under GLP (good laboratory practice) conditions and has been published only as an abstract. This result is therefore not very reliable. Various acrylic acid esters, which are hydrolyzed to acrylic acid, produced no carcinogenic effects after inhalation. A carcinogenicity study with ethyl acrylate and administration by gavage produced forestomach tumours in rats and mice.

In animal studies, acrylic acid caused no adverse effects on reproduction or developmental toxicity at not parentally toxic doses. Delayed development in the pups was observed only at doses that produced adverse effects on the parent animals.

2 Mechanism of Action

The irritant and clastogenic effects are probably the result of the acidic character and reactivity of the substance's double bond.

The reactivity of the double bond of acrylic acid is, however, lower than that of acrylates, which is evident for example in the less potent clastogenic effect of acrylic acid compared with that of methyl acrylate and ethyl acrylate (Moore *et al.* 1988).

3 Toxicokinetics and Metabolism

3.1 Absorption, distribution, elimination

The studies available for [^{14}C]acrylic acid with rats and mice indicate that acrylic acid is rapidly absorbed from the gastrointestinal tract after oral administration. Within a few hours, most of it is exhaled as CO_2 via the lungs (60–80 %) and a small amount is eliminated via the kidneys and faeces (2–9 %). The highest levels were determined in the stomach shortly after administration (e.g. 37 % in rats after 65 minutes). After 72 hours, between 10 % and 25 % of the radioactivity was still detected in fat, muscles, liver, blood, blood plasma, skin and stomach (deBethizy *et al.* 1987; Black *et al.* 1995; Rohm and Haas Company 1985; Winter and Sipes 1993; Winter *et al.* 1992).

Studies with non-occlusive application of [1-^{14}C]acrylic acid to the skin of rats and mice show that acrylic acid penetrates the skin in both species. Between 12 % and 19 % to 26 % was absorbed by mice and rats, respectively. However, most of it evaporated because of the open form of application. In both species, about 75 % to 80 % of the absorbed amount was metabolized to CO_2 and eliminated within 24 hours. Elimination via the urine and faeces amounts to a mere 0.5 %. Elimination of the radioactivity from the plasma, liver and kidneys was very rapid (Black *et al.* 1995; Winter and Sipes 1993; Winter *et al.* 1992).

Studies of the dermal penetration of acrylic acid with human skin *in vitro* showed that absorption is considerably dependent on the pH and solvent. After a dose of 1 mg the values varied up to 600-fold. The highest absorption rate was obtained with acetone as a vehicle (absorption rate in acetone > phosphate buffer pH 6.0 > ethylene glycol > phosphate buffer pH 7.4). The decrease in the absorption rate with increasing pH correlated well with the decrease in the octanol/water partition coefficient with increasing pH (D'Souza and Francis 1988).

After one-minute inhalation exposure of rats to [^{11}C]acrylic acid vapour in a head-only system, 18.3 % had been absorbed 1.5 minutes after the end of exposure; 28.4 % of the radioactivity absorbed was found in the rats' snouts and another large amount was detected in the upper respiratory tract. Sixty-five minutes after inhalation, only 8.1 % of the radioactivity measured in the body was found in the snouts and about 60 % of the radioactive dose had been exhaled as $^{11}CO_2$. Elimination followed a biphasic course with a half-life of the α phase of 30.6 minutes. Details of the β phase could not be given because of the short half-life of the isotope (20 minutes). About 15 % (of the 18.3 % absorbed) of the radioactivity was eliminated via the urine and faeces by the 65th minute. The remaining fraction in the liver, adipose tissue and stomach increased considerably between 1.5 and 65 minutes. This was attributed to ingestion during grooming of the fur around the snouts (Kutzman *et al.* 1982). The relevance of the quantitative data is unclear because of the short half-life of the ^{11}C isotope.

Investigations with an anatomical model of the nasal cavity of an adult rat showed that only 7 % to 10 % of the airflow inhaled passes over the olfactory epithelium (Keyhani *et al.* 1995). Similar investigations with a rat nose model showed that between 13.2 % and 21.7 % of the inspiratory airflow reaches the affected olfactory epithelium in the dorsal medial meatus, depending on the flow rate and the direction of the nostrils (upwards or downwards). At the normal inspiratory rate of 288 ml/min for the F344 rat examined, the rates were 14.6 % if the nostrils pointed downwards and 17 % if they pointed upwards (Kimbell *et al.* 1997).

After an acrylic acid concentration of 131 ml/m^3 was drawn through the upper respiratory passages of tracheotomized F344 rats at a flow rate of 200 ml/min, 97 % of the substance was found to be deposited in the respiratory tract (Morris and Frederick 1995).

The local tissue dose of inhaled acrylic acid in the nasal cavity of humans and rats was estimated by means of a CFD-PBPK model (computational fluid dynamics and physiologically based pharmacokinetic dosimetry model) by way of comparison. Steady-state simulations of the anterior nasal cavity of rats and the human nasal cavity were carried out in a 3-dimensional model. The release of acrylic acid vapour onto the mucous membranes of the upper respiratory tract was determined region for region in the simulations. Calculations were carried out with respiratory minute volumes of 0.1, 0.3 and 0.5 l/minute for rats and 11.4 and 18.9 l/minute for humans. Two compartments with olfactory epithelia (dorsal meatus and ethmoid region) were included for rats while only one compartment was included for the human model since humans have no ethmoid. The model included the effect of the buffer capacity of the mucus on the degree of ionization of the acid and the diffusion of the ionized and non-ionized species. The model also incorporated local physicochemical processes, partition coefficients, the metabolism of acrylic acid, flow conditions and respiratory parameters. Deposition of 50 % was calculated for the human nasal cavity using this model; compared with the 97 % measured in rats (Morris and Frederick 1995) this is plausible in view of the somewhat wider respiratory passages of the human nose (Frederick *et al.* 1998). The amounts deposited of a number of substances were also well predicted for rats (Frederick *et al.* 2001). In further calculations using this model, similar effects on the olfactory epithelium were estimated after acrylic acid concentrations of 4.4 and 25 ml/m^3 for rats (respiratory minute volume 0.250 l/minute) and humans (respiratory minute volume 13.9 l/minute) (Andersen *et al.* 2000; Frederick *et al.* 2001). However, since the values of important model parameters (e.g. gas phase diffusivity, diffusivity in mucus, diffusivity in squamous epithelium, tissue diffusivity and blood flow through the nasal epithelium) are not based on measurements, but on model simulations and assumptions, the conclusions drawn from the model are not reliable.

Assuming uniform distribution over the surface of the nasal cavity and the parameters listed in Table 1, the ratio for the amounts of acrylic acid deposited in the olfactory epithelium of rats and humans per time unit is about 1 to 3. Taking into account the different airflow that passes over the olfactory epithelium of both species, the effect on the olfactory epithelium is about 1 to 1.6 (Table 1). This calculation considers the respiratory minute volume increased under workplace conditions as the worst-case assumption. In view of the log P_{ow} of 0.46 (EU 2002) for acrylic acid, it must be assumed, however, that a larger fraction is deposited in the anterior region of the nose than in the posterior region. Since the olfactory epithelium of the rat is located in the posterior region of the nasal cavity and accounts for almost 50 % of the surface area of the nasal cavity (in humans only 4 %; Frederick *et al.* 1998), i.e. most of the olfactory epithelium is closer to the portal of entry in rats, more acrylic acid will reach the olfactory epithelium in rats than in humans. It can therefore be assumed that the concentrations to which the olfactory epithelium is exposed are not higher in humans than in rats.

Within 48 hours, 48 % of the intravenous dose of [^{14}C]acrylic acid was exhaled by mice as CO_2. Most of it was exhaled within the first 6 hours. Altogether, 5 % was eliminated with the urine and faeces, and 4 % of the dose administered could still be detected in the tissues after 48 hours (Rohm and Haas Company 1988).

Table 1. Comparison of the exposure level of the olfactory epithelium in rats and humans; exposure to acrylic acid concentrations of 25 ml/m^3 (75 µg/l)

	Rat	Human
respiratory minute volume	0.175 l/minute for 250 g rat (EPA 1988)	10 m^3/8 hours = 20.8 l/minute
inhaled acrylic acid amount/minute	13.1 µg	1563 µg
surface of the nasal cavity	13.79 cm^2 (Frederick *et al.* 1998)	245.9 cm^2 (Frederick *et al.* 1998)
surface of the olfactory epithelium	6.72 cm^2 (Frederick *et al.* 1998)	13.2 cm^2 (Frederick *et al.* 1998)
deposition in the nasal cavity	97 % (Morris and Frederick 1995)	about 50 % (Frederick *et al.* 1998)
average dose on the nasal surface (assuming uniform distribution in the nose)	13.1 µg/minute/13.79 cm^2 × 0.97 = 0.92 µg/cm^2/minute	1563 µg/minute/245.9 cm^2 × 0.5 = 3.18 µg/cm^2/minute
airflow over the olfactory epithelium	about 15 % (Frederick *et al.* 1998)	about 7 % (Frederick *et al.* 1998)
ratio of the exposure of the olfactory epithelium taking into account the different airflow over the olfactory epithelium (rat = 1)	1	3.18/0.92 × 7 %/15 % = 1.6

3.2 Metabolism

In vitro

After incubation of rat liver microsomes with [2,3-^{14}C]acrylic acid, no epoxidized metabolites were found. Unchanged acrylic acid was again isolated from the exposure mixture (deBethizy *et al.* 1987).

After incubation of rat liver homogenates with (1-^{14}C, 2,3-^{14}C or 1,2,3-^{14}C)acrylic acid, the substance was rapidly oxidized to CO_2. HPLC analysis revealed several metabolites in the culture medium, but only 3-hydroxypropionic acid was identified unequivocally. The authors concluded that acrylic acid enters intermediary metabolism after cleavage of CO_2 via acetyl-CoA (Ziegler *et al.* 1992).

In a comparative study of *in vitro* metabolism, [^{14}C]acrylic acid and acrylates were tested in tissue homogenates of male F344 rats and B6C3F$_1$ mice. Unlike acrylates, acrylic acid did not react with reduced glutathione in liver, kidney and lung homogenates or in the blood of rats in the presence or absence of soluble enzymes. No depletion of non-protein-bound sulfhydryl groups (NPSH) was detected (Miller *et al.* 1981b).

The oxidation rate for the formation of $^{14}CO_2$ from [^{14}C]acrylic acid was investigated *in vitro* using rat liver homogenates. Rapid oxidation of acrylic acid to CO_2 was observed. Mitochondria isolated from the liver homogenate and tested under the same conditions yielded higher oxidation rates. HPLC analysis of the mitochondria solution showed 3-hydroxypropionic acid to be the main metabolite (Finch and Frederick 1992).

In mice, the oxidation of [1-^{14}C or 2,3-^{14}C]acrylic acid was investigated *in vitro* in a total of 13 tissues (liver, kidneys, forestomach and glandular stomach, large and small intestines, spleen, brain, heart, lungs, skeletal muscles, adipose tissue and skin). The maximum oxidation rates were 48, 616 and 2890 nmol/hour and gram for skin, liver and kidneys, respectively. Acrylic acid was oxidized in all tissues examined, but, compared with the kidneys and liver, only in a relatively small proportion. No sex differences were observed. 3-Hydroxypropionic acid was the only metabolite found in the culture medium. The end products of the metabolism are acetyl-CoA, which enters the tricarboxylic acid cycle, and CO_2 (Black *et al.* 1993).

The oxidation rates and the tissue partition coefficients of [1-^{14}C]acrylic acid were measured in tissue sections from 13 organs (liver, kidneys, forestomach and glandular stomach, large and small intestines, spleen, brain, heart, lungs, skeletal muscles, adipose tissue and skin) of two rat strains (F344 and Sprague-Dawley). The kidneys and liver were the organs with the highest conversion rates of a maximum of 4 and 2 µmol/hour and gram. Acrylic acid oxidation in the remaining organs was only 40 % of the liver value or even lower. The conversion rates were similar in both rat strains. The oxidation rates in the rat organs were 2 to 3 times higher, relative to the absolute oxidation rates per gram tissue, than those in the mice organs (Black and Finch 1995; Black *et al.* 1993).

In vivo

CO_2 is the main metabolite (see Section 3.1). After rats were given single oral doses of [^{14}C]-labelled acrylic acid of 4 to 1000 mg/kg body weight, 3-hydroxypropionic acid was identified in the urine (depending on the dose in amounts of 0.4 % to 1.4 %). It was not possible to identify a second metabolite that was eliminated with the urine (3 % of the dose administered). 2,3-Epoxypropionic acid and *N*-acetyl-*S*-(2-carboxy-2-hydroxyethyl)cysteine were not found. After one hour, a significant depletion of NPSH was detected in the glandular stomach after doses of 40 mg/kg body weight and above and in the forestomach after 1000 mg/kg body weight. The NPSH concentrations in the blood and liver were not impaired. The authors concluded that acrylic acid enters the normal cell metabolism via 3-hydroxypropionic acid and acetyl-CoA (deBethizy *et al.* 1987; Rohm and Haas Company 1985).

The metabolite spectrum of [1,2,3-^{13}C]acrylic acid in the urine of rats given single gavage doses (400 mg/kg body weight) was investigated by means of ^{13}C NMR analysis. 3-Hydroxypropionic acid was detected as the main metabolite of acrylic acid in the urine. Metabolic degradation takes place by means of a β-oxidative pathway of mitochondrial propionate metabolism via acrylyl-CoA to 3-hydroxypropionyl-CoA, 3-hydroxypropionic acid, malonsemialdehyde, then to acetyl-CoA and CO_2. This metabolic pathway is of subordinate relevance in rats and humans, and quantitative differences between the two species have not been investigated. *N*-Acetyl-*S*-(2-carboxyethyl)cysteine and *N*-acetyl-*S*-(2-carboxyethyl)cysteine-*S*-oxide were detected as further metabolites. At the high acrylic acid concentrations used here, these metabolites are probably formed as soon as the main metabolic pathway is saturated. Unchanged acrylic acid and epoxidized compounds were not detected. The authors assume that non-ionized acrylic acid leads to glutathione conjugation or that coupling with CoA and the resulting activation of the double bond facilitates glutathione conjugation (Winter *et al.* 1992; Winter and Sipes 1993). In this way, acrylic acid could also react with other nucleophiles, such as DNA.

Other research groups also detected 3-hydroxypropionic acid as the main metabolite in the urine of rats and mice. Up to seven further metabolites were found, some of them in very small proportions, but they could not be characterized in detail. Neither unchanged acrylic acid nor the metabolites were detected in plasma or liver after one hour (Black *et al.* 1995; Rohm and Haas Company 1985).

After dermal application of [^{14}C]acrylic acid, 3-hydroxypropionic acid and a metabolite that could not be identified (after a high dose) were detected only in the urine of rats. No acrylic acid was found in the liver of mice at any time (Black *et al.* 1995).

4 Effects in Humans

A total of six occupational accidents with acrylic acid in which the persons affected were admitted to hospital for further treatment were reported between 1967 and 1995 and in

1998. Skin and eye burns were the main effects. Irritation of the respiratory tract was additionally observed in only one case (BASF AG 1993, 1994, 1996, 1999).

No sensitization was found among any of the 450 workers employed in acrylic acid production during the occupational health examinations that have regularly been carried out since 1989 (EU 2002).

According to an inadequately documented study, one of 45 patients with proven sensitization to one of a total of 64 tested substances possibly occurring in shoe materials also reacted to acrylic acid (no other details) (Grimalt and Romaguera 1975). One female patient with sensitization presumably caused by an implant made of a methyl methacrylate/methyl acrylate copolymer, who reacted to the polymer, some methacrylates, 0.1 % acetonitrile and 0.1 % methacrylic acid, also produced a 2+ reaction to 0.1 % acrylic acid in petrolatum (Romaguera *et al.* 1985).

A male worker who developed pronounced, generalized urticarial reactions after prolonged contact with acrylate resins and acrylic acid, produced an immediate reaction to 2 % acrylic acid in olive oil, but not to several acrylates and methacrylates tested. The authors did not consider the oedematous, erythematous reaction (2+) detected in the test area after 48 hours to be a clear indication of a delayed allergic reaction as they also believed an inflammatory reaction to be a possible result of the immediate reaction (Fowler 1990). One woman with a generalized urticarial reaction to an adhesive tape containing polyacrylate on the basis of acrylic acid also produced an immediate reaction to 2 % acrylic acid in petrolatum in an epicutaneous test (Daecke *et al.* 1993).

None of the six workers who had been sensitized by contact with anaerobic sealants and had reacted, for example, to 2 % hydroxyethyl methacrylate in an epicutaneous test, reacted to 0.1 % acrylic acid in petrolatum (Condé-Salazar *et al.* 1988). Nor did a worker with sensitization to hydroxypropyl acrylate react to 0.1 % acrylic acid in petrolatum in an epicutaneous test (Lovell *et al.* 1985).

The odour threshold of acrylic acid, at which a specially trained cohort definitely assigned an odour to acrylic acid, was specified to be 1.04 ml/m^3 (Hellman and Small 1974).

The chemosensory thresholds of acrylic acid were determined with psychophysical methods in a random sample of persons stratified for gender and age (n = 72) (van Thriel *et al.* 2006). The median for the odour thresholds was 0.5 ml/m^3 (5th percentile: 0.001 ml/m^3; 95th percentile: 2 ml/m^3). The median of the lateralization threshold as an index of unambiguous trigeminal stimulation was 31 ml/m^3 (5th percentile: 13 ml/m^3; 95th percentile: 82 ml/m^3). No conclusions about irritation after 8-hour exposures can be drawn from these values.

5 Animal Experiments and *in vitro* Studies

5.1 Acute toxicity

The results of acute toxicity tests show that acrylic acid has adverse effects after oral administration and dermal application (Table 2). The 4-hour LC_{50} for rats was found to be more than 1700 ml/m^3 in a well documented study. The oral LD_{50} values vary in accordance with the concentration of acrylic acid administered and are in a range of 150 to 2750 mg/kg body weight for rats. The dermal LD_{50} for rabbits is between 295 mg/kg body weight and 1000 mg/kg body weight.

Local irritation was the main symptom of toxicity. Inhalation of acrylic acid concentrations of 1000 ml/m^3 for 4 hours led in rats to marked NPSH depletion in the liver, lungs and kidneys and, to a lesser extent, also in the blood (Silver and Murphy 1981).

Table 2. Acute toxicity of acrylic acid

Species (strain)	LD_{50}/LC_{50}	Exposure	References
inhalation			
rat (Wistar)	♂: 3600 mg/m^3 (1200 ml/m^3)	4 hours	Majka *et al.* 1974
rat (Sprague-Dawley)	mortality: 0/12 (30 minutes) 1/6 (1 hour) 6/6 (3 hours)	atmosphere saturated at 20°C; concentration not specified	BASF AG 1979a
rat (Sprague-Dawley)	> 5100 mg/m^3 (1700 ml/m^3) no mortality	4 hours; whole-body exposure; vapour	BASF AG 1980
rat (Holtzman)	mortality: 0/6 (1300 ml/m^3) 6/6 (1600 ml/m^3) 5/5 (2100 ml/m^3)	4 hours	Silver and Murphy 1981
monkey (*Macaca fascicularis*)	mortality: 0/3 (3 hours) 0/3 (6 hours) no clinical symptoms and no gross-pathological abnormalities; lesions in the proximal olfactory epithelium, increasing with the period of exposure; maximum affected area of the olfactory epithelium: about 40 %	75 ml/m^3 nose-only inhalation for 3 or 6 hours	BAMM 2001

Table 2. continued

Species (strain)	LD_{50}/LC_{50}	Exposure	References
oral			
rat (Sprague-Dawley)	♂: 1337 mg/kg body weight ♀: 718 mg/kg body weight	undiluted	Dow Chemical Co 1992
rat (CDF)	♂: 151 mg/kg body weight ♀: 526 mg/kg body weight		
rat (Wistar)	♂: 1350 mg/kg body weight	10 % in water	Majka *et al.* 1974
rat (strain not specified)	1500 mg/kg body weight	10 % in 0.9 % NaCl solution	BASF AG 1958
rat (albino)	2.59 ml/kg body weight (2750 mg/kg body weight)	undiluted	Smyth *et al.* 1962
rat (albino)	360 mg/kg body weight	undiluted	Carpenter *et al.* 1974
mouse (strain not specified)	1200 mg/kg body weight	10 % in 0.9 % NaCl solution	BASF AG 1958
dermal			
rabbit (albino)	1000 mg/kg body weight	occlusive; 24-hour exposure	Smyth *et al.* 1962
rabbit (albino)	295 mg/kg body weight	anhydrous acrylic acid	Carpenter *et al.* 1974
rabbit (Vienna White)	640 mg/kg body weight	occlusive; 24-hour exposure	BASF AG 1979b

5.2 Subacute, subchronic and chronic toxicity

5.2.1 Inhalation

In a 13-week inhalation study (6 hours/day for 5 days/week), groups of 15 male and 15 female F344 rats and $B6C3F_1$ mice were exposed to acrylic acid concentrations of 5, 25 or 75 ml/m^3. Ten animals per group were examined histopathologically (nose: 4 planes of section). The NOAECs (no observed adverse effect concentrations) for systemic effects were 5 ml/m^3 in female mice and 75 ml/m^3 in male mice and rats. Only lower body weight gains were observed, but no systemic organ or tissue damage. As a result of slight histological damage to the olfactory epithelium, the LOAEC (lowest observed adverse effect concentration) was given as 5 ml/m^3 in mice. No effects on the nasal mucosae were observed in the two low concentration groups in rats. The NOAEC for local irritation was thus 25 ml/m^3 (Table 3; Miller *et al.* 1981a).

Table 3. Histological observations in the nasal mucosa of rats and mice after 13-week exposure to acrylic acid (Miller *et al.* 1981a)

	Concentration [ml/m³]							
	0	5	25	75	0	5	25	75
	male rat				female rat			
slight focal degeneration in the olfactory epithelium	0/10	0/10	0/10	7/10	0/10	0/10	0/10	10/10
slight inflammation characterized by infiltration of mononuclear cells into the mucosa and submucosa	0/10	0/10	0/10	0/10	1/10	0/10	0/10	0/10
	male mouse				female mouse			
focal degeneration of the olfactory epithelium (dorso-medial); in some cases, replacement by epithelium similar to the respiratory epithelium								
slight to moderate	1/10	1/10	0/11	10/10	0/10	0/10	0/10	10/12
focal degeneration of the olfactory epithelium (dorso-medial)								
slight	0/0	0/10	10/11	0/10	0/10	0/10	9/10	1/12
very slight	0/10	1/10	1/11	0/10	0/10	4/10	0/10	0/12
ungraded because of autolysis	0/10	0/10	0/11	0/10	0/10	0/10	0/10	1/12
focal infiltrations of inflammatory cells into the mucosa and submucosa in regions with degeneration of the mucosa								
slight	0/10	0/10	0/11	0/10	0/10	0/10	2/10	0/12
very slight	0/10	0/10	1/11	10/10	0/10	0/10	0/10	10/12
focal hyperplasias of the submucousal glands in regions with degeneration of the mucosa								
very slight	0/10	0/10	0/11	10/10	0/10	0/10	0/10	10/12

In a study carried out as a result of these findings, groups of 15 female $B6C3F_1$ mice were exposed to acrylic acid vapour concentrations of 0, 5 or 25 ml/m³ for 15 days for 6 or 22 hours a day and also to 25 ml/m³ for 4.4 hours a day. Ten animals were sacrificed at the end of exposure and the remaining 5 after a 6-week recovery phase. Clinical parameters were investigated but only the nasal cavity was examined histopathologically. Exposure to 5 ml/m³ for 6 hours caused no changes. Both 5 ml/m³ for 22 hours and 25 ml/m³ for 4.4, 6 and 22 hours led to concentration and time-related changes to the olfactory epithelium consisting of atrophy, basal cell hypertrophy, necrosis and degeneration of Bowman's glands. The findings obtained after 22-hour exposure to 5 ml/m³ and after 4.4-hour and 6-hour exposure to 25 ml/m³ were completely reversible after 6 weeks. In contrast, the animals of the 25 ml/m³ group were found to have localized areas in which the olfactory epithelium had turned into respiratory-like epithelium (respiratory metaplasia) after 22-hour exposure (Lomax *et al.* 1994; Rohm and Haas Company 1994). The findings from this 15-day study show that at most a slight increase in the effects with time must be assumed as regards irritation of the olfactory epithelium. With the exception of the very slight changes to the olfactory epithelium at 5 ml/m^3 in the medium-term study, the effects observed were similar after 15 and 90 days.

In a study with male F344 rats and B6C3F$_1$ mice designed to detect differences between the species, the animals were exposed to acrylic acid concentrations of 75 ml/m^3 in whole body exposure chambers for 6 hours a day for 4 days. On day 5, nose-only exposure was carried out for 6 hours. The respiratory rate decreased by about the same extent in both species (rats between 16 % and 23 %; mice between 32 % and 37 %). The effect on the inhaled volume was only very slight in rats (93 % to 103 % of the values of the control animals); in mice, it was also only slightly affected. The minute volume decreased by about 23 % in rats and 27 % to 34 % in mice. For mice, an 88 % higher tissue dose per time unit (3.5–3.8 mg/min and cm^2 compared with 1.8–2.1 mg/min and cm^2 in rats) was calculated from the concentration, the surface area of the nasal cavity and the respiratory minute volume. Lesions of the respiratory tract, which were limited to the nasal section and mainly affected the olfactory epithelium, were found in both species. The lesions were more pronounced in mice than in rats. The animals were treated with tritium-labelled thymidine 18 hours after the last exposure to acrylic acid to determine the effects on olfactory cell proliferation. Cell division was increased 17 times in mice and 4 times in rats (Barrow 1984, 1986; Swenberg *et al.* 1987).

In a range-finding study for a prenatal toxicity study, acrylic acid led to lesions of the nasal mucosa also in rabbits after concentrations of 30 ml/m^3 and above (see Section 5.5.2; Neeper-Bradley *et al.* 1997).

5.2.2 Ingestion

F344 rats were given acrylic acid in doses of 83, 250 or 750 mg/kg body weight with the drinking water for 3 months. 15 animals aged between 41 and 43 days were used per test group. No animals died during the study. Significantly reduced body weights and reduced feed consumption were observed in both sexes of the highest dose group. At the end of the study, body weights were significantly reduced in the female rats of the middle dose. Various clinico-chemical parameters such as alkaline phosphatase were changed after doses of 250 mg/kg body weight and above, and the absolute liver, spleen and heart weights were reduced after 750 mg/kg body weight. The red and white blood counts showed no abnormalities. No significant gross-pathological or histopathological changes were found in any group. The NOAEL (no observed adverse effect level) was 83 mg/kg body weight and day (Inter-Company Acrylate Study Group 1980a).

Groups of 10 male and 10 female Wistar rats were given aqueous acrylic acid by gavage in doses of 0, 150 or 375 mg/kg body weight and day for 3 months. Both doses were lethal and led to the premature death of 6/10 male and 9/10 female rats after 375 mg/kg body weight and day and of 5/10 males and 5/10 females in the low dose group. Pronounced symptoms of irritation in the forestomach and glandular stomach and also changes in the kidneys and nasal mucosa were the main histopathological findings. This form of administration was therefore found unsuitable for repeated treatment at the doses selected (BASF AG 1987a; Hellwig *et al.* 1993).

In a subsequent study, Wistar rats were exposed to acrylic acid concentrations of 0, 120, 800, 2000 or 5000 mg/l (about 0, 6/10, 40/66, 100/150, 200/375 mg/kg body weight and day; ♂/♀) in the drinking water for either 3 months (10 ♂/10 ♀) or 12 months (20 ♂/20 ♀). Body weight gain was slightly impaired in males after doses of 2000 mg/l, and more markedly so at the 5000 mg/l dose. There were no changes in the females. Drinking water consumption was slightly reduced at the 2000 mg/l dose and markedly reduced at 5000 g/l in both sexes. Reduced feed consumption was observed only in the male rats of the 5000 mg/l group. Although ingestion of the substance at the two high doses corresponded to the amount taken in by gavage, no specific organ or tissue damage was detected in the comprehensive haematological and clinico-pathological examinations and urinalyses or in the detailed gross-pathological and histopathological examinations (BASF AG 1987b; Hellwig *et al.* 1993).

The results of the above-mentioned studies can be seen in detail in Table 4.

5.2.3 Dermal absorption

Acrylic acid was applied to the shaved dorsal skin of groups of 30 female ICR and $B6C3F_1$ mice and 30 male C3H mice as 1 % or 4 % solutions in acetone 3 times a week for 13 weeks. There were no clinical symptoms of toxicity or any effects on body weight (McLaughlin *et al.* 1995; see also Section 5.3.1).

Table 4. Studies with repeated oral administration of acrylic acid

Species, strain, number, sex/group	Exposure	Dose: findings	References
rat, Fischer-344, groups of 15♂/15 ♀	3 months; 0, 83, 250, 750 mg/kg body weight and day; drinking water	**83 mg/kg body weight**: NOAEL **250 mg/kg body weight and above**: ♂/♀: water consumption decreased; relative kidney weights and protein content of the urine increased; ♂: relative testis weights increased; ♀: body weight gains and cholesterol decreased; urea nitrogen and alkaline phosphatase increased **750 mg/kg body weight**: ♂/♀: feed consumption, body weight gains and absolute liver, spleen and heart weights decreased; relative kidney weights increased; ♂: relative liver and spleen weights and urea nitrogen increased; ♀: absolute kidney weights, glucose and aspartate aminotransferase increased; urine pH decreased	Inter-Company Acrylate Study Group 1980a
rat, Wistar, groups of 10 ♂/10 ♀	3 months; 0, 150, 375 mg/kg body weight and day; 5 days/week; gavage	**150 mg/kg body weight**: mortality (5/10 ♂ and 5/10 ♀) **150 mg/kg body weight and above**: erosions/ulcerations of the gastric mucosa, oedema, hyperaemia, dystelectasis in the lungs, rhinitis and tubular nephrosis **375 mg/kg body weight**: mortality (♂: 6/10; ♀: 9/10)	BASF AG 1987a; Hellwig *et al.* 1993
rat, Wistar, groups of 10 ♂/10 ♀	12 months; 0, 120, 800, 2000, 5000 mg/l (0, 6/10, 40/66, 100/150, 200/375 mg/kg body weight and day; ♂/♀); drinking water	**40 mg/kg body weight**: ♂: NOAEL **100 mg/kg body weight and above**: ♂: body weight gains decreased; ♂/♀: water consumption decreased **200 mg/kg body weight**: ♂: feed consumption decreased **375 mg/kg body weight**: ♀: NOAEL	BASF AG 1987b; Hellwig *et al.* 1993

5.3 Local effects on skin and mucous membranes

5.3.1 Skin

Acrylic acid is corrosive to the skin (EU 2002).

The corrosive potential of undiluted acrylic acid was determined on the unshaved dorsal skin of New Zealand White rabbits in a study carried out according to OECD Test Guideline 404. A single topical application of 0.5 ml under semi-occlusive conditions over a period of 3 minutes led to brownish skin discoloration after only one hour. The histopathological examination revealed deep skin necrosis after an observation period of 14 days (BASF AG 1998).

Acrylic acid was applied to the shaved dorsal skin of groups of 30 female ICR and $B6C3F_1$ mice and 30 male C3H mice as 1 % or 4 % solutions in acetone three times a week for 13 weeks to evaluate differences in sensitivity after dermal exposure and to establish a tolerable concentration for a long-term study. About 24 hours after the 3rd, 6th, 12th and 24th treatments (1, 2, 4 and 8 weeks), 5 mice per group were sacrificed; the remaining animals were sacrificed after the 39th treatment (13 weeks). In all three mouse strains, the 4 % solution induced marked skin irritation beginning after 1 to 2 weeks and continuing up to week 13. The maximum irritation was observed between weeks 3 and 5. The histopathological examination revealed proliferative, degenerative and inflammatory changes of dermis and epidermis, the severity of which remained more or less the same throughout the study period. ICR mice were somewhat less sensitive than the two other strains. The authors considered the effects of 1 % acrylic acid to be tolerable for long-term exposure (McLaughlin *et al.* 1995).

5.3.2 Eyes

Acrylic acid is corrosive to the eyes (EU 2002).

After instillation of 0.1 ml into the conjunctival sac of albino rabbits, acrylic acid neutralized with potassium hydroxide solution (60 % aqueous potassium acrylate) led to damage to the eyes which was very much dependent on the exposure duration. If the eyes were not rinsed with lukewarm water, or were rinsed only after 20 or 60 seconds, corneal opacity, in some cases linked with ulceration, was observed. If the eyes were rinsed after 2 or 4 seconds, the cornea was clear again after only 7 days and only mild irritation of the conjunctiva remained after 18 days (Hoechst Celanese Corp 1992).

5.4 Allergenic effects

The results of the sensitization studies are shown in Table 5. In various adjuvant tests, the repeated application of pure or distilled acrylic acid did not lead to dermal hypersensitivity reactions in guinea pigs, irrespective of the induction or challenge procedure. If, however, contaminated or commercial samples not characterized in detail were used, pronounced sensitization reactions were observed on the skin. They were due to α,β-diacryloxypropionic acid (DAPA; 2,3-di(acryloxy)propionic acid), which was identified in the acrylic acid in amounts of up to 7 % and found to be a potent sensitizing agent (Parker and Turk 1983; Rao *et al.* 1981; Waegemaekers and van der Walle 1984; van der Walle *et al.* 1982).

Table 5. Animal sensitization studies with acrylic acid

Type of test, species	Conduct of the study	Remarks	Result	References
Freund's complete adjuvant test, guinea pig	induction: 0.1 ml (0.5 M) emulsion in FCA; intradermal; days 0, 2, 4, 7, 9 challenge: topical; 0.025 ml; days 21, 35	8/8 produced skin reactions to an unpurified sample (purity 55 %) 0/8 reacted to the distillate (purity > 98 %) 8/8 reacted to the residue (45 %)	not sensitizing	van der Walle *et al.* 1982
modified Freund's complete adjuvant test, guinea pig	induction: 0.1 ml (0.17 M = 1.2 %) emulsion in FCA; intradermal; days 0, 5, 9 challenge: topical; 0.025 ml; (1 M = 7.2 %), days 21, 35, 49	8/8 produced skin reactions to a commercial sample 0/8 reacted to the distillate DAPA (up to 7 % in the product) was found to be the sensitizing impurity	not sensitizing	Waegemaekers and van der Walle 1984
modified split adjuvant test, guinea pig	induction: topical: 0.1 ml aliquot; 4× in 10 days; 0.2 ml FCA for 3rd application, intradermal on adjacent site challenge: topical after 2 weeks	0/10 produced skin reactions	not sensitizing	Rao *et al.* 1981
Polak test, guinea pig	induction: intradermal: day 0, in ethanol/NaCl solution (1:4) in FCA challenge: topical; weekly from day 7 for 12 weeks; 0.02 ml 5 % acrylic acid in acetone/olive oil (4:1)	average reaction: 0.5	questionably sensitizing (no details of DAPA level)	Parker and Turk 1983

DAPA: α,β-diacryloxypropionic acid; FCA: Freund's complete adjuvant

5.5 Reproductive toxicity

5.5.1 Fertility

In a study not carried out in conformity with test guidelines, groups of 10 male and 20 female F344 rats were given acrylic acid in the drinking water for 13 weeks. The concentrations used corresponded to the ingestion of 0, 83, 250 and 750 mg/kg body weight and day. One male was mated with two females in each case, and exposure continued during pregnancy up to the end of lactation. Symptoms of toxicity in the form of reduced water and feed consumption, reduced body weight gains and organ weight changes in the parent animals were observed mainly at the highest dose and, to a lesser extent, at the middle dose. The fertility and gestation indices were numerically reduced in this group, but the corresponding indices in the controls were also very low. A reduced number of born and weaned pups, and significantly reduced body weight gains with organ weight changes, were further effects found at the high dose. In spite of systemic toxicity, there was no clear evidence of substance-induced effects on fertility because the differences between the control group and the treated animals were not significant. No substance-induced changes were detected histopathologically (DePass *et al.* 1983; Inter-Company Acrylate Study Group 1980b).

In a 2-generation study carried out according to OECD Test Guideline 416, acrylic acid was administered to groups of 25 Wistar rats per sex and group in concentrations of 0, 500, 2500 or 5000 mg/l (corresponding to 53, 240 or 460 mg/kg body weight and day) with the drinking water for 10 weeks before and during mating and during gestation and lactation. One litter was formed per generation. Selected pups were reared and after weaning treated in an identical way. General clinical parameters, fertility parameters and toxicity and developmental parameters were assessed in the pups. The parent animals were also subjected to a comprehensive histopathological examination in which the reproductive organs were considered in particular. No impairment of fertility was detected either clinically or histopathologically up to and including the highest concentration. As symptoms of systemic toxicity in the parent animals, reduced body weights and reduced feed and drinking water consumption were observed at the 460 mg/kg body weight dose in the F_0 generation and at both the 460 mg/kg and 240 mg/kg body weight doses in the F_1 generation. Minimal hyperkeratosis in the limiting ridge of the forestomach and minimal submucosal oedema in the glandular stomach were the only treatment-induced histopathological findings observed in both parent generations. Developmental toxicity was manifest as reduced body weight gains in the F_1 and F_2 pups of the middle and high dose groups. Further evidence of possibly delayed development were a lower number of F_2 pups with open eyes on day 15 after birth in the 460 mg/kg body weight dose group and with open auditory canals in the 240 mg/kg body weight group. These findings were, however, within the range of the historical control data. No morphological changes whatsoever occurred in the pups of the two parent generations. Thus, a NOAEL of 460 mg/kg body weight and day was obtained for reproductive performance/fertility. The NOAEL for systemic toxicity was 240 mg/kg body weight and day for the F_0 generation and 53 mg/kg body weight and day for the F_1 parent animals. The NOAEL for developmental toxicity was 53 mg/kg body weight and day (Hellwig *et al.* 1997).

Details of the fertility studies are summarized in Table 6.

Table 6. Fertility studies with acrylic acid after administration with the drinking water

Species, strain, number of animals, sex/group	Exposure (dose, route, duration)	Dose: findings	References
rat, F344, 20 ♂/10 ♀	**1-generation study**, 0, 83, 250, 750 mg/kg body weight and day; drinking water **F_0:** from week 13 before mating up to weaning of the 1st litter (F_1)	**83 mg/kg body weight**: NOAEL systemic toxicity for F_0 animals **250 mg/kg body weight**: NOAEL developmental toxicity **F_0**: ♂/♀: water consumption decreased; ♀: body weight gains decreased; relative liver and absolute kidney weights increased **750 mg/kg body weight**: NOAEL fertility **F_0**: ♂/♀: water consumption, body weight gains and feed consumption decreased; absolute liver weights decreased; relative kidney weights increased; ♂: relative spleen and relative testis weights increased; ♀: absolute spleen weights decreased; **F_1:** body weight gains decreased; ♂: absolute and relative liver weights, absolute kidney and heart weights decreased; ♀: absolute and relative spleen weights and absolute liver weights decreased	DePass *et al.* 1983; Inter-Company Acrylate Study Group 1980b
rat, Wistar, groups of 25 ♂/25 ♀	**2-generation study**, 0, 500, 2500, 5000 mg/l (about 53, 240, 460 mg/kg body weight and day); drinking water **F_0:** from 10 weeks before mating; ♂: up to the end of mating; ♀: up to weaning of litter (F_1) **F_1**: from weaning and during mating; ♂: up to the end of mating; ♀: during gestation up to the end of lactation	**53 mg/kg body weight**: NOAEL systemic toxicity (F_1) and developmental toxicity **240 mg/kg body weight**: NOAEL systemic toxicity (F_0) **F_0/F_1 parent animals**: ♂/♀: water consumption and body weight gains not significantly decreased; ♀: feed consumption not significantly decreased (gestation and lactation) **F_1 pups**: body weight gains decreased **F_2 pups:** body weight gains decreased; delayed opening of auditory canal **460 mg/kg body weight**: NOAEL fertility **F_0/F_1 parent animals**: ♂/♀: water consumption and body weight gains decreased; histopathological findings in the forestomach wall (hyperkeratosis) and in the glandular stomach (submucosal oedema); ♀: feed consumption decreased (gestation and lactation) **F_1 pups**: body weight gains decreased **F_2 pups:** body weight gains decreased; delayed opening of eyes	Hellwig *et al.* 1997

5.5.2 Developmental toxicity

Groups of 30 Sprague-Dawley rats were exposed to acrylic acid vapour in concentrations of 0, 40, 120 or 360 ml/m^3 (0, 120, 360 and 1080 mg/m^3) in whole body exposure chambers for 6 hours a day from days 6 to 15 of gestation. In the dams, treatment led to reduced corrected body weight gains (body weight from days 0 to 20 minus uterus weight) after concentrations of 40 ml/m^3 and above, and also to reduced corrected body weights (body weight on day 20 minus uterus weight) and reduced feed consumption after 120 ml/m^3 and above. The highest concentration induced both irritation of the respiratory tract and eyes and clear reductions in body weights and feed consumption. No adverse effects on gestation parameters or evidence of embryotoxicity or foetotoxicity were observed in any group. The number of implantations, live foetuses per litter and resorptions per litter, the sex ratio of the foetuses and the number and form of malformations were similar in all groups and not related to the concentration. A NOAEC for maternal toxicity was not obtained; the highest concentration investigated of 360 ml/m^3 was the NOAEC for developmental toxicity. The study was carried out in compliance with OECD Test Guideline 414 from 1981 (BASF AG 1983; Klimisch and Hellwig 1991).

In a study carried out with a prolonged treatment interval according to the current OECD Test Guideline 414, 20 to 24 pregnant Sprague-Dawley rats were exposed to acrylic acid vapour concentrations of 0, 50, 100, 200 or 300 ml/m^3 in whole body exposure chambers for 6 hours a day from days 6 to 20 of gestation. Reduced feed consumption occurred in all groups: during the first half of exposure after concentrations of 50 ml/m^3 and 100 ml/m^3, and permanently at higher concentrations. Body weight gains were reduced in the first half of exposure after concentrations of 200 ml/m^3 and throughout the exposure period after 300 ml/m^3. No clinical findings were described. No abnormalities were found for gestation parameters. Maternal toxicity was accompanied by reduced foetal weights only at 300 ml/m^3. Otherwise, no evidence of embryotoxicity or foetotoxicity was observed at any concentration. In particular, the number and form of variations, retardations and malformations were similar in all groups. The NOAEC for maternal toxicity was 100 ml/m^3, and that for developmental toxicity 200 ml/m^3 (Saillenfait *et al.* 1999).

In a pilot study for another study with rabbits and acrylic acid concentrations of 0, 30, 60, 125 or 250 ml/m^3 (days 10 to 22 of gestation; 6 hours/day), the animals had blepharospasms at concentrations of 125 ml/m^3 and above. The histopathological examination of the nasal area revealed concentration-related lesions of the nasal mucosa (erosion, ulceration, and in some cases squamous metaplasia in the nasal epithelium) in all groups immediately after exposure. During the examination on day 29, these histopathological changes were less pronounced with a NOAEC of 30 ml/m^3. In the main study, groups of 16 New Zealand White rabbits were exposed to acrylic acid vapour in concentrations of 0, 25, 75 or 225 ml/m^3 in whole body exposure chambers for 6 hours daily from days 6 to 18 of gestation. The treatment led to reduced body weight gains and reduced feed consumption in the dams after concentrations of 75 ml/m^3 and above. When the body weight parameters were corrected for the uterus weight, however, they did not differ from those of the controls. Symptoms of slight irritation occurred at

75 ml/m³. The highest concentration led to marked irritation of the respiratory tract including gross-pathological ulceration of the nasal mucosa in the turbinate region in one dam. No adverse effects on gestation parameters or evidence of embryotoxicity or foetotoxicity were observed in any group. The number of implantations, live foetuses per litter and resorptions per litter, the sex ratio of the foetuses, and the number and form of variations and malformations were similar in all groups. The NOAEC for maternal toxicity was 25 ml/m³; the NOAEC for developmental toxicity was the highest concentration investigated of 225 ml/m³. The study was carried out in compliance with OECD Test Guideline 414 from 1981 (Neeper-Bradley *et al.* 1997).

Details of the above-mentioned studies are shown in Table 7.

Table 7. Studies of the developmental toxicity in rats and rabbits after inhalation of acrylic acid vapour

Species, strain, number/group	Exposure	Concentration: findings	References
rat, Sprague-Dawley, groups of 30 pregnant ♀	GD 6–15 0, 40, 120, 360 ml/m³; 6 hours/day; vapour; whole body exposure	**40 ml/m³ and above**: dams: body weight gains minus uterus weights decreased; LOAEC **120 ml/m³ and above**: dams: body weights on day 20 minus uterus weights decreased **360 ml/m³**: NOAEC developmental toxicity; dams: irritation of the eyes/respiratory tract; body weight gains decreased in accordance with OECD Test Guideline 414 from 1981	BASF AG 1983; Klimisch and Hellwig 1991
rat, Sprague-Dawley, groups of 20–24 pregnant ♀	GD 6–20 0, 50, 100, 200, 300 ml/m³; 6 hours/day; vapour; whole body exposure	**50 ml/m³ and above**: feed consumption decreased **100 ml/m³**: dams: NOAEC **200 ml/m³**: NOAEC developmental toxicity **200 ml/m³ and above**: dams: body weight gains decreased **300 ml/m³**: foetal weights decreased	Saillenfait *et al.* 1999
rabbit, New Zealand White, groups of 16 pregnant ♀	GD 6–18 0, 25, 75, 225 ml/m³; 6 hours/day; vapour; whole body exposure	**25 ml/m³**: dams: NOAEC **75 ml/m³ and above**: dams: feed consumption and body weight gains decreased; irritation of the respiratory tract **225 ml/m³**: NOAEC developmental toxicity in accordance with OECD Test Guideline 414 from 1981	Neeper-Bradley *et al.* 1997

GD: day of gestation

5.6 Genotoxicity

5.6.1 *In vitro*

In an *in vitro* study of DNA binding, acrylic acid induced 2-carboxyethyl adducts with all four DNA bases after 40-day incubation with calf thymus DNA and the individual nucleosides at pH 7.0. The authors considered the reaction to be slow (Cote *et al.* 1986b; Segal *et al.* 1987). Since the very long incubation time makes it difficult to interpret the result, the DNA reactivity of acrylic acid is considered to be very slight in accordance with the authors' opinion.

The results of the *in vitro* studies of genotoxicity can be seen in Table 8.

Table 8. Genotoxicity of acrylic acid *in vitro*

Test system		Concentration	Result		Remarks	References
			– m. a.	+ m. a.		
BMT	*Salmonella typhimurium* TA98, TA100, TA1535, TA1537	3.15–1000 nl/plate (3.15–1000 µg/plate)	negative	negative	highest concentration not toxic; test result cannot be assessed	BASF AG 1977
BMT	*Salmonella typhimurium* TA98, TA100, TA1535, TA1537	0–3333 µg/plate	negative	negative	bacteriotoxicity at 3333 µg/plate	Zeiger *et al.* 1987
BMT	*Salmonella typhimurium* TA98, TA100, TA1535, TA1537, TA1538; preincubation test and plate incorporation test	0–1000 µg/plate	negative	negative	maximum non-bacteriotoxic dose: preincubation test: 250 µg/plate plate incorporation test: 1000 µg/plate	Lijinski and Andrews 1980
BMT	*Salmonella typhimurium* TA98, TA100, TA1535, TA1537; plate incorporation test	33–5000 µg/plate	negative	negative	bacteriotoxicity at 10000 µg/plate	Cameron *et al.* 1991
Mammalian cells						
UDS	primary rat hepatocytes	0.01–0.6 µl/ml (about 0.01–0.6 mg/ml)	negative	not carried out	severe cytotoxicity at the highest dose	BAMM 1988b; McCarthy *et al.* 1992
UDS	SHE cells	1–300 µg/ml	negative	not carried out	no details of cytotoxicity; test result could not be evaluated	Wiegand *et al.* 1989

Table 8. continued

Test system		Concentration	Result		Remarks	References
			– m. a.	+ m. a.		
CA	CHO cells	– m. a.: 2.846–5 µl/ml (about 2.8–5 mg/ml) + m. a.: 1.615–3.769 µl/ml (about 1.6–3.7 mg/ml)	positive after 3.769 µl/ ml (3.769 mg/ml) and above	positive after 1.615 µl/ml (1.615 mg/ ml) and above	< 68 % surviving cells at positive concentrations	Celanese Corp 1986a; McCarthy *et al.* 1992
CA	CHL cells	0.25–0.75 mg/ml	positive at 0.75 mg/ ml	not carried out	no details of cytotoxicity	Ishidate 1979; Ishidate *et al.* 1988
MNT	SHE cells	0.5–10 µg/ml	negative	not carried out	no details of cytotoxicity; test result could not be evaluated	Wiegand *et al.* 1989
HPRT	CHO cells	+ m. a.: 1.0–2.8 µl/ml (about 1–2.8 mg/ml) – m. a.: 0.3–1.9 µl/ml (about 0.3–1.9 mg/ml); test medium pH unchanged	negative	negative	cytotoxicity – m. a. at 2.8 µl/ ml; + m. a. at 6.08 µl/ml; concentrations > 2.8 µl/ml caused too high osmolality of the medium	BAMM 1988a; McCarthy *et al.* 1992
$TK^{+/-}$	mouse lymphoma cells (L5178Y)	– m. a.: 1.62–5.44 mM + m. a.: 4.41–26.5 mM	positive after 2.65 mM (190 µg/ ml) and above	positive after 16.2 mM (1116 µg/ ml) and above	< 55 % surviving cells at positive concentrations; colony count not specified	Cameron *et al.* 1991
$TK^{+/-}$	mouse lymphoma cells (L5178Y)	– m. a.: 300–600 µg/ml	positive after about 300 µg/ml and above	not carried out	< 40 % surviving cells at positive concentrations; almost exclusive-ly small colonies induced (chromo-some aberrations)	Moore and Doerr 1990; Moore *et al.* 1988
$TK^{+/-}$	mouse lymphoma cells (L5178Y)	– m. a.: 0.05–0.5 µl/ml (about 0.05–0.5 mg/ml) + m. a.: 0.3–3.0 µl/ml (about 0.3–3 mg/ml)	positive after about 0.2 µl/ml (0.2 mg/ ml) and above	positive after about 1 µl/ml (1 mg/ml) and above	< 50 % surviving cells at positive concentrations; colony count not recorded	National Cancer Institute 1982

BMT: bacterial mutagenicity test; CA: test for structural chromosome aberrations; CHL: Chinese hamster lung; CHO: Chinese hamster ovary; HPRT: hypoxanthine guanine phosphoribosyl transferase; m. a.: metabolic activation; MNT: micronucleus test; SHE: Syrian hamster embryo; TK: thymidine kinase mutation assay; UDS: test for DNA repair synthesis

Acrylic acid was not found to be mutagenic in bacterial test systems (Cameron *et al.* 1991; Lijinski and Andrews 1980; Zeiger *et al.* 1987). In mammalian cells, no induction of gene mutations was observed in CHO cells in an HPRT gene mutation assay (BAMM 1988a; McCarthy *et al.* 1992). Negative results described in a micronucleus test and with SHE (Syrian hamster embryo) cells in a UDS assay cannot be evaluated since there is no information on cytotoxicity (Wiegand *et al.* 1989). Positive results were obtained in three $TK^{+/-}$ gene mutation assays with L5178Y mouse lymphoma cells and in a chromosome aberration test with CHO cells and another with CHL (Chinese hamster lung) cells (Cameron *et al.* 1991; Celanese Corp 1986a; Ishidate 1979; Ishidate *et al.* 1988; McCarthy *et al.* 1992; Moore *et al.* 1988; Moore and Doerr 1990; National Cancer Institute 1982). Effects occurred in the chromosome aberration test after concentrations of 750 µg/ml and above (about 10 mM) and are thus in a relatively high concentration range. In a mouse lymphoma test, small colonies were preferably formed, also at high concentrations (about 2.5 mM); this is regarded as evidence of a clastogenic potential *in vitro* that corresponds with the effects found in the chromosome aberration tests with CHO and CHL cells. Clastogenicity in the mouse lymphoma test was in the range of moderate to severe cytotoxicity (< 55 % surviving cells), which according to OECD Test Guideline 473 should, however, be reached experimentally, whereas no clastogenic effect was found at weak cytotoxicity.

The acrylates methyl acrylate and ethyl acrylate tested for comparison caused effects in a mouse lymphoma test at about 10 times lower concentrations (Moore *et al.* 1988). Comparison of the results obtained for several substances in mouse lymphoma tests with those from other genotoxicity test systems showed that most of the DNA-damaging substances in the mouse lymphoma test produced positive results not only at cytotoxic concentrations but also at concentrations which induced no pronounced cytotoxicity (> 60 % cell growth) (Galloway 2000). This is regarded as evidence that the positive findings for acrylic acid are due to a cytotoxic mechanism rather than a primarily genotoxic mechanism.

5.6.2 *In vivo*

The results of the *in vivo* studies on genotoxicity can be seen in Table 9.

No SLRL mutations were induced in fruit flies after feeding or injection. In mammals, no increased chromosome aberration rates were found in the bone marrow of rats. An *in vivo* chromosome aberration test was carried out after single doses of 100, 333 or 1000 mg/kg body weight and another one after administration of acrylic acid in the drinking water (2000 or 5000 mg/l) for 5 days. A dominant lethal test after single or multiple doses in mice also produced negative results (Celanese Corp 1986b; McCarthy *et al.* 1992).

Table 9. Genotoxicity of acrylic acid *in vivo*

Test system	Species, strain	Exposure	Remarks	Result	References
SLRL	*Drosophila melanogaster*	intrathoracic injection; once; 0.3 µl/fly		negative	McCarthy *et al.* 1992
		feeding; 3 days; 2 % acrylic acid in 5 % sucrose solution		negative	
CA, bone marrow	rat, Sprague-Dawley, 15 ♂/15 ♀ per dose	gavage; once; 0, 100, 333, 1000 mg/kg body weight; 5 ♂/5 ♀ examined 6, 12 and 24 hours after administration	333 mg/kg body weight: 1 ♀ died 1000 mg/kg body weight: body weight gains decreased (♂, ♀); irregular, laboured respiration; 1 ♀ died no significant effects on mitosis index	negative	Celanese Corp 1986b; McCarthy *et al.* 1992
CA, bone marrow	rat, Sprague-Dawley, 5 ♂/5 ♀ per dose	drinking water; 5 days; 0, 2000, 5000 mg/l (about 180 and 450 mg/kg body weight)	body weight gains decreased	negative	Celanese Corp 1986b; McCarthy *et al.* 1992
dominant lethal test	mouse, CD-1, mated with 15–59 ♀ per 4-day interval	gavage; once; 0, 32, 108, 324 mg/kg body weight	maximum tolerated dose: 324 mg/kg body weight (=LD_1) and 162 mg/kg body weight ($LD_1/2$); mortality increased after 800 mg/kg body weight and above (acute)	negative	McCarthy *et al.* 1992
		gavage; 5 days; 0, 16, 54, 162 mg/kg body weight and day		negative	

CA: test for structural chromosome aberrations; LD_1: lethal dose for 1 % of the animals; SLRL: test for recessive lethal mutations on the X chromosome

5.7 Carcinogenicity

5.7.1 Short-term studies

Acrylic acid induced no cell transformations in SHE cells at concentrations of 5 µg/ml to 50 µg/ml (Wiegand *et al.* 1989). This result cannot be evaluated since there is no information on cytotoxicity.

5.7.2 Long-term studies

Inhalation

Carcinogenicity studies after long-term inhalation exposure to acrylic acid are not available. There are, however, studies with ethyl acrylate in rats and mice and with methyl and butyl acrylate in rats. None of the esters was carcinogenic after inhalation (see the previous MAK documentation in the present series: "*n*-Butyl acrylate", Volumes 5 and 12; "Ethyl acrylate", Volume 6; "Methyl acrylate", Volume 6). Since the esters are very rapidly degraded to acrylic acid and their corresponding alcohol in the organism and particularly in the nasal mucosa (deBethizy *et al.* 1987; Miller *et al.* 1981b; Stott and McKenna 1985), it can be assumed that acrylic acid is not carcinogenic after exposure by inhalation.

Ingestion

In a carcinogenicity study, groups of 50 male and 50 female Wistar rats were given acrylic acid concentrations of 0, 120, 400 or 1200 mg/l (0, 8, 27 and 78 mg/kg body weight and day) in the drinking water. Drinking water consumption was not significantly reduced in the rats of either sex given 1200 mg/l. In the 400 mg/l group, it was, however, increased. According to the original study report, any association with administration of the substance may therefore be incidental. There were no substance-induced systemic findings, nor an impairment in body weights or the survival period. In particular, the tumour incidences were not increased in either the male or in the female rats (see Table 10; BASF AG 1989; Hellwig *et al.* 1993). The MTD (maximum tolerated dose) was not reached in this study. In the 12-month study, drinking water consumption was reduced in the males and females and body weight gains were reduced in the males after concentrations of 2000 and 5000 mg/l. In the carcinogenicity study, a concentration very much higher than 1200 mg/l would, therefore, not have been possible. No tumours or their precursors occurred in the 12-month study (see Section 5.2.2) at concentrations up to 5000 mg/l.

Table 10. Study of the carcinogenicity of acrylic acid after oral administration

Authors:	BASF AG 1989; Hellwig *et al.* 1993
Substance:	acrylic acid
Species:	rat, Wistar
Administration route:	drinking water
Concentration:	0, 120, 400 and 1200 mg/l (about 0, 8, 27 and 78 mg/kg body weight and day)
Duration:	♂: 26 months; ♀: 28 months
Toxicity:	78 mg/kg body weight: slightly reduced drinking water consumption; body weight gains and survival rates not different from those of the controls
Tumours:	no increased tumour incidences induced by the substance

Dermal absorption

In a lifetime study, 25 µl of a 1 % acrylic acid solution in acetone (0.25 mg acrylic acid/day; about 8 mg/kg body weight for a mouse with a body weight of 30 g) or acetone were applied to the shaved dorsal skin of groups of 40 male C3H/HeJ mice three times a week. 5 % acrylic acid led to desquamation of the skin. The histopathological examination included the dorsal skin regions and gross-pathological abnormalities. The survival rate of the acrylic acid group (515 days) did not significantly differ from that of the acetone control group (484 days). No tumours of the skin or subcutis developed in any of the animals treated with acrylic acid or acetone. No signs of skin irritation were found. Epidermal hyperplasia was observed in one mouse after treatment with acrylic acid. In the positive control group (0.1 % 3-methylcholanthrene in acetone), however, 39/40 animals developed tumours, 33 of which were malignant epitheliomas (DePass *et al.* 1984; Inter-Company Acrylate Study Group 1982).

In a study published only as an abstract and poster, acrylic acid in acetone was applied to the shaved dorsal skin of groups of 30 female ICR/Ha mice (identical with Hsd:(ICR)BR) with and without dimethylbenzanthracene (DMBA) pretreatment three times a week for 18 months. The dose used was specified in the abstract to be 25 µl of 4 % acrylic acid and in the poster to be 4 mg in 0.1 ml acetone. It had been established as the maximum tolerated dose in a preliminary study. While no skin tumours occurred in the control group, two squamous cell carcinomas ($p < 0.07$ compared with the historical controls) were found in the animals treated with acrylic acid, and 3 papillomas (in 2 mice) and one squamous epithelial carcinoma ($p < 0.05$ compared with the historical controls) in those pretreated with DMBA. The incidence of leukaemia increased significantly to 86 % in the acrylic acid group compared with 30 % in the controls (Cote *et al.* 1986a, 1986b). In a re-evaluation of the study with a follow-up examination of the sections, "1 % solution in 0.1 ml acetone" was the dose according to written communications from the authors, "10 µg" was given by another author in an interim report and "72 µg" was specified in a laboratory journal. The housing of the animals and test conditions were regarded as "completely acceptable" in the re-evaluation, but there were no records of clinical observations of the animals which may possibly have contained statements about infections, nor was a detailed histopathological report or the laboratory journal to be found. The repeated analysis confirmed the two keratinizing squamous cell carcinomas in the animal group treated only with acrylic acid (re-examined: 29 of originally 30 animals). The control group treated with acetone showed no pathological changes. However, skin samples could be localized from only 21 animals. The skin areas treated with acrylic acid produced no changes outside the tumour regions. Although no significant difference in the tumour incidence was calculated when Fisher's exact test was used, it must be borne in mind that a keratinizing squamous cell carcinoma of the mouse skin very rarely occurs after treatment with acetone alone. The re-evaluation confirmed this. The 180 control animals of the institute that were suitable for comparison produced no skin carcinomas. If the 2/30 animals that developed tumours are compared with the 180 animals of the control group without carcinomas, there is a significant difference in the development of tumours induced by acrylic acid. For the tumour promotion study, the following proliferative changes were diagnosed based on the histopathological preparations: one animal with a keratinizing squamous cell carcinoma, one animal with a keratoacanthoma and one animal with one papilloma and one with two papillomas. The findings initially published with a total of four animals with tumours were thus confirmed. Although the reported incidence level of leukaemia/lymphoma was not confirmed, it was significantly increased in the acrylic acid group, but, according to the authors, remained within the range for that of the historical controls. Since the organotropy of these tumours was inconsistent among the treatment groups, the re-evaluators did not assume that there was an association between lymphoma and treatment. These inadequacies show that the study was not carried out according to GLP regulations. The re-evaluators concluded that acrylic acid (presumably as a 1 % solution) induces benign and malignant skin tumours, but the interpretation of the results is problematical since there was no written protocol and no definitive documentation of the dose actually used (Celanese Corp 1986c).

To clarify the results of the above-mentioned study, 25 µl and 100 µl of a 1 % acrylic acid solution in acetone (acrylic acid doses of about 0.25 and 1 mg/day; about 8 and 33 mg/kg body weight for an animal/mouse with a body weight of 30 g) were applied to the shaved dorsal skin of groups of 50 male and 50 female C3H/HeN Hsd BR and Hsd:(ICR)BR mice three times a week for 21 months. The dorsal skin, internal organs and gross-pathological changes were examined histopathologically. There was no evidence of skin irritation in either mouse strain. The mortality rates and body weight gains did not differ from those of the solvent controls. No clinical signs of systemic toxicity or the development of skin tumours were observed in any group. A tumour incidence of 95 % was obtained for the animals of the positive control group (0.1 % benzo(a)pyrene in acetone). Gross-pathological examination did not yield any treatment-related findings induced by acrylic acid. No histopathological changes were found in the ICR mice or the male C3H mice. A significant increase in the incidence of lympho-sarcomas was observed only in the female C3H animals of the 100 µl group (BAMM 1990). A differentiated re-evaluation of all test groups including an updated classification showed that although the incidence of haematopoietic neoplasms was increased to a total of 8/50 (1/50 histiocytic sarcoma, 1/50 lymphoblastic lymphoma and 6/50 follicular centre cell lymphomas), the incidence in the control groups (25 µl: 2/50;

100 µl: 1/50) was very low compared with that in the historical controls (BAMM 1991), which was specified to be about 10 % in female C3H mice older than 18 months (Frith and Wiley 1981). Thus the increased incidence of lymphomas cannot be attributed with certainty to acrylic acid. The tumourigenic effects on the skin could not be confirmed.

Since an increased skin tumour incidence was not observed in this adequately carried out and documented study, which used the same strain of mice and presumably the same dose as Cote *et al.* (1986a, 1986b) as well as a longer period of application and a higher number of animals, more importance is attached to the result obtained by BAMM (1990).

The results of the dermal carcinogenicity studies with mice can be seen in detail in Table 11.

Table 11. Studies of the dermal carcinogenicity of acrylic acid in mice

Authors:	DePass *et al.* 1984; Inter-Company Acrylate Study Group 1982
Substance:	acrylic acid
Species:	mouse, C3H/HeJ 40 ♂ and 40 ♂ control animals
Application:	dermal (shaved dorsal skin)
Concentration:	25 µl; 1 % acrylic acid solution in acetone (about 0.25 mg acrylic acid/day; 8 mg/kg body weight for a mouse with a body weight of 30 g) solvent controls: 25 µl acetone positive controls: 0.1 % 3-methylcholanthrene
Duration:	lifetime (3 times a week)
Toxicity:	no mortality; no effects on the survival rate; no irritation; 1/40 epidermal hyperplasia
Tumours:	none
Authors:	BAMM 1990, 1991
Substance:	acrylic acid
Species:	mouse, C3H/HeN Hsd BR (groups of 50 ♂ and 50 ♀)
Application:	dermal (shaved dorsal skin)
Concentration:	25 µl or 100 µl; 1 % acrylic acid solution in acetone (about 0.25 mg or 1 mg acrylic acid/day; 8 or 33 mg/kg body weight for a mouse with a body weight of 30 g) solvent controls: 25 µl or 100 µl acetone positive controls: 0.1% benzo(a)pyrene
Duration:	21 months (3 times a week)
Toxicity:	no mortality; no effects on the survival rate or body weight gains; no irritation
Tumours:	none

Table 11. continued

Authors:	BAMM 1990, 1991
Substance:	acrylic acid
Species:	mouse, Hsd:(ICR)BR (identical with ICR/Ha) (groups of 50 ♂ and 50 ♀)
Application:	dermal (shaved dorsal skin)
Concentration:	25 µl or 100 µl; 1 % acrylic acid solution in acetone (about 0.25 or 1 mg acrylic acid/day; 8 or 33 mg/kg body weight for a mouse with a body weight of 30 g) solvent controls: 25 µl or 100 µl acetone positive controls: 0.1 % benzo(a)pyrene
Duration:	21 months (3 times a week)
Toxicity:	no mortality; no effects on the survival rate or body weight gains; no irritation
Tumours:	none

Subcutaneous injection

Of 30 female Hsd:(ICR)BR mice which were given subcutaneous injections of 1.4 mg (20 µmol) acrylic acid in 0.05 µl trioctanoin once a week for 52 weeks and which were observed for a further 93 days, 2 developed sarcomas at the application site. The first tumour was detected after 323 days. No tumours were found in the untreated control animals and in those treated with trioctanoin. For comparison, the tumour incidence was 17/30 using the unequivocally carcinogenic β-propiolactone (4 µmol) at the same application frequency with the same strain of mice (Segal *et al.* 1987). In laboratory rodents, sarcomas can be induced also after repeated subcutaneous application of large amounts of glucose, distilled water and sodium chloride (Grasso 1987). The incidence of sarcomas observed with acrylic acid may, therefore, be due to a non-specific effect.

6 Manifesto (MAK value, classification)

Irritation of the skin, eyes and nose is the critical effect of acrylic acid.

Positive findings in *in vitro* chromosome aberration tests and mouse lymphoma tests indicate clastogenic effects in cultivated mammalian cells. No clastogenic effects were detected in *in vivo* chromosome aberration tests, but no reliable statement can be made about the extent to which the target cells were exposed. A dominant lethal test also yielded a negative result.

An inhalation study of the carcinogenicity of acrylic acid itself is not available, but studies with methyl acrylate, ethyl acrylate and butyl acrylate yielded negative results. Since the ester is rapidly cleaved in all tissues (also in the nasal mucosa), it can be assumed that acrylic acid is not carcinogenic after exposure by inhalation. A GLP drinking water study in rats, in which the MTD was not reached, provided no evidence of carcinogenicity. A carcinogenicity study which yielded positive results after administration of ethyl acrylate by gavage was described and assessed in the 1985 MAK documentation. The relevance of the tumours observed in the forestomach was qualified because of the oral administration and the local irritation of the target organ (see "Ethyl acrylate", Volume 6, present series). Well documented dermal carcinogenicity studies with mice did not produce either local or systemic tumours. Another dermal carcinogenicity study (Cote *et al.* 1986a, 1986b) with positive results and inadequate documentation has been published only as an abstract.

Therefore, when all data are considered, there does not seem to be an adequate basis for classifying acrylic acid as a carcinogen or tumour promoter.

Since there are no reports of effects at the workplace, a threshold value cannot be derived. In a 90-day study, a NOAEC of 25 ml/m^3 was obtained for rats after 6-hour inhalation per day, whereas slight damage to the olfactory epithelium was observed in mice even at 5 ml/m^3. It can be estimated (see Section 3) that the exposure level of the olfactory epithelium in humans at an increased respiratory minute volume (about 21 l/minute = 10 m^3/8 hours) is not higher than that in rats (resting tidal volume) and is lower than that in mice. Since irritation of the olfactory epithelium does not substantially increase with time and since interindividual differences are probably slight as acrylic acid does not have to be metabolized, the MAK value for acrylic acid has been established at 10 ml/m^3. Because irritation is the critical effect, acrylic acid has been classified in Peak Limitation Category I, and since there are no studies to establish an excursion factor, a basic excursion factor of 1 has been fixed.

Prenatal inhalation toxicity studies with rats and rabbits provided no evidence of developmental toxicity after concentrations of up to about 200 ml/m^3. Acrylic acid is therefore classified in Pregnancy Risk Group C.

Pure acrylic acid was found to have no sensitizing potential in animal studies. Skin reactions in animals were very probably induced by a sensitizing impurity, α,β-diacryloxypropionic acid. The findings obtained in humans do not show acrylic acid to have a substantial contact sensitizing potential. There is no evidence of sensitization of the respiratory tract. Acrylic acid is therefore not designated with either an "Sa" or with an "Sh".

Ready dermal penetration of about 26 % in rats and 12 % in mice was detected when a non-irritant concentration of 1 % was applied (Black *et al.* 1995). The dermal LD_{50} in rabbits is 295 to 1000 mg/kg body weight, but is of only limited suitability for assessing dermal penetration because of the corrosive effect. When handling non-irritant concentrations (1 %) and taking into account that 25 % of the dose is absorbed, 16000 mg/kg body weight of a 1 % acrylic acid solution corresponds to the systemic oral NOAEL of 40 mg/kg body weight in rats obtained from the one-year study (BASF AG 1987b). This means a tolerable exposure to 1120 g of a 1 % acrylic acid solution for a man with a body weight of 70 kg. This level is so high that systemic toxicity from dermal exposure to acrylic acid is not to be expected and the substance is not designated with an "H".

No data are available which would justify classification in one of the categories for germ cell mutagenicity.

7 References

Andersen M, Sarangapani R, Gentry R, Clewell H, Covington T, Frederick CB (2000) Application of a hybrid CFD-PBPK nasal dosimetry model in an inhalation risk assessment: an example with acrylic acid. *Toxicol Sci 57*: 312–325

BAMM (Basic Acrylic Monomer Manufacturers) (1988a) *CHO/HGPRT mutation assay, acrylic acid.* Microbiological Associates Inc, Laboratory Study No. T5372.332, BAMM, Washington, DC, USA, unpublished report

BAMM (1988b) *Test for chemical induction of unscheduled DNA synthesis in primary cultures of rat hepatocytes (by autoradiography).* Microbiological Associates Inc, Laboratory Study No. T5372.380, BAMM, Washington, DC, USA, unpublished report

BAMM (1990) *Chronic dermal oncogenicity study with acrylic acid in [C3H/HeN Hsd BR] and [Hsd: (ICR)BR] mice.* Project Report 52–619, 05.12.1990, NTIS/OTS0510541-3, DOC ID 8EHQ-0191-0592S, NTIS, Springfield, VA, USA

BAMM (1991) *Report – Histopathological review of tissues for hematopoietic neoplasms from a chronic dermal oncogenicity study with acrylic acid in female [C3H/HeN Hsd BR] mice.* NTIS/OTS0510541-2, NTIS, Springfield, VA, USA

BAMM (2001) *Single dose inhalation study of ethylacrylate and acrylic acid in nonhuman primates.* Histopathology report, JE Harkema, Michigan State University, unpublished report

Barrow CS (1984) *Species differences in toxicology of the nasal passages: acrylic acid and dimethylamine.* CIIT Act 4: 1–5

Barrow CS (1986) Quantitation of nasal "dose" with formaldehyde, acrylic acid, and dimethylamine. In: Barrow CS (Ed.) *Toxicology of the nasal passages*, Hemisphere Publishing Corporation, Washington, 113–122

BASF AG (1958) *Bericht über die biologische Prüfung der reinen und rohen Acrylsäure* (Report on the biological study of pure and crude acrylic acid) (German). Report No. VII/365-366, 21.07.1958, BASF AG, Ludwigshafen, unpublished

BASF AG (1977) *Ames-Test an den Substanzen Acrylsäure, Methylacrylat, Butylacrylat* (Ames test on the substances acrylic acid, methyl acrylate and butyl acrylate) (German). Mainz University, undated, unpublished report

BASF AG (1979a) *Bericht über die Prüfung der akuten Inhalationsgefahr (akutes Inhalationsrisiko) von Acrylsäure rein an Sprague-Dawley-Ratten* (Report on the study of the acute inhalation hazard of pure acrylic acid in Sprague-Dawley rats) (German). 22.05.1979, BASF AG, Ludwigshafen, unpublished report

BASF AG (1979b) *Bericht über die Prüfung der akuten dermalen Toxizität von Acrylsäure rein an der Rückenhaut weißer Kaninchen (78/520)* (Report on the study of the acute dermal toxicity of pure acrylic acid to the dorsal skin of white rabbits (78/520)) (German). 08.01.1979, BASF AG, Ludwigshafen, unpublished report

BASF AG (1980) *Bestimmung der akuten Inhalationstoxizität LC_{50} von Acrylsäure rein als Dampf bei 4stündiger Exposition (79/62)* (Determination of the acute inhalation toxicity LC_{50} of pure acrylic acid as a vapour after 4-hour inhalation (79/62)) (German). 31.07.1980, BASF AG, Ludwigshafen, unpublished report

BASF AG (1983) *Prenatal toxicity of acrylic acid after inhalation in Sprague-Dawley rats.* Department of Toxicology, Project No. 37R0386/8017, BASF AG, Ludwigshafen, unpublished report

BASF AG (1987a) *Bericht über die Prüfung der Toxizität von Acrylsäure an Ratten nach 3monatiger Gabe per Schlundsonde* (Report on the study of the toxicity of acrylic acid in rats after 3-month administration by gavage) (German). Department of Toxicology, Project No. 35C0380/8250, BASF AG, Ludwigshafen, unpublished report

BASF AG (1987b) *Bericht über die Prüfung der Toxizität von Acrylsäure an Ratten bei 12monatiger Gabe über das Trinkwasser* (Report on the study of the toxicity of acrylic acid in rats after 12-month administration via the drinking water) (German). Department of Toxicology, Project No. 74C0380/8239, BASF AG, Ludwigshafen, unpublished report

BASF AG (1989) *Study of a potential carcinogenic effect of acrylic acid in rats after long-term administration in the drinking water.* Department of Toxicology, Project No. 72C0380/8240, BASF AG, Ludwigshafen, unpublished report

BASF AG (1993) Occupational Medical Service, communication, 15.02.1993, BASF AG, Ludwigshafen, unpublished

BASF AG (1994) Occupational Medical Service, communication, 02.09.94, BASF AG, Ludwigshafen, unpublished

BASF AG (1996) Occupational Medical Service, communication, 16.02.96, BASF AG, Ludwigshafen, unpublished

BASF AG (1998) *Report – acrylic acid glacial: acute dermal irritation/corrosion in rabbits.* Department of Toxicology, Laboratory Project Identification 18H0007/982003, BASF AG, Ludwigshafen, unpublished report

BASF AG (1999) Occupational Medical and Health Protection Department, communication, 02.12.99, BASF AG, Ludwigshafen, unpublished

deBethizy JD, Udinsky JR, Scribner HE, Frederick CB (1987) The disposition and metabolism of acrylic acid and ethyl acrylate in male Sprague-Dawley rats. *Fundam Appl Toxicol 8*: 549–561

Black KA, Finch L (1995) Acrylic acid oxidation and tissue-to-blood partition coefficients in rat tissues. *Toxicol Lett 78*: 73–78

Black KA, Finch L, Frederick CB (1993) Metabolism of acrylic acid to carbon dioxide in mouse tissues. *Fundam Appl Toxicol 21*: 97–104

Black KA, Beskitt JL, Finch L, Tallant MJ, Udinsky JR, Frantz SW (1995) Disposition and metabolism of acrylic acid in C3H mice and Fischer 344 rats after oral or cutaneous administration. *J Toxicol Environ Health 45*: 291–311

Cameron TP, Rogers-Back AM, Lawlor TE, Harabell JW, Seifried HE, Dunkel VC (1991) Genotoxicity of multifunctional acrylates in the *Salmonella*/mammalian-microsome assay and mouse lymphoma TK+/– assay. *Environ Mol Mutagen 17*: 264–271

Carpenter CP, Weil CS, Smyth HF (1974) Range-finding toxicity data: list VIII. *Toxicol Appl Pharmacol 28*: 313–319

Celanese Corp (1986a) *Cytogenicity study – Chinese hamster ovary (CHO) cells in vitro*, test article CJP-60. Microbiological Associates, Study No. T4901.338, NTIS/OTS0000367-4, NTIS, Springfield, VA, USA

Celanese Corp (1986b) *Cytogenicity study – rat bone marrow in vivo*, test article CJP-60. Microbiological Associates, Study No. T4901.106, NTIS/OTS0000367–4, NTIS, Springfield, VA, USA

Celanese Corp (1986c) *Evaluation of acrylic acid mouse skin tumor bioassay and DNA adduct study performed at the New York University medical center, New York, New York*, Arthur D Little, Inc, NTIS/OTS 0510540, NTIS, Springfield, VA, USA

Condé-Salazar L, Guimaraens D, Romero LV (1988) Occupational allergic contact dermatitis from anaerobic acrylic sealants. *Contact Dermatitis 18*: 129–132

Cote I, Hochwalt AE, Seidman I, Budzilovich GN, Solomon JJ, Segal A (1986a) Acrylic acid: skin carcinogenesis in ICR/HA mice. *Toxicologist 6*: 235

Cote I, Hochwalt AE, Seidman I, Budzilovich GN, Solomon JJ, Segal A (1986b) *Acrylic acid: skin carcinogenesis in ICR/HA mice*, Poster, SOT meeting, 1986

Daecke C, Schaller S, Schaller J, Goos M (1993) Contact urticaria from acrylic acid in Fixomull® tape. *Contact Dermatitis 29*: 216–217

DePass LR, Woodside MD, Garman RH, Weil CS (1983) Subchronic and reproductive toxicology studies on acrylic acid in the drinking water of the rat. *Drug Chem Toxicol 6*: 1–20

DePass LR, Fowler EH, Meckley DR, Weil CS (1984) Dermal oncogenicity bioassays of acrylic acid, ethyl acrylate, and butyl acrylate. *J Toxicol Environ Health 14*: 115–120

Dow Chemical Co (1992) *A comparison of single-dose oral LD50's for SPB: (Sprague-Dawley) rats and CD F (Fischer 344-derived) rats.* NTIS/OTS 0537283, Old Doc ID 8EHQ-0592-3831, New Doc ID 88-920002473, NTIS, Springfield, VA, USA

D'Souza WR, Francis WR (1988) Vehicle and pH effects on the dermal penetration of acrylic acid: *in vitro-in vivo* correlations. *Toxicologist 8*: 209

EPA (Environmental Protection Agency) (1988) *Reference physiological parameters in pharmaco-kinetic modeling*. EPA/600/6-88/004, US Environmental Protection Agency, Washington, DC, USA

EU (European Union) (2002) *Risk assessment report, acrylic acid*, 1st priority list, Volume 28, Office for Official Publications of the European Communities, Luxemburg

Finch L, Frederick CB (1992) Rate and route of oxidation of acrylic acid to carbon dioxide in rat liver. *Fundam Appl Toxicol 19*: 498–504

Fowler JF (1990) Immediate contact hypersensitivity to acrylic acid. *Dermatol Clin 8*: 193–195

Frederick CB, Bush ML, Lomax LG, Black KA, Finch L, Kimbell JS, Morgan KT, Subramaniam RP, Morris JB, Ultman JS (1998) Application of a hybrid computational fluid dynamics and physiologically based inhalation model for interspecies dosimetry extrapolation of acidic vapors in the upper airways. *Toxicol Appl Pharmacol 152*: 211–231

Frederick CB, Gentry PR, Bush ML, Lomax LG, Black KA, Finch L, Kimbell JS, Morgan KT, Subramaniam RP, Morris JB, Ultman JS (2001) A hybrid computational fluid dynamics and physiologically based pharmacokinetic model for comparison of predicted tissue concentrations of acrylic acid and other vapors in the rat and human nasal cavities following inhalation exposure. *Inhal Toxicol 13*: 359–376

Frith CH, Wiley LD (1981) Morphologic classification and correlation of incidence of hyperplastic and neoplastic hematopoietic lesions in mice with age. *J Gerontol 36*: 534–545

Galloway SM (2000) Cytotoxicity and chromosome aberrations *in vitro*: experience in industry and the case for an upper limit on toxicity in the aberration assay. *Environ Mol Mutagen 35*: 191–201

Grasso P (1987) Persistent organ damage and cancer production in rats and mice. *Arch Toxicol, Suppl 11*: 75–83

Grimalt F, Romaguera C (1975) New resin allergens in shoe contact dermatitis. *Contact Dermatitis 1*: 169–174

Hellman TM, Small FH (1974) Characterization of the odor properties of 101 petrochemicals using sensory methods. *J Air Pollut Control Assoc 24*: 179–182

Hellwig J, Deckardt K, Freisberg KO (1993) Subchronic and chronic studies of the effects of oral administration of acrylic acid to rats. *Food Chem Toxicol 31*: 1–18

Hellwig J, Gembardt C, Murphy SR (1997) Acrylic acid: two-generation reproduction toxicity study in Wistar rats with continuous administration in the drinking water. *Food Chem Toxicol 35*: 859–868

Hoechst Celanese Corp (1992) *Potassium acrylate – 60% neutralized solution*. Project No. 3859, NTIS/OTS0536927, Old Doc ID 8EHQ-0792-5862, New Doc ID 88-920004507, NTIS, Springfield, VA, USA

Inter-Company Acrylate Study Group (1980a) *Acrylic acid – subchronic toxicity study: inclusion in the drinking water of rats for three months*. Union Carbide, Bushy Run Research Center, Project Report 43-529, 30.04.1980, Inter-Company Acrylate Study Group, unpublished

Inter-Company Acrylate Study Group (1980b) *Acrylic acid – inclusion in the drinking water of rats for one generation of reproduction*. Union Carbide, Bushy Run Research Center, Project Report 43-528, NTIS/OTS 0534940, NTIS, Springfield, VA, USA

Inter-Company Acrylate Study Group (1982) *Acrylic acid – Lifetime dermal carcinogenicity study in male C3H/HeJ mice*. Union Carbide, Bushy Run Research Center, Project-Report 45-512, NTIS/OTS0520806, NTIS, Springfield, VA, USA

Ishidate M (Ed.) (1979) *Chromosomal aberration test in vitro*, Realize Inc, Tokyo, 9

Ishidate M, Harnois MC, Sofuni T (1988) A comparative analysis of data on the clastogenicity of 951 chemical substances tested in mammalian cell cultures. *Mutat Res 195*: 151–213

Keyhani K, Scherer PW, Mozell MM (1995) Numerical simulation of airflow in the human nasal cavity. *J Biomech Eng 117*: 429–441

Kimbell JS, Godo MN, Gross EA, Joyner DR, Richardson RB, Morgan KT (1997) Computer simulation of inspiratory airflow in all regions of the F344 rat nasal passages. *Toxicol Appl Pharmacol 145*: 388–398

Klimisch H-J, Hellwig J (1991) The prenatal inhalation toxicity of acrylic acid in rats. *Fundam Appl Toxicol 16*: 656–666

Kutzman RS, Meyer G-J, Wolf AP (1982) The biodistribution and metabolic fate of [^{11}C]acrylic acid in the rat after acute inhalation exposure or stomach intubation. *J Toxicol Environ Health 10*: 969–979

Lijinski W, Andrews AW (1980) Mutagenicity of vinyl compounds in *Salmonella typhimurium*. *Teratogen Carcinogen Mutagen 1*: 259–267

Lomax LG, Brown DW, Frederick CB (1994) Regional histopathology of the mouse nasal cavity following two weeks of exposure to acrylic acid for either 6 or 22 hours per day. *Toxicologist 14*: 312

Lovell CR, Rycroft RJG, Williams DMJ, Hamlin JW (1985) Contact dermatitis from the irritancy (immediate and delayed) and allergenicity of hydroxypropyl acrylate. *Contact Dermatitis 12*: 117–118

Majka J, Knobloch K, Sterkiewicz J (1974) [Evaluation of acute and subacute toxicity of acrylic acid] (Polish). *Med Pr 25*: 427–435

McCarthy KL, Thomas WC, Aardema MJ, Seymour JL, Putman DL, Yang LL, Curren RD, Valencia R (1992) Genetic toxicology of acrylic acid. *Food Chem Toxicol 30*: 505–515

McLaughlin JE, Parno J, Garner FM, Clary JJ, Thomas WC, Murphy SR (1995) Comparison of the maximum tolerated dose (MTD) dermal response in three strains of mice following repeated exposure to acrylic acid. *Food Chem Toxicol 33*: 507–513

Miller RR, Ayres JA, Jersey GC, McKenna MJ (1981a) Inhalation toxicity of acrylic acid. *Fundam Appl Toxicol 1*: 271–277

Miller RR, Ayres JA, Rampy LW, McKenna MJ (1981b) Metabolism of acrylate esters in rat tissue homogenates. *Fundam Appl Toxicol 1*: 410–414

Moore MM, Amtower A, Doerr CL, Brock KH, Dearfield KL (1988) Genotoxicity of acrylic acid, methyl acrylate, ethyl acrylate, methyl methacrylate, and ethyl methacrylate in L5178Y mouse lymphoma cells. *Environ Mol Mutagen 11*: 49–63

Moore MM, Doerr CL (1990) Comparison of chromosome aberration frequency and small-colony TK-deficient mutant frequency in L5178Y/TK$^{+/-}$-3.7.2C mouse lymphoma cells. *Mutagenesis 5*: 609–614

Morris JB, Frederick CB (1995) Upper respiratory tract uptake of acrylate esters and acid vapors. *Inhal Toxicol 7*: 557–574

National Cancer Institute (1982) *Mouse lymphoma mutagenesis assay with #46145 (ML-NCI #41)*, Microbiological Associates, Contract N01-CP-15739, 20.05.1982, unpublished report

Neeper-Bradley T, Fowler E, Pritts I, Tyler T (1997) Developmental toxicity study of inhaled acrylic acid in New Zealand White rabbits. *Food Chem Toxicol 35*: 869–880

Parker D, Turk JL (1983) Contact sensitivity to acrylate compounds in guinea pigs. *Contact Dermatitis 9*: 55–60

Rao KS, Betso JE, Olson KJ (1981) A collection of guinea pig sensitization test results – grouped by chemical class. *Drug Chem Toxicol 4*: 331–351

Rohm and Haas Company (1985) *Pharmacokinetic and metabolism studies of acrylic acid and ethyl acrylate in young adult male Sprague-Dawley rats.* Toxicology Department; Report No. 84R-154, unpublished

Rohm and Haas Company (1988) *Acrylic acid and ethyl acrylate: range-finding i.v. kinetic study in male mice.* NTIS/OTS 0520855, New Doc ID 86-8900013556, NTIS, Springfield, VA, USA

Rohm and Haas Company (1994) *Acrylic acid: regional histopathology of the mouse nasal cavity following 15 days of exposure to acrylic acid either 6 or 22 hours per day.* Toxicology Department, Report No. 93R-199, unpublished study

Romaguera C, Grimalt F, Vilaplana J (1985) Methyl methacrylate prosthesis dermatitis. *Contact Dermatitis 12*: 172–183

Saillenfait AM, Bonnet P, Gallissot F, Protois JC, Peltier A, Fabriès JF (1999) Relative developmental toxicities of acrylates in rats following inhalation exposure. *Toxicol Sci 48*: 240–254

Segal A, Fedyk J, Melchionne S, Seidman I (1987) The isolation and characterization of 2-carboxyethyl adducts following *in vitro* reaction of acrylic acid with calf thymus DNA and bioassay of acrylic acid in female Hsd:(ICR)Br mice. *Chem Biol Interact 61*: 189–197

Silver EH, Murphy SD (1981) Potentiation of acrylate ester toxicity by prior treatment with the carboxylesterase inhibitor triorthotolyl phosphate (TOTP). *Toxicol Appl Pharmacol 57*: 208–219

Smyth HF, Carpenter CP, Weil CS, Pozzani UC, Striegel JA (1962) Range-finding toxicity data: list VI. *Am Ind Hyg Assoc J 23*: 95–107

Stott WT, McKenna MJ (1985) Hydrolysis of several glycol ether acetates and acrylate esters by nasal mucosal carboxylesterase *in vitro*. *Fundam Appl Toxicol 5*: 399–404

Swenberg JA, Gross EA, Randall HW (1987) Localization and quantitation of cell proliferation following exposure to nasal irritants. In: Barrow CS (Ed.) *Toxicology of the nasal passages*, Hemisphere Publishing Corporation, Washington, 291–300

van Thriel C, Schäper M, Kiesswetter E, Kleinbeck S, Juran S, Blaszkewicz M, Fricke H-H, Altmann L, Berresheim H, Brüning T (2006). From chemosensory thresholds to whole body exposures – experimental approaches evaluating chemosensory effects of chemicals. *Int Arch Occup Environ Health 79*: 308–321

Waegemaekers THJM, van der Walle HB (1984) alpha,beta-Diacryloxypropionic acid, a sensitizing impurity in commercial acrylic acid. *Dermatosen Beruf Umwelt 32*: 55–58

van der Walle HB, Klecak G, Geleick H, Bensink T (1982) Sensitizing potential of 14 mono (meth) acrylates in the guinea pig. *Contact Dermatitis 8*: 223–235

Wiegand HJ, Schiffmann D, Henschler D (1989) Non-genotoxicity of acrylic acid and *n*-butyl acrylate in a mammalian cell system (SHE cells). *Arch Toxicol 63*: 250–251

Winter SM, Sipes IG (1993) The disposition of acrylic acid in the male Sprague-Dawley rat following oral or topical administration. *Food Chem Toxicol 31*: 615–621

Winter SM, Weber GL, Gooley PR, MacKenzie NE, Sipes IG (1992) Identification and comparison of the urinary metabolites of [1,2,3-$^{13}C_3$]acrylic acid and [1,2,3-$^{13}C_3$]propionic acid in the rat by homonuclear ^{13}C nuclear magnetic resonance spectroscopy. *Drug Metab Dispos 20*: 665–672

Zeiger E, Anderson B, Haworth S, Lawlor T, Mortelmans K, Speck W (1987) *Salmonella* mutagenicity tests: III. Results from the testing of 255 chemicals. *Environ Mutagen 9*: 1–110

Ziegler TL, Winter SM, MacKenzie NE, Sipes IG (1992) Metabolism of acrylic acid in rat liver homogenates. *Toxicologist 12*: 61

completed 09.03.2005

Aniline[1]

Supplement 2002

MAK value (1981)	**2 ml/m³ (ppm) ≙ 7.7 mg/m³**
Peak limitation (2002)	**Category II, excursion factor 2**
Absorption through the skin (1981)	**H**
Sensitization	**–**
Carcinogenicity (1989)	**Category 3B**
Prenatal toxicity (1990)	**Pregnancy Risk Group D**
Germ cell mutagenicity	**–**
BAT value (1985)	**1 mg aniline (free)/l urine 100 µg aniline (released from haemoglobin conjugate)/l blood**
Synonyms	aminobenzene C. I. 76000 phenylamine
Chemical name (CAS)	benzenamine
CAS number	62-53-3

Peak Limitation Category

In addition to central nervous stimulation, the critical effects are the formation of methaemoglobin and Heinz bodies and the sequelae hypoxaemia and haemolytic anaemia. Methaemoglobin formation is ascribed to the formation of the metabolites phenylhydroxylamine and aminophenol. The higher sensitivity of humans compared with that in animals is associated with the increased formation of phenylhydroxylamine and its further metabolism. A single oral dose of 45 mg is considered to be the critical limit at which a healthy person shows no symptoms of methaemoglobinaemia (Jenkins *et al.* 1972). A total intake of 92 mg was calculated for 8-hour exposure to aniline at the

[1] A more recent supplement follows

The MAK-Collection Part I: MAK Value Documentations, Vol. 26. DFG, Deutsche Forschungsgemeinschaft

ISBN: 978-3-527-32306-7

level of the MAK value ("Aniline," Volume 6, present series). The half-life for aniline in humans is 3.5 hours (Piotrowski 1972).

Aniline has moderately strong irritant effects on the skin and in the eyes of rabbits (Czajkowska *et al.* 1977).

Aniline is classified in Peak Limitation Category II because of its predominantly systemic effects.

As the half-life for aniline is 3.5 hours, the critical concentration for methaemoglobin formation would not be reached even with an excursion factor of 8. However, as the relevance of methaemoglobin formation for a carcinogenic effect has not been clarified, an excursion factor of 2 is established.

References

Czajkowska T, Krysiak B, Stetkiewicz J (1977) [Comparative evaluation of the acute and subacute toxic action of aniline and *o*-isopropoxyaniline] (Polish). *Med Pracy 28*: 157–174

Jenkins FP, Robinson JA, Gellatly JB, Salmond GW (1972) The no-effect dose of aniline in human subjects and a comparison of aniline toxicity in man and the rat. *Food Cosmet Toxicol 10*: 671–679

Piotrowski J (1972) Certain problems of exposure tests for aromatic compounds. *Pracov Lék 24*: 94–97

completed 28.02.2002

Aniline

Supplement 2007

MAK value (1981)	**2 ml/m^3 (ppm) ≙ 7.7 mg/m^3**
Peak limitation (2002)	**Category II, excursion factor 2**
Absorption through the skin (1981)	**H**
Sensitization (2006)	**Sh**
Carcinogenicity (2006)	**Category 4**
Prenatal toxicity (2006)	**Pregnancy Risk Group C**
Germ cell mutagenicity	**–**
BAT value (1985)	**1 mg aniline (free)/l urine** **100 µg aniline (released from haemoglobin conjugate)/l blood**
Chemical name (CAS)	benzenamine
CAS number	62-53-3

Since the MAK documentation from 1992 ("Aniline", Volume 6, present series), numerous studies of the toxicity of aniline after inhalation and ingestion, its genotoxicity and in particular the mechanism of splenic toxicity in rats have been carried out; the substance must therefore be re-evaluated.

1 Toxic Effects and Mode of Action

Damage to the haematopoietic system was detected in animal studies as a result of the toxic effects on the erythrocytes caused by the methaemoglobin-forming activity of aniline. The spleen in particular can be damaged by the increased degradation of damaged erythrocytes and the resulting overload with cell debris, haemoglobin (Hb) and redox-active iron. Another important aspect in animal studies is the increased extramedullary haematopoiesis in the spleen.

As a result of physiological differences, species differences are to be expected for the secondary effects of methaemoglobin (metHb) formation in the spleens of humans and rats; this makes it unlikely that the toxic effects on rat spleens that produce severe extramedullary haematopoiesis and lead to tumours can be quantitatively applied to humans.

Reproductive studies in rats and mice show that aniline-induced effects on the offspring occur only at doses that are markedly higher than those that can be absorbed if the MAK value is observed.

Studies of the genotoxicity of aniline indicate low genotoxic potential for the substance, which is found only in the toxic/erythrotoxic dose range.

2 Mechanism of Action

2.1 Primary effects: erythrocytes

The effects of aniline are manifest primarily in the erythrocytes in the form of the dose-related formation of methaemoglobin. Methaemoglobin formation, mainly induced by the metabolite phenylhydroxylamine (PHA), is reversible via the activity of methaemoglobin reductase (NADH-dependent) and does not necessarily lead to erythrocyte damage. However, massive oxidation of iron can lead to the formation of superoxide anions and thus to intra-erythrocytic oxidative stress (Dickinson and Forman 2002; Grossman *et al.* 1992; Jarolim *et al.* 1990; Jollow and McMillan 2001). In addition, in erythrocytes that are heavily loaded with methaemoglobin, redox-active iron is released, which, as a catalyst of auto-oxidation reactions, contributes via Fenton and Haber-Weiss reactions to the further formation of oxygen radicals (Papanikolaou and Pantopoulos 2005). The redox equilibrium of the erythrocyte is impaired. This can lead to membrane changes via lipid peroxidation (Wilhelm and Herget 1999), disturbance of the reversible interactions between the thiol groups of haemoglobin and glutathione and the binding of haemoglobin to the erythrocyte membrane (Mawatari and Murakami 2004). Irreversible haemichromes, which precipitate and form Heinz bodies, can be formed from methaemoglobin (Jarolim *et al.* 1990). This can lead to destabilization of the membrane skeleton and a change in erythrocyte deformability and surface properties (Dickinson and Forman 2002; Grossman *et al.* 1992; Jarolim *et al.* 1990; Wilhelm and Herget 1999).

There are clear species differences in the methaemoglobin reductase activity, which are responsible for the regeneration of the functioning haem from methaemoglobin. Markedly lower activities were detected in the erythrocytes of dogs than in those of humans (Rockwood *et al.* 2003) or rodents (Srivastava *et al.* 2002). In mice and hamsters, the methaemoglobin reductase activity is much higher than that in rats (Srivastava *et al.* 2002).

2.2 Secondary effects in rats: spleen and haematopoietic system

The secondary effects of erythrocyte toxicity are manifest in rats mainly in the spleen. The spleen, which is involved in haematopoiesis by up to 5% under normal conditions in adult rats, assumes up to 80% of the haematopoietic activity under hypoxic conditions (Seifert and Marks 1985). The bone marrow and liver are also stimulated to form blood, but to a lesser degree. In addition, toxic effects occur in rat spleens that can even lead to the formation of splenic tumours. The causes of splenic toxicity in rats are described below.

Mechanical disturbance

Erythrocytes with reduced elasticity and deformability can pass the splenic sinuses of the red pulp only with difficulty or not at all (Harvey 1989). This leads to an accumulation of cell debris in the splenic vessels of rats that obstructs the splenic sinuses, resulting in swelling, inflammatory reactions, and hyperplastic and fibrotic changes of the splenic stroma (CIIT 1982).

Oxidative stress

Degradation of the damaged erythrocytes and haem leads to a drastic dose and time-related accumulation of iron (Khan *et al.* 1993, 1995a, 1997; Pauluhn 2004; Wu *et al.* 2005) or of free, redox-active iron (Wu *et al.* 2005) in the splenic red pulp of rats, mainly in the macrophages of the splenic reticulum (Khan *et al.* 1993, 1997, 2003a). The redox equilibrium is impaired, resulting in oxidative stress. In addition, the rat spleen tries to compensate the oxygen deficiency in the body via increased haematopoiesis—the iron cycle and cell proliferation are activated, which in turn promote the formation of radicals. The following observations demonstrate this effect cascade:

Lipid peroxidation and the formation of malondialdehyde–protein adducts (Khan *et al.* 1997, 1999a, 1999b; Pauluhn 2004) and protein oxidation (Khan *et al.* 1997, 1999a, 1999b), which is a marker of oxidative stress, increase in rat spleens as a function of dose and time after treatment with aniline.

The iron deposits in the splenic red pulp are found close to the malondialdehyde–protein adducts that are formed (Khan *et al.* 2003a). Malondialdehyde is a by-product of lipid oxidation and binds covalently to proteins (Khan *et al.* 2003a) and DNA (Stone *et al.* 1990) and can lead to damage there. Feed containing iron potentiates the typical effects of aniline on the spleen as described above (Khan *et al.* 1999a).

After oral treatment of rats with aniline for seven days, immunohistochemical examinations revealed the marked formation of tyrosine-nitrated proteins and increased nitric oxide synthase activity in the macrophages and sinusoidal cells of the red pulp (Khan *et al.* 2003b, 2003c), more evidence of oxidative stress caused by nitrogen and oxygen radicals.

The development of fibrosis in rats treated with aniline is closely related to the overloading of the spleen with iron (Khan *et al.* 1993, 1995b, 1997) and the activation of fibrinogenous or inflammatory cytokines and transcription factors in the spleen (Khan *et al.* 2000b; Wang *et al.* 2005).

The interaction of mechanical disturbance, oxidative stress triggered by the activation of extramedullary erythropoiesis, inflammatory reactions and proliferative processes in the spleen can be regarded as the cause of the species-specific splenic tumours induced by aniline in rats.

2.3 Transferability of the secondary effects from rats to humans

As a result of physiological differences, species differences are to be expected for the secondary effects of methaemoglobin formation in the spleens of humans and rats.

1) Unlike in rats, in which about 5% of erythropoiesis takes place in the spleen throughout the animals' lifetime, haematopoiesis stops after birth in the human spleen (Seifert and Marks 1985) and takes place in the spleen only if the function of the bone marrow is impaired because of illness (Freedman and Saunders 1981; Seifert and Marks 1985).

2) Whereas haematopoiesis takes place in rat spleens via the proliferation of local erythroblasts (Pauluhn 2004), in human spleens (patients with osteopetrosis) it takes place via the immigration of erythroid precursors from the blood (Freedman and Saunders 1981) and subsequent proliferation in the splenic sinus.

3) Under hypoxic conditions, extramedullary haematopoiesis is initially activated in rat spleens (Pepelko 1970), whereas no haematopoiesis is detected in human spleens even if there are anaemic conditions, such as sickle-cell anaemia and thalassaemia (Freedman and Saunders 1981).

4) As a result of severe haemolytic anaemia, rat spleens can take on up to 80% of erythropoiesis, with the splenic red blood cells being able to undergo five or six additional cell divisions and those of the bone marrow only two or three (Pauluhn 2004). These physiological differences make it unlikely that the effects on rats can be applied to humans. Instead, the high sensitivity of male rats to the erythrotoxic effects of aniline on the spleen seems to be species and sex-specific (see Section 5.7.1).

Splenic tumours are extremely rare in humans. Despite decades of experience with the substance, no abnormalities have been described in human spleens after exposure to aniline or aniline intoxication; however, no systematic epidemiological studies have been carried out.

Splenic tumours have also not been reported for human conditions accompanied by methaemoglobinaemia (congenital methaemoglobinaemia), haemolysis (sickle cell anaemia and thalassaemia) or the increased incorporation of iron (haemochromatosis).

2.4 Effects of the metabolites

In the spleen, the toxicity spectrum of nitrosobenzene and phenylhydroxylamine, the two *N*-oxidized metabolites of aniline, is similar to that of aniline. Thus, these metabolites are not only mainly responsible for the methaemoglobin-forming effect, but may also contribute to the splenic toxicity of aniline (Khan *et al.* 1998, 2000a).

In vitro incubation with 0.1 mM phenylhydroxylamine for 2 hours led to the formation of methaemoglobin in rat erythrocytes together with an increase in the level of free iron and a decrease in the level of glutathione. Aniline produced no significant effects under the same conditions (Ciccoli *et al.* 1999). The maximum increase in methaemoglobin formation in isolated rat erythrocytes was reached within the first 10 minutes (Singh and Purnell 2005a).

The dose-dependent degradation of rat erythrocytes pre-treated *in vitro* with phenylhydroxylamine was faster than that of untreated erythrocytes when re-injected into rats (Singh and Purnell 2005b).

Phenylhydroxylamine and nitrosobenzene have a toxicological profile comparable to that of aniline. Rats received 0.093% phenylhydroxylamine with the diet (about 70 mg/kg body weight and day) for 13 days. Loss of weight, lower feed consumption and Heinz bodies in the blood were observed in the animals. The spleen was enlarged and erythropoietic activity was increased. Haemosiderin deposits were observed in both the splenic medulla and the splenic follicles, and also in the parenchymal tissue and Kupffer's cells of the liver and in the proximal tubular epithelium of the kidneys (Jenkins *et al.* 1972).

Treatment of groups of 5 male Sprague-Dawley rats with phenylhydroxylamine (Khan *et al.* 1998) and nitrosobenzene (Khan *et al.* 2000a) with gavage doses of 0.025, 0.05, 0.1 and 0.5 mmol/kg body weight and day for 4 days led to the toxic effects typical of aniline, such as a dose-dependent increase in methaemoglobin, an enlarged spleen and increased spleen weights, iron deposits, congestion and oxidative stress in the spleen. Doses of 0.5 and 1 mmol/kg body weight with the same treatment pattern were required for aniline hydrochloride to induce lipid peroxidation, iron accumulation or protein oxidation (Khan *et al.* 1997), whereas phenylhydroxylamine and nitrosobenzene caused these effects even at 10 to 20 times lower doses (Khan *et al.* 1998, 2000a). In the phenylhydroxylamine study, 3 of 4 animals of the high dose group had necrotic changes in the heart, which were assumed to be caused by severe oxygen deficiency induced by phenylhydroxylamine (Khan *et al.* 1998). Aniline did not lead to such cardiotoxic effects (see Section 5.2). After administration of *p*-aminophenol (0.093%) to rats with the diet, there were no organ changes or any other general toxic effects (Jenkins *et al.* 1972).

3 Toxicokinetics and Metabolism

The toxicokinetics and metabolism of aniline were described in detail in the 1992 MAK documentation ("Aniline", Volume 6, present series). The results are summarized briefly here and supplemented by recent studies.

In humans and in the investigated animal species, aniline is rapidly absorbed, distributed to the tissues and after metabolism eliminated mainly via the kidneys.

Absorption after ingestion is 89% to 96% in rats and 72% in mice (1992 MAK documentation). Aniline is absorbed dermally from the liquid and gaseous phases (1992 MAK documentation). Absorption from the gaseous phase was confirmed in an inhalation study with dogs (Pauluhn 2005). After identical external exposure to about 240 mg/m^3, the methaemoglobin level was about three times higher in dogs after whole-body exposure than in those after head-only exposure.

After single oral doses of radioactively labelled aniline, the highest aniline concentration was found in the erythrocytes, followed by the plasma, spleen, kidneys, lungs, heart, brain and fat. Repeated administration of aniline led to accumulation in the spleen, whereas the radioactivity in the other tissues rapidly decreased (Khan *et al.* 1995b).

As a representative of the aromatic amines, aniline is metabolized primarily by ring and *N*-hydroxylation (reaction dependent on cytochrome P450) or *N*-acetylation (via *N*-acetyltransferase). These conversions take place mainly in the liver. The position of ring hydroxylation (ortho, meta or para) seems to vary considerably from species to species (McCarthy *et al.* 1985). The *N*-phenylhydroxylamine formed by *N*-hydroxylation is metabolized to nitrosobenzene in the erythrocytes, while the Fe^{2+} in haemoglobin is converted in a linked reaction to trivalent iron, leading to methaemoglobin formation. Reactive phenylhydroxylamine can in turn be formed from nitrosobenzene in a redox cycle with NADPH. The higher level of exposure to this metabolite limits its own formation from nitrosobenzene because of the consumption of NADPH in the erythrocytes. Part of the nitrosobenzene binds covalently to the SH group of cysteine in globin (Zwirner-Baier *et al.* 2003). The methaemoglobin-forming and haemolytic properties of aniline are attributed mainly to phenylhydroxylamine and nitrosobenzene (Bradshaw *et al.* 1995; Harrison and Jollow 1987; Khan *et al.* 1998, 2000a; Kiese 1974). Other methaemoglobin-forming metabolites, such as aminophenols, are probably of subordinate relevance since they are effectively detoxified via conjugation with sulfate or glucuronic acid (Harrison and Jollow 1987). Aniline itself induced no methaemoglobin formation in isolated rat erythrocytes (Ciccoli *et al.* 1999) and is thus not a direct methaemoglobin inducer.

Acetanilide is formed by *N*-acetylation of aniline and is then hydroxylated to *p*-hydroxyacetanilide. The conjugates of *p*-hydroxyacetanilide with glucuronic acid and sulfate are the main metabolites of aniline in urine (1992 MAK documentation).

Aniline and its metabolites are eliminated relatively rapidly via the urine. The half-life for aniline is 3.5 hours in humans (Piotrowski 1972) and 177 minutes in rats (1 mmol/kg body weight; intraperitoneal) (Kim and Carlson 1986). At lower doses, the main metabolic pathway in rats is via sulfation. This reaction pathway seems to be saturated at higher doses. In mice, conjugation with glucuronic acid is the main route of elimination, irrespective of the dose (McCarthy *et al.* 1985). Therefore, mice can eliminate aniline more effectively than rats and should be less sensitive to long-term effects (1992 MAK documentation). In addition, the haemoglobin binding index (HBI) for aniline is ten times lower in mice than in female rats (Albrecht and Neumann 1985; Birner and Neumann 1988).

Rats and humans react to intoxication with aniline by forming methaemoglobin. Methaemoglobin formation can already be detected shortly after the absorption of aniline. In the blood of rats, the concentration of methaemoglobin reached its peak of

37.2% after only 0.5 hours after single oral aniline hydrochloride doses of 2 mmol/kg body weight (259 mg/kg in water; about ½ LD_{50}) and then continuously decreased to about 3% after 6 to 48 hours (Khan *et al.* 1997). The half-life of methaemoglobin after inhalation is 75 minutes in rats (Kim and Carlson 1986) and 100 minutes in dogs (Pauluhn 2002). The half-life of methaemoglobin in humans after intoxication with nitro/amino compounds with methaemoglobin concentrations of 50% to 72% is specified to be 3 to 10 hours (Anonymous 2001). For aniline, the half-life of methaemoglobin in humans is 3.5 hours (1992 MAK documentation).

4 Effects in Humans

Since the 1992 MAK documentation, no new data have been published about the effects of aniline in humans which would modify the results known to date.

Acute intoxication with aniline is primarily apparent in the formation of methaemoglobin, mainly caused by the aniline metabolite phenylhydroxylamine, which in the erythrocytes induces oxidation of the iron atom in haemoglobin. The physiological methaemoglobin level of 1% (ECB 2004) to 2% (Smith 1986) in the blood of humans and most mammals is low. Up to 5% methaemoglobin is regarded as safe and tolerable on the basis of decades of experience from occupational medicine (Henschler and Lehnert 1986). Methaemoglobin values exceeding these levels lead to an impairment in oxygen transport. Various intra-erythrocytic mechanisms reduce methaemoglobin to haemoglobin, methaemoglobin reductase being responsible for 95% (Anonymous 2001). Since *N*-acetyltransferase activity is low in about 50% of Europeans ("slow acetylators"), the reaction from aniline to acetanilide can be shifted in favour of the formation of phenylhydroxylamine and nitrosobenzene, resulting in increased methaemoglobin formation. In a study by Lewalter and Korallus (1985), 7 of 14 workers who had been industrially exposed to aniline concentrations of less than 2 ml/m^3 (no other details) had methaemoglobin values of 0.9% (0.7%–1.2%), levels of aniline released from haemoglobin conjugates of less than 10 µg/l blood and acetanilide concentrations of 340 µg/g creatinine in the urine at the end of a working day (mean values). Average acetanilide values of 27 µg/g creatinine and average levels of aniline released from haemoglobin conjugates of 123 µg/l blood were detected for the remaining 7 workers with a mean methaemoglobin level of 1.4% (1.0%–1.5%). The parameters additionally determined (*p*-aminophenol, *p*-acetaminophenol and aniline) in the urine of the two groups hardly differed.

Studies with groups of 20 healthy volunteers showed that oral aniline doses of 5 and 15 mg did not lead to increased methaemoglobin formation, whereas a dose of 25 mg caused an increase in the methaemoglobin level of 2.5%. In groups of 5 persons, doses of 35, 45 and 65 mg led to methaemoglobin values that were increased by 3.7% to 7.1%. The highest methaemoglobin level (16%) in one volunteer who had ingested 65 mg aniline was detected 2 hours after administration; one hour later, the level had returned to normal. The formation of Heinz bodies was not observed at any of these doses. A slight increase in serum bilirubin occurred after 45 and 65 mg (Jenkins *et al.* 1972).

Despite decades of experience with aniline, no findings in human spleens have been described after contact or poisoning with the substance; however, no systematic epidemiological studies have been carried out.

Occupational medical monitoring programmes have shown that no adverse effects on health occur when the MAK value or BAT value is observed (Lewalter *et al.* 2002).

Sensitization of the skin

There are several studies available in which aniline was investigated in patch tests with a large number of patients at low test concentrations (Enders 1986; IVDK 2005; Moriearty *et al.* 1978) (Table 1). However, in these studies, as in studies with higher test concentrations (Calas *et al.* 1978; Malten 1969; Rudzki and Kleniewska 1970, 1971; Scarpa and Ferrea 1966; Schultheiss 1959; Zündel 1936), hardly any data exist as to whether the reactions observed were evidence of sensitization after previous exposure to aniline. This also applies to the studies of patients with *ulcus cruris* or dermatitis on the lower leg (Angelini *et al.* 1975; Eberhartinger 1984; Ebner and Lindemayr 1977; Malten *et al.* 1973).

Reactions to aniline occur often in association with cross-reactions in patients with sensitization to aromatic amines substituted in para position (Czarnecki 1977; Düngemann and Borelli 1966; Kleniewska 1975; Meneghini *et al.* 1967; Schulz 1962a, 1962b) (Table 2). In some other, mostly earlier reports and reviews, there is also mention of occasional cross-reactions to aniline in patients with sensitization to *p*-phenylenediamine or 4-aminophenol (Bonnevie 1939; Mayer 1928, 1954; Pambor 1971; Paschoud 1963; Zündel 1936); *o*-aminoazotoluene (reactions to 2% aniline in wool wax alcohol/petrolatum in 4 of 11 persons tested (Foussereau *et al.* 1973)); and Pellidol® (diacetylaminoazotoluene; 2 of 20 patients with reactions to 10% aniline in olive oil (Jirásek *et al.* 1966)); 2 patients with reactions to 5% aniline in petrolatum (Mayer 1929)), but the severity and course of the reactions are not documented in detail. However, 6 patients who had been sensitized to *o*-aminoazotoluene by the patch test produced no reaction to aniline (Foussereau *et al.* 1973). "Clearly positive" reactions to aniline in patients with sensitization to local anaesthetics of the 4-aminobenzoic acid ester type were also reported, but no details given (Flandin *et al.* 1936; Meltzer and Baer 1949). In another study, sensitization to anhydroformaldehyde aniline (hexahydro-1,3,5-triphenyl-1,3,5-triazine; CAS No. 91-78-1) was frequently associated with a reaction to 5% aniline (in wool wax alcohol ointment) and also to *p*-toluidine (Schultheiss 1959). There are no details of the severity of the reactions, however.

Table 1. Reports of patch test reactions to aniline in patients with contact dermatitis or suspected contact allergy

Persons tested	Concentration (vehicle)	Results	Comments	References
113 patients with dermatitis on the lower leg	5% (petrolatum)	10/113 positive (8.9%) (no other details)	test period: 1970–1973; 113 of a total of 306 patients were tested with aniline; reactions to 1% *p*-phenylenediamine in petrolatum in 49 of 306 patients	Angelini *et al.* 1975
139 patients with contact eczema	25% (wool wax alcohol ointment)	0/139 positive (no other details)	test period: not specified; no data for exposure; reaction to aniline in 1 of 7 additional patients with sensitization to *p*-phenylene-diamine or 4-aminophenol	Bonnevie 1939
500 patients	10% (almond oil)	4.9% positive (no other details)	test period: not specified; high test concentration; unclear whether all patients were tested with aniline; reactions to 2% *p*-phenylene-diamine in petrolatum in 6.2% of the persons tested	Calas *et al.* 1978
206 patients with eczema on the hand; 205 patients with eczema on the lower leg	not specified	4/206 positive (1.9%); 6/205 positive (2.9%) (no other details)	test period: 1969–1975; no data for reactions during the test period 1976–1983 or no longer tested with aniline; reactions to *p*-phenylenediamine in 19 of 206 (9.2%) and 10 of 205 (4.9%) patients	Eberhartinger 1984
200 patients with leg ulcers	5% (petrolatum)	8/200 positive (4%) (no other details)	test period: not specified	Ebner and Lindemayr 1977
111 patients	1% (not specified)	0/111 positive	test period: 1977–1983; testing with “printer’s series”	Enders 1986
148 patients	1% (not specified)	0/148 positive	testing with “leather series”	
191 patients	1% (not specified)	1/191 positive (0.5%) (no other details)	testing with “painter’s series”	
1119 patients	1% (petrolatum or wool wax alcohol ointment)	20× 1+, 2× 2+, 3× 3+ (after 72 hours)	test period: 1992–6/2005; some patients also tested with higher concentrations and other vehicles; however, all reactions to 1% aniline in petrolatum or wool wax alcohol ointment	IVDK 2005
1000 patients	5% (olive oil)	1.9% positive (no other details)	test period: not specified	Malten 1969

Table 1. continued

Persons tested	Concentration (vehicle)	Results	Comments	References
100 patients with leg ulcers	0.5% (olive oil)	10/100 positive (10%) (no other details)	test period: not specified; in the 10 patients, also reactions to *p*-phenylenediamine; in addition, reactions to 1% *p*-phenylenediamine in petrolatum in 17 other patients	Malten *et al.* 1973
717 workers from the building sector	not specified	3/717 positive (0.4%)	test period: 1960–1962	Meneghini *et al.* 1963
392 domestic employees		14/392 positive (3.6%)		
242 mechanics		4/242 positive (1.7%)		
196 varnishers and painters		15/196 positive (7.7%)	in 5/196 patients reactions to *p*-phenylenediamine	
129 chemists		10/129 positive (7.8%)	in 2/129 patients reactions to *p*-phenylenediamine	
80 workers in the synthetic resin industry		3/80 positive (3.8%)	in 1/80 patients reaction to *p*-phenylenediamine	
77 workers in the textiles industry		9/77 positive (11.7%)	in 3/77 patients reactions to *p*-phenylenediamine	
70 carpenters and varnishers		1/70 positive (1.4%)	in 1/70 patients reaction to *p*-phenylenediamine	
56 hairdressers and cosmeticians		5/56 positive (8.9%)	in 2/5 patients with reactions to aniline also reactions to *p*-phenylenediamine; in addition, reactions to *p*-phenylenediamine in 26/56 patients	
51 workers in rubber production		3/51 positive (5.9%)		
45 workers in the graphics industry		3/45 positive (6.7%)		
43 workers in leather processing		1/43 positive (2.3%) (no other details)	in 2/43 patients reactions to *p*-phenylenediamine	

Table 1. continued

Persons tested	Concentration (vehicle)	Results	Comments	References
536 patients	2% (paraffin oil)	1/536 positive (0.2%) (no other details)	test period: 1976; reaction in 1/271 men; reactions to 2% *p*-phenylene-diamine in 1.1% of the persons tested	Moriearty *et al.* 1978
70 patients	2% (not specified)	3/70 positive (no other details)	test period: 1970; patients used for control testings; 12 of the patients tested sensitized to *p*-phenylene-diamine	Pambor 1971
93 patients with atopic dermatitis	5% (not specified)	0/93 positive (no other details)	test period: not specified	Rudzki and Grzywa 1975
600 patients	5% (yellow petrolatum)	3.6% positive (no other details)	test period: 9/1967–1/1970; reactions to 2% *p*-phenylenediamine in 15.2% of 1205 patients; probably cohort overlap with Rudzki and Kleniewska (1971)	Rudzki and Kleniewska 1970
1057 patients	5% (petrolatum)	32/1057 positive (3.0%) (no other details)	test period: not specified; reactions to 2% *p*-phenylenediamine in 144/1565 (9.2%) patients; probably cohort overlap with Rudzki and Kleniewska (1971)	Rudzki and Kleniewska 1971
3105 patients	10% (almond oil)	269/3105 positive (8.7%) (no other details)	test period: 1956–1965; in 1956–1959, 1959–1962 and 1962–1965, reactions in 102/781 (13.0%), 94/886 (10.6%) and 73/1438 (5.1%); reactions to *p*-phenylenediamine in these periods: in 49/571 (8.6%), 38/430 (9.9%) and 11/263 (4.1%) persons tested	Scarpa and Ferrea 1966
737 patients	5% (wool wax alcohol ointment)	62/737 positive (8.4%)	testing with aniline in the standard series; application 24 hours; 62 reactions at the least clearly positive, reading after 48 hours; in addition, 11 weak reactions or reactions with decrescendo course and 28 questionable reactions; reactions to 2% *p*-phenylenediamine in 28/737 (3.8%) patients	Schultheiss 1959
1000 patients	not specified	reactions in 3.2%	test period: 8/1970–12/1971; reactions to *p*-phenylenediamine in 5.2%	Schwarz and Gottmann-Lückerath 1982
1000 patients	not specified	reactions in 1.9% (no other details)	test period: 5/1976–1/1979; reactions to *p*-phenylenediamine in 7%; testing with aniline as a component of the standard series	

Table 1. continued

Persons tested	Concentration (vehicle)	Results	Comments	References
1529 patients	10% (not specified)	154/1529 positive (10.1%) (no other details)	test period: 1933–4/1935; no data for severity of reaction; reactions to aniline in 50/242 and 47/146 patients with reactions to 0.5% *p*-phenylenediamine or 10% 4-aminophenol and in 22/70 patients with reactions to 25% Pellidol® (diacetylaminoazo-toluene)	Zündel 1936

Reactions to aniline in patients with sensitization to diaminodiphenylmethane were reported in some studies: one patient with a marked reaction to 10% aniline in olive oil after 72 hours (van Joost *et al.* 1987); reactions to aniline in 5 of 8 patients as well as reactions to *p*-toluidine in all 8 patients; no data for the test concentration (Rudzki *et al.* 1995).

Between 1992 and June 2005, in the clinics of the Information Network of German Departments of Dermatology (IVDK: Informationsverbund Dermatologischer Kliniken) 25 of 1119 patients tested reacted to 1% aniline in petrolatum or anhydrous Eucerin®, including five 2+ or 3+ reactions. A reaction to *p*-phenylenediamine or *p*-aminoazobenzene was observed in 23 of these 25 patients. In one of the two patients who did not react to *p*-phenylenediamine or *p*-aminoazobenzene, a painter with a 1+ reaction to aniline, a simultaneous reaction to *p*-aminodiphenylamine was recorded (IVDK 2005).

A specific test preparation for aniline is not available. The identity and purity of the test preparations used in the listed studies are therefore not known.

Aniline was found to have mildly sensitizing effects on the skin of volunteers in maximization tests. Reactions were observed in 7 of 25 persons after induction with a 20% aniline preparation in petrolatum and challenge with a 10% preparation (Kligman 1966). There are no data for the purity of the preparation used in this study. In addition, it is not clear whether the volunteers experimentally exposed to aniline in this study were previously or simultaneously exposed to other aromatic amines, such as *p*-phenylenediamine.

Sensitization of the respiratory tract

There are no data available for sensitization of the respiratory tract induced by aniline.

Table 2. Reports of patch test reactions to aniline in patients with existing sensitization to *p*-phenylenediamine or similar aromatic (di-)amino compounds

Persons tested	Concentration (vehicle)	Results	Contact/comments	References
74 patients with hairdressers' dermatitis	25% (olive oil)	16/74 positive (no other details)	very high test concentration; also reactions to 1% *p*-phenylenediamine in all 16 patients	Czarnecki 1977
181 patients with sensitization to aromatic amines	1% (not specified)	24/181 positive (13.3%)	also contact with mafenide-sulfanilamide powder/weak to moderate reactions	Düngemann and Borelli 1966
160 patients with initial reactions to *p*-phenylene-diamine	5% (yellow petrolatum)	reactions in 60/160 ("37.3%") (no other details)	no reactions in 40 persons tested with 5% *N*-methylaniline and 40 tested with *N,N*-dimethylaniline; in addition, reactions to 2% *p*-toluidine in 37/58 tested patients sensitized to *p*-phenylenediamine	Kleniewska 1975
30 patients with sensitization to "para substances"	not specified	18/30 positive	at least 2+ reactions in all patients	Meneghini *et al*. 1967
30 patients	1% (wool wax alcohol ointment/olive oil)	15/30 positive	reactions in 3/3 patients "with a broad spectrum of group sensitization to amino and nitro compounds" and in 6/12 patients with primary sensitization to *p*-phenylenediamine and 7/15 patients with primary sensitization to *p*-phenylenediamine or azo dyes; reactions in 0/12 patients to 1% *N,N*-dimethylaniline; allocation of the reactions to the individual cohorts unclear; 2+ to 3+ reactions to aniline in a total of 9/15 cases	Schulz 1962a, 1962b

5 Animal studies

5.1 Acute toxicity

5.1.1 Inhalation

The LC_{50} values for aniline are listed in Table 3.

Table 3. Acute inhalation toxicity of aniline

Species (number)	Type of exposure	Exposure period [hours]	Results	References
rat (10/concentration)	head only	4	LC_{50} 839 ml/m^3	Anonymous 2001
rat (10/concentration)	whole body	4	LC_{50} 479 ml/m^3	Anonymous 2001
rat (6)	whole body	4	250 ml/m^3 mortality 2–4/6	Carpenter *et al.* 1949
rat (6)	whole body	1	545 ml/m^3 mortality 0/6	Bio-Fax 1969
dog (4/concentration)	head only	4	155 or 174 mg/m^3 (40 or 45 ml/m^3) transient cyanotic appearance of mucosa	Pauluhn 2002
dog (4/concentration)	head only	4	$\geq$ 58.6 mg/m^3 (15.2 ml/m^3) transient cyanotic appearance of mucosa	Pauluhn 2005
dog (4)	whole body	4	241 mg/m^3 (62.6 ml/m^3) cyanotic appearance of mucosa lasting for up to 24 hours	Pauluhn 2005

Cyanosis, tremor, weakness, hair loss on the head, reddish-brown discharge from the nose, eyes and mouth, and corneal opacity were observed in the studies with rats (Anonymous 2001).

Two groups of 4 dogs were exposed head only to aniline vapour concentrations of 174 mg/m^3 or 155 mg/m^3 (45.2 and 40.3 ml/m^3) for 4 hours (group 1) or 4 one-hour periods (group 2). The aniline dose absorbed was established via serum aniline adduct formation (group 1) and the respiratory minute volume (group 2). Except for the transient cyanotic appearance of the visible oral mucosa in 4 of 8 dogs on the day of exposure, 7 of 8 dogs tolerated the treatment without any specific clinical symptoms. The methaemoglobin levels of these animals, which were below 0.5% before treatment, increased to a maximum of 3% to 7% as a result of exposure. An absorbed cumulative aniline dose of 12 to 18 mg/kg body weight was established on the basis of the respiratory minute volumes determined in group 2. The half-life of methaemoglobin was 100 minutes (Pauluhn 2002). Groups of 4 dogs were exposed head only to aniline concentrations of 15.8, 30.3, 58.6, 116.1, 243.4 and 493.6 mg/m^3 (4.1, 7.9, 15.2, 30.2, 63.2 and 128.2 ml/m^3) for a period of four hours. A slight cyanotic appearance of the oral mucosa was observed after concentrations of 58.6 and 116.1 mg/m^3, but this had disappeared one hour after the end of exposure. The signs of cyanosis intensified after concentrations of 243.4 mg/m^3. Comparison of the methaemoglobin values before, during and after exposure to 15.8 mg/m^3 revealed a minimal increase in the methaemoglobin level (0.8% compared with 0.5% before exposure). The values were, however, in the range of the methaemoglobin levels in the blood of dogs not exposed to the substance (0.8%). The concentration of 30.3 mg/m^3 resulted in a statistically significant increase in the methaemoglobin levels, but with a proportion of less than 1% they were only of borderline significance. No consistent concentration-dependent or time-dependent changes were found in the other haematological parameters examined. However, an increase in the α-tocopherol concentration was observed in the erythrocytes after concentrations of 243.4 mg/m^3 and above, evidence of intra-erythrocytic oxidative stress. During the treatment, marked, stress-induced hyperventilation (immobilization stress; respiratory minute volume up to 10 l/min) was seen in a few animals of both dog studies (Pauluhn 2002, 2005). The methaemoglobin levels determined in these animals were markedly higher (a maximum of 45%) and persisted longer than in the animals of the same concentration group which were breathing normally (average of 4 l/min) (a maximum of 10%). These results suggest that the severity of methaemoglobinaemia depends more on the rate of dose delivery rather than on the total dose. By means of a benchmark analysis, taking into account the dog study (Pauluhn 2002) and the results of the hyperventilating animals, a threshold was calculated at which aniline led to a methaemoglobin formation of 0.8% in dogs exposed head only. This was found to be 23.6 mg/m^3 (6.1 ml/m^3) after four hours exposure and 20.6 mg/m^3 (5.4 ml/m^3) after eight hours exposure (Pauluhn 2005).

The methaemoglobin reductase activity, responsible for the regeneration of the functioning haemoglobin from methaemoglobin, is lower in dog blood than in human blood (Rockwood *et al.* 2003). The dog can thus be regarded as a species particularly sensitive to methaemoglobinaemia.

5.1.2 Ingestion

Recently published data for the acute oral toxicity of aniline lend further support to the 1992 MAK assessment, in which an LD_{50} of about 440 mg/kg body weight was described for rats. The LD_{50} established in the additional studies was 442 mg/kg body weight (Bio-Fax 1969) to 930 mg/kg body weight (Ethyl Corporation 1980) for male rats and 780 mg/kg body weight (Ethyl Corporation 1980) for female rats. The clinical symptoms described included tremor, convulsion, salivation, deeper or accelerated breathing, cyanosis and lethargy. Autopsy of the animals revealed inflammation and haemorrhage in the gastrointestinal tract, hyperaemia of the lungs and in one study distended bladders.

In a study of methaemoglobin formation in rats, 20 mg/kg body weight was determined to be the highest single oral aniline dose that does not cause relevant methaemoglobin formation in rats. Aniline doses of 40 to 300 mg/kg body weight led to methaemoglobin levels of up to 18%, while a maximum of 48% methaemoglobin was detected after 1000 mg/kg body weight. The maximum values were generally observed one to four hours after the beginning of treatment (Jenkins *et al.* 1972).

The cat is regarded as a species relatively sensitive to aniline intoxication on account of the high levels of methaemoglobin formed in this animal. An oral dose of 0.1 ml/kg body weight (102.2 mg/kg body weight) administered to cats led to severe cyanosis, staggering gait and weakness, and to the death of 1 of 2 animals (BASF AG 1970). In another study, 2 cats had methaemoglobin levels of more than 80% four hours after oral administration of 0.05 ml/kg body weight (51.1 mg/kg body weight). These values returned to normal within two to three days. The death of 1 of the 2 animals on day 4 may be due to acute pneumonia resulting from an administration error (BASF AG 1971). After treatment with 1 µl/kg body weight (1 mg/kg body weight), cyanosis was not found in any of 10 cats. Slight cyanosis was observed one day after administration of 2.5 µl/kg body weight (2.6 mg/kg body weight) (BASF AG 1970). Oral treatment of groups of 2 cats with 10 and 50 mg/kg body weight caused a 60% increase in methaemoglobin levels and a 55% increase in Heinz bodies after 24 hours. Loss of body weight was the only clinical symptom reported after treatment with 50 mg/kg body weight (Bayer AG 1984).

Single oral doses of 15 mg/kg body weight administered to 4 dogs led to a transient cyanotic appearance of the visible oral mucosa and to a maximum increase in the methaemoglobin levels from below 0.5% to 19% to 29% three hours after treatment (Pauluhn 2002). A comparative investigation with inhalation exposure of the dogs to an almost identical systemically available aniline dose (described in Section 5.1.1) showed that a cumulative inhaled aniline dose of 15 mg/kg body weight, determined via the respiratory minute volume, induces methaemoglobin levels between 3% and 7% in dogs. Compared with this, ingestion of the same dose was 5 to 6 times more potent as regards methaemoglobin formation. Hepatic and intestinal first-pass activation of aniline is assumed to be the cause of this discrepancy (Pauluhn 2002). In addition, the rate of dose delivery of the aniline during inhalation seems to be of great importance for the severity of methaemoglobinaemia (see Section 5.1.1; Pauluhn 2005).

5.1.3 Dermal absorption

The acute dermal toxicity of aniline for rats was specified to be 670 mg/kg body weight (1992 MAK documentation). A dermal LD_{50} of 1540 mg/kg body weight was established in rabbits. Doses of 2150 mg/kg body weight and above led to the death of all treated animals. Hyperactivity, hypersensitivity and salivation were the symptoms of toxicity observed. Severe irritation was seen on the treated skin (Bio-Fax 1969). An LD_{50} of 820 mg/kg body weight was established on the abraded skin (doses not specified) in another study with rabbits (4/dose) (Roudabush *et al.* 1965).

The dermal LD_{50} for guinea pigs after occlusive application was specified to be 1290 mg/kg body weight. The LD_{50} determined in parallel on the abraded skin was 2150 mg/kg body weight (Roudabush *et al.* 1965).

Application of aniline to the skin of cats resulted in a dermal LD_{50} of 254 mg/kg body weight (no other details) (Kondrashov 1969).

A comparative inhalation study in dogs with whole-body exposure to aniline concentrations of 243.1 mg/m^3 (63.1 ml/m^3) and head-only exposure to 241 mg/m^3 (62.6 ml/m^3) demonstrated that dermal absorption contributes greatly to the severity and course of aniline intoxication by inhalation. The methaemoglobin levels of 35% in the dogs' blood induced by aniline were 3.5 times higher and persisted for longer than after head-only exposure (Pauluhn 2005).

5.2 Subacute, subchronic and chronic toxicity

5.2.1 Inhalation

The most important results from the studies of subacute and subchronic inhalation toxicity in rats are listed in Table 4.

Table 4. Effects of aniline (A) and aniline hydrochloride (AH) after repeated inhalation exposure in animal studies

Strain, number of animals, sex	Exposure	Findings	References
rat, Crl:Cd, groups of 16 ♂	**2 weeks** 6 hours/day, 5 days/week 0, 17, 45, 87 ml/m³ (0, 65.9, 174.4, 337 mg/m³) 13-day observation	concentration-dependent: **17 ml/m³ and above**: histology*: effects on spleen (minimal at 17 ml/m³): RES hypertrophy, haemosiderosis (not reversible), haematopoiesis increased **45 ml/m³ and above**: metHb increased, reticulocytes increased, RBC decreased, Hb decreased, Hc decreased, MCV increased, MCHC decreased; urine: bilirubin increased; histology*: spleen: weight increased, congestion; bone marrow: myeloid/erythroid ratio decreased **87 ml/m³**: cyanosis, diarrhoea, lymphocytes increased, PMN decreased, platelets decreased; histology*: liver: haematopoiesis increased	ECB 2004
rat, Wistar, groups of 30 ♂ (interim necropsies of 5 rats per group on days 0, 4, 11, 14)	**2 weeks** 6 hours/day, 5 days/week 0, 9.2, 32.6, 96.7, 274.9 mg/m³ (2.4, 8.5, 25.1, 71.4 ml/m³) 2-week observation	**2.4 ml/m³**: NOAEL **8.5 ml/m³ and above**: extramedullary haematopoiesis and haemosiderosis (at 8.5 ml/m³ marginal) **25.1 ml/m³ and above**: clinical smptoms, body weights decreased, spleen weights increased, metHb increased, RBC decreased, reticulocytes increased, Heinz bodies increased; histology*: iron in the spleen increased, dose-related effects on spleen (minimal at 25.1 ml/m³): RES hypertrophy, haemosiderosis (not reversible), haematopoiesis increased **71.4 ml/m³**: liver: haemosiderosis	Pauluhn 2004

* histopathological examination of selected organs; Hc: haematocrit; MCHC: mean corpuscular Hb concentration; MCV: mean corpuscular volume; metHb: methaemoglobin; NOAEL: no observed adverse effect level; PMN: polymorphonuclear neutrophils; RBC: red blood cells; RES: reticuloendothelial system

Exposure to aniline 8 or 12 hours a day for 4 to 5 days and 3, 6 or 9 hours a day for up to 2 weeks (5 days/week) consistently resulted in a NOAEC (no observed adverse effect concentration) of 10 ml/m^3 (1992 MAK documentation). These results match those obtained by Pauluhn (2004), who observed initial marginal signs of the onset of extramedullary haematopoiesis and haemosiderosis in rats after aniline concentrations of 8.5 ml/m^3 and above for two weeks. However, since there were no significant signs of erythrocyte toxicity or associated splenic sequestration, iron accumulation or lipid peroxidation at this concentration, the NOAEC for these effects was specified to be 8.5 ml/m^3.

Initial histopathological effects were observed in the spleen in the two-week studies after aniline concentrations of 30 ml/m^3 (1992 MAK documentation). Apart from the spleen, also the bone marrow (ECB 2004) and, at high concentrations ($\geq$ 71.4 ml/m^3), the liver (ECB 2004; Pauluhn 2004) were affected.

5.2.2 Ingestion

The most important results from recent studies of repeated oral administration of aniline or aniline hydrochloride in rats and mice are listed in Table 5.

Although the database for the oral toxicity of aniline after repeated administration is large, its fundamental shortcoming is that methaemoglobin formation as a sensitive parameter of aniline intoxication was determined too late or not at all. Since the half-life of methaemoglobin in rats was specified to be about 75 minutes (Kim and Carlson 1986) and 100 minutes in dogs (Pauluhn 2002), a delay in examination may yield a false-low methaemoglobin level (Pauluhn and Mohr 2001). In addition, no NOAEL (no observed adverse effect level) was obtained in the studies with a treatment period of more than one week which included haematological and histopathological examinations. Most of the studies were carried out only with the more sensitive male rats.

Treatment of male SD rats with aniline doses of 23 mg/kg body weight and day for 4 days did not lead to any adverse effects (Khan *et al.* 1997). In a feeding study with male F344 rats given aniline hydrochloride doses of 10, 30 or 100 mg/kg body weight and day for one or four weeks (Mellert *et al.* 2004; Zwirner-Baier *et al.* 2003), the concentrations of the test substance in the diet were adjusted weekly because of stability problems. The actual amount of aniline hydrochloride absorbed was therefore 6, 17 and 57 mg/kg body weight and day, corresponding to 4, 12 and 41 mg/kg body weight and day of aniline. Clinico-chemical parameters were not examined, and only selected organs were weighed and subjected to histopathological examination. No NOEL (no observed effect level) was obtained in this study. The LOEL (lowest observed effect level) of 4 mg/kg body weight and day established in the 28-day study was based on a dose-dependent and linear increase in haemoglobin–aniline adducts in the blood of the animals (0.8, 84.7, 198.4 and 352.4 nmol/g haemoglobin), the occurrence of Heinz bodies and vascular congestion in the spleen. At the next highest dose of 12 mg/kg body weight and day, clear signs of a haematological disorder (decreased erythrocyte count and haemoglobin level, increased incidence of Heinz bodies in all animals, increased thrombocyte count, etc.) and changes in the spleen (weight gain and increased congestion) were observed. These and other effects (e.g. hypochromasia) increased in the highest dose group and are evidence of regenerative haemolytic anaemia induced by methaemoglobin formation. Although the methaemoglobin concentration was slightly increased in the high dose group, the increase was not statistically significant at any of the doses tested. This discrepancy can be explained by the rapid regeneration of methaemoglobin by methaemoglobin reductase. Since blood sampling in the test animals was carried out only after a fasting period of 16 to 20 hours, the regeneration of methaemoglobin had already been completed at the time of blood sampling. Histopathological examination of the spleen revealed slight to moderate multifocal perisplenitis in all animals of the high dose group treated for four weeks. There were no significantly increased haemosiderin deposits in the animals' spleens compared with those of the controls although the high dose animals were severely anaemic and haemosiderin deposits were observed in the livers of these animals. Nor was extramedullary haematopoiesis observed in the spleen. A dose-related increase in aniline–haemoglobin adducts was observed after treatment for one week (0.8, 45.0, 130.7 and 350.3 nmol/g haemoglobin after aniline doses of 0, 4, 12 and 41 mg/kg body weight), whereas prolonged treatment for 28 days did not lead to a further increase in adduct formation (0.8, 84.7, 198.4 and 352.4 nmol/g haemoglobin). This suggests that a saturation process was involved (Mellert *et al.* 2004; Zwirner-Baier *et al.* 2003).

In a 2-year feeding study, aniline hydrochloride doses of 10, 30 or 100 mg/kg body weight and day (corresponding to aniline doses of 7, 22 and 72 mg/kg body weight and day) were administered to groups of 130 male and 130 female F344 rats. Interim necropsies of 10 animals per sex were carried out after 26 and 52 weeks, followed by further interim necropsies of 20 animals per sex in week 78 of the study. The remaining 90 animals per sex and group were treated with the test substance until the end of the study (see also Sections 5.5 and 5.7). In this study, the lowest observed adverse effect level (LOAEL) was about 7 mg/kg body weight and day for the aniline dose, based on initial signs of haemosiderosis and haematopoiesis in the animals' spleens, which increased with the exposure duration, and haematological changes (reduced: erythrocyte count, haemoglobin level and haematocrit; increased: methaemoglobin levels, mean corpuscular volume and reticulocyte count) (CIIT 1982).

Effects on organs other than the spleen are generally induced only after higher aniline doses. Thus, increased bone marrow activity and pigment accumulation in pancreatic lymph nodes of the adrenal gland and liver were observed in the high dose group of the long-term study described above (100 mg/kg body weight corresponding to an aniline dose of 72 mg/kg body weight and day) (CIIT 1982). In the 30-day range-finding study for the 2-year feeding study, pigment deposits were observed in the adrenal cortex at the lethal dose of 1000 mg/kg body weight and day in addition to the known effects on the blood count and spleen (CIIT 1982).

In another long-term study, tubular haemosiderosis was found in the kidneys of the rats treated with 174 mg/kg body weight and day, while Kupffer's cell haemosiderosis was observed in addition in the livers of the high dose group (350 mg/kg body weight and day) (NCI 1978). Haemosiderin deposits in the liver and increased erythropoiesis in the bone marrow of the high dose animals were also detected by Mellert *et al.* (2004).

Table 5. Effects of aniline (A) and aniline hydrochloride (AH) after repeated oral exposure of rats and mice

Strain, number of animals, sex	Exposure	Findings	References
Rat			
SD, groups of 5 ♂	**4 days** 0, 0.25, 0.5, 1, 2 mmol/kg body weight and day (AH in water; aniline doses of about 23, 46, 93, 186 mg/kg body weight and day) gavage	**23 mg/kg body weight**: no adverse effects **46 mg/kg body weight and above**: iron in the spleen increased (72%, 172%, 325%; redox status not determined); lipid peroxidation increased (24%, 32%, 44%); histology*: expansion of the splenic red pulp increased **93 mg/kg body weight and above**: relative and absolute spleen weights increased; Hb and Hc decreased; histology*: spleen: splenic red pulp cellularity increased, cellular fragmentation, congestion, iron deposits (Fe^{3+}) in phagocytes, erythrophagocytosis **186 mg/kg body weight**: RBC decreased, WBC increased	Khan *et al.* 1997
SD, 5 ♂	**7 days** 0, 1 mmol/kg body weight and day (AH in water; aniline doses of about 93 mg/kg body weight and day) gavage	only spleen examined: immunohistochemical analysis of nitrotyrosine, inducible nitrotyrosine and nitrated proteins: mainly red pulp	Khan *et al.* 2003b
F344, 12 ♂ (after necropsy 6 animals for haematology and histology and 6 for cell proliferation, etc.; see text)	**1 week** 0, 10, 30, 100 mg/kg body weight and day in the diet (AH; aniline doses of about 4, 12, 41 mg/kg body weight and day**)	**4 mg/kg body weight and above**: Hb adducts increased**,** Heinz bodies in 2/6 increased **12 mg/kg body weight and above**: absolute and relative spleen weights increased; Hb decreased, Heinz bodies increased, thrombocytes increased, MCHC increased, iron in serum increased (not significant); histology*: spleen: congestion (grade 1) in 6/6 **41 mg/kg body weight**: splenomegaly in 6/6; metHb increased (not significant), RBC and Hc decreased, anisocytosis, hypochromasia, polychromasia, WBC increased, reticulocytes increased, transferrin increased, iron-binding capacity increased; histology*: spleen: vascular congestion (grade 3) in 6/6, focal perisplenitis in 1/6	Mellert *et al.* 2004; Zwirner-Baier *et al.* 2003
Colworth-Wistar, 12 ♂/12 ♀	**13 days** 0, 0.093% (A; about 70 mg/kg body weight and day) diet	**70 mg/kg body weight**: spleen weight increased; spleen: congestion, erythropoiesis, haemosiderosis; liver: slight erythropoiesis	Jenkins *et al.* 1972

Table 5. continued

Strain, number of animals, sex	Exposure	Findings	References
SD, 15 ♂, 5 per each necropsy time	**14 days** 0, 0.7 mmol/kg body weight and day (AH in mineral oil; aniline doses of about 65 mg/kg body weight and day) gavage; necropsy 1, 7 and 28 days after end of exposure	**65 mg/kg body weight**: spleen weight increased (absolute: all time periods; relative: days 1 and 7); metHb increased (day 1), RBC decreased (days 1 and 7), Hb and Hc decreased (day 1), WBC increased (day 1); histology*: spleen: congestion (decreasing from day 1 and reversible), increased iron deposits (haemosiderin) in macrophages/pulp (all time periods), focal fibrosis pronounced in the capsule (day 28)	Khan *et al.* 1995b
SD, 15 ♂, 5 per each necropsy time	**30, 60 or 90 days** 0, 600 mg/l drinking water (AH; aniline doses of about 43 mg/kg body weight and day)	**43 mg/kg body weight**: spleen weight (relative) increased (+56%, 61%, 53%); metHb increased (+89%, 59%, 45%), RBC decreased (all time periods), Hb and Hc decreased (days 30 and 90), MCV and MCH increased (days 60 and 90), WBC increased (day 30); histology*: spleen (all time periods; increase with duration of treatment): expansion of the splenic red pulp in sinusoids, fibroblasts, increased iron deposits (haemosiderin) in macrophages/pulp	Khan *et al.* 1993
SD, 3–6 ♂	**30, 60 or 90 days** 0, 600 mg/l drinking water (AH; aniline doses of about 43 mg/kg body weight and day	only specific parameters examined (metHb, iron in spleen and liver) **43 mg/kg body weight**: relative spleen weights increased, no change in liver weights; metHb increased; free iron in RBC and spleen increased; free iron in liver (only at 60 days)	Ciccoli *et al.* 1999
groups of 10 ♂ (strain not specified)	**28 days** 0, 25, 150, 600 mg/kg diet (about 2.22, 13.5, 48.5 mg/kg body weight and day)	no haematology; no histology **2.22 mg/kg body weight**: NOAEL **13.5 mg/kg body weight and above**: enlarged, black spleen	Bio-Fax 1969
F344, 12 ♂ (after necropsy 6 animals for haematology and histology and 6 for cell proliferation etc.; see text)	**4 weeks** 0, 10, 30, 100 mg/kg body weight and day in the diet (AH; aniline doses of about 4, 12, 41 mg/kg body weight and day**)	**4 mg/kg body weight and above**: Hb adducts increased, Heinz bodies in 4/6 increased; histology*: spleen: congestion (grade 1) in 4/6 **12 mg/kg body weight and above**: absolute and relative spleen weights increased; RBC and Hb decreased, Heinz bodies increased, thrombocytes increased, MCH and MCV increased, hypochromasia, iron in serum increased; histology*: spleen: congestion (grade 2) in 6/6	Mellert *et al.* 2004; Zwirner-Baier *et al.* 2003

Table 5. continued

Strain, number of animals, sex	Exposure	Findings	References
		41 mg/kg body weight: splenomegaly in 6/6; metHb increased (not significant), Hc decreased, polychromasia, MCHC decreased, WBC increased, transferrin increased, iron-binding capacity increased; histology*: spleen: congestion (grade 3) in 6/6, slight to moderate multifocal perisplenitis in 6/6; bone marrow: slight hypercellularity in 6/6; liver: minimal haemosiderin deposits in 6/6; minimal extramedullary haematopoiesis in 1/6	
F344, groups of 10 ♂/10 ♀	**30 days** 0, 30, 100, 300, 1000 mg/kg body weight and day in the diet (AH; aniline doses of about 21, 72, 210 mg/kg body weight and day)	no histology **30 mg/kg body weight and above**: metHb increased, Heinz bodies decreased, reticulocytes increased **100 mg/kg body weight and above**: spleen and liver enlarged and with irregular surface; spleen and kidney discoloured **300 mg/kg body weight and above**: cyanosis **1000 mg/kg body weight**: 10 premature deaths in ♀ on days 24–27, feed consumption and body weight gains reduced, black renal cortex and enlarged pancreatic lymph nodes	CIIT 1977
F344, groups of 5 ♂/5 ♀	**8 weeks** 0%, 0.01%, 0.03%, 0.3%, 1% in the diet (AH; aniline doses of about 5, 15, 151, 504 mg/kg body weight and day; calculation based on relative feed consumption of 7% of body weight) range-finding study	no haematology; no histology **up to 15 mg/kg body weight**: NOAEL **151 mg/kg body weight and above**: spleen: black, granular, enlarged **504 mg/kg body weight**: body weight gains reduced	NCI 1978
F344, groups of 50 ♂/50 ♀ (25 ♂/25 ♀ as controls)	**103 weeks** 0%, 0.3%, 0.6% in the diet (AH; aniline doses of about 174.4, 350.5 mg/kg body weight and day; calculation based on relative feed consumption of 7% of body weight) 4-week observation	no haematology **174.4 mg/kg body weight and above**: histology: spleen: capsular and trabecular fibrosis, fatty metamorphosis, papillary hyperplasia of the capsule; kidney: tubular haemosiderosis **350.5 mg/kg body weight**: Kupffer's cell haemosiderosis	NCI 1978

Table 5. continued

Strain, number of animals, sex	Exposure	Findings	References
F344, 130 ♂/130 ♀ (interim necropsies after 26 and 52 weeks with groups of 10/sex and after 78 weeks with groups of 20/sex)	**104 weeks** 0, 10, 30, 100 mg/kg body weight and day in the diet (AH; aniline doses of about 7, 22, 72 mg/kg body weight and day)	**7 mg/kg body weight and above**: spleen: pigment accumulation (haemosiderosis) and haematopoiesis with increasing severity (from week 52); RBC, Hb, Hc decreased, MCV increased, reticulocytes increased **22 mg/kg body weight and above**: spleen weight increased, dose-dependent increase in extramedullary haematopoiesis (week 26); metHb increased **72 mg/kg body weight**: proportion of animals that died increased; spleen: solid, enlarged, irregular surface; liver weights increased; Heinz bodies and MCH increased; histology: spleen: chronic capsulitis (from week 26), stromal hyperplasia and fibrosis (week 104), lymphoid atrophy (week 104), fatty degeneration, tumour formation; bone marrow: erythroid and myeloid hyperplasia increased; pigment accumulation in pancreatic lymph nodes, adrenals and liver	CIIT 1982
Mouse			
B6C3F1, groups of 5 ♂/5 ♀	**8 weeks** 0%, 0.01%, 0.03%, 0.3%, 1% in the diet (AH; aniline doses of about 10.8, 32.4, 324, 1080 mg/kg body weight and day; calculation based on relative feed consumption of 15% of body weight) range-finding study	no haematology; no histology **up to 32.4 mg/kg body weight**: NOAEL **324 mg/kg body weight and above**: black, granular, enlarged spleen	NCI 1978
$B6C3F_1$, groups of 50 ♂/49–50 ♀	**103 weeks** 0%, 0.6%, 1.2% in the diet (AH; aniline doses of about 733–737, 1510–1560 mg/kg body weight and day) 4-week observation	no haematology **733 mg/kg body weight and above**: ♂: bile duct inflammation **1510 mg/kg body weight**: body weight gains reduced	NCI 1978

Hb: haemoglobin; Hc: haematocrit; MCHC: mean corpuscular Hb concentration; MCH: mean corpuscular Hb level; MCV: mean corpuscular volume; RBC: red blood cells; WBC: white blood cells

* histopathological examination of selected organs; ** calculated on the basis of the limited stability of AH (56.6%) in the diet (6, 17, 56 mg/kg body weight and day); AH concentration was adjusted weekly; aniline level was 72% of the AH level

Inflammation of the bile duct in the livers of the males and reduced body weight gains in all animals of the high dose group were observed in mice given 0.6% or 1.2% aniline hydrochloride with the diet for 103 weeks, corresponding to aniline doses of about 735 and 1540 mg/kg body weight and day (NCI 1978).

In the 8-week range-finding study, the spleen was black and enlarged after aniline hydrochloride concentrations of 0.3% and above (corresponding to aniline doses of about 320 mg/kg body weight and day) (NCI 1978). This may be indirect evidence of toxic effects on the haematopoietic system. Studies of haematological effects, methaemoglobin formation and clinical chemistry in mice are not available. Aniline was not found to be tumourigenic in mice (see Section 5.7).

5.3 Local effects on skin and mucous membranes

5.3.1 Skin

In two studies carried out before the introduction of test guidelines, the application of undiluted aniline to rabbit skin led only to slight erythema (no other details; Bio-Fax 1969; ECB 2004).

Marked irritation with subdermal haemorrhage and severe erythema was reported in an acute dermal toxicity study with aniline on rabbit skin (probably 24-hour exposure) (Bio-Fax 1969).

5.3.2 Eyes

Aniline led to severe, long-lasting damage to the rabbit eye. Severe irritation to the conjunctivae and opacity or corneal lesions were observed in various studies (ECB 2004; Sziza and Podhragyai 1957). A pannus developed in one study eight days after treatment (ECB 2004). Corneal opacity was observed in inhalation studies in rats after exposure to aniline concentrations of up to 896 ml/m^3 for four hours (Anonymous 2001).

5.4 Allergenic effects

5.4.1 Skin

The skin sensitizing potential of aniline was investigated in a study with guinea pigs in three different tests (Goodwin *et al.* 1981). The maximization test according to Magnusson and Kligman produced a reaction in 1 of 10 animals with a 1.5% intracutaneous and 25% epicutaneous induction concentration and a 10% challenge concentration. In the single injection adjuvant test), 10 guinea pigs were given a single intracutaneous injection of a 1.5% aniline solution in 0.9% saline. After 12 to 14 days, the challenge was carried out occlusively (occluded chamber application) with 20% aniline for six hours. Five of ten animals reacted. In a modified Draize test, induction was carried out by means of four simultaneous intracutaneous injections of 2.5% aniline solution above the axillary and inguinal lymph nodes. Intracutaneous challenge treatment was carried out on the opposite shaved flanks 14 days later using a 1% concentration. No sensitization was observed in this test.

In another maximization test, the intradermal and epicutaneous induction treatments were carried out with 0.5% aniline in physiological saline and undiluted aniline, respectively. 90% of the Dunkin-Hartley guinea pigs (number of animals not specified) reacted to the 24-hour occlusive challenge treatment with undiluted aniline at the readings after 48 and 72 hours (Basketter and Scholes 1992).

Negative (Haneke *et al.* 2001) and questionable or weakly positive findings (Basketter and Scholes 1992; Basketter *et al.* 2003) were described for the local lymph node assay. The concentration necessary for tripling lymphocyte proliferation (EC_3 value), given as more than 50% or 89% (in acetone/olive oil), indicates a very small sensitization potential (Basketter *et al.* 2003; Gerberick *et al.* 2005).

In the popliteal lymph node test in mice, aniline, *p*-aminophenol, *N*-acetyl-*p*-aminophenol and nitrobenzene did not lead to cell proliferation, whereas nitrosobenzene and phenylhydroxylamine induced marked and slight cell proliferation, respectively (Wulferink *et al.* 2001).

The results from studies of structure–effect relationships, according to which aniline was identified as a (weak) contact allergen (Cronin and Basketter 1994; Fedorowicz *et al.* 2005; Li *et al.* 2005), are not included in the present assessment.

Nor can incompletely documented findings for the sensitizing effects of 5% aniline in cooking oil in rabbits (Mierzecki and Miklaszewska 1958) be used for assessment.

5.4.2 Respiratory tract

There are no data available for airway sensitization induced by aniline.

5.5 Reproductive and developmental toxicity

5.5.1 Fertility

Multi-generation studies to determine the effects of aniline on the fertility of rodents are not available. Long-term feeding studies in rats and mice, in which the weight, pathology and histopathology of the reproductive organs were determined, can, however, provide evidence of such effects.

In a long-term study, F344 rats were given aniline hydrochloride in the diet for 104 weeks. The daily doses were 0, 10, 30 or 100 mg/kg body weight, corresponding to aniline doses of 7, 22 and 72 mg/kg body weight and day. 130 animals were used per group and sex. Interim necropsies of 10 animals per sex were carried out after 26 and 52 weeks, followed by another interim necropsy of 20 animals per sex in week 78 of the study. The remaining 90 animals per sex and group were treated with the test substance until the end of the study (see also Sections 5.2.2 and 5.7). No substance-induced changes were found in the reproductive organs of the male rats examined (testes, epididymides and prostate). In the females, the ovary weights were slightly increased (not significantly) at the interim necropsies, but at the final necropsy were significantly lower (0.095 g) than those of the control animals (0.190 g). The increased incidence of endometrial uterine polyps only in the treated animals sacrificed after 78 weeks is not regarded as substance-induced (CIIT 1982).

In another feeding study, groups of 50 F344 rats and 50 B6C3F$_1$ mice were treated with aniline hydrochloride for 103 weeks. The doses used corresponded to aniline doses of about 174.4 and 350.5 mg/kg body weight and day in rats and 737 and 1510 mg/kg body weight and day in mice. It is not reported whether the weights of the reproductive organs were determined. Histopathological examinations revealed an increased, but not dose-related incidence of endometrial uterine polyps in female rats. After 103 weeks, such polyps were found in 31% of the low dose animals, 14% of the high dose animals and 8% of the controls. No abnormal findings were reported for the reproductive organs of male rats or of male and female mice (NCI 1978).

The treatment of mice with five daily, intraperitoneally injected aniline doses between 17 and 200 mg/kg body weight did not lead to increased incidences of sperm-head abnormalities five weeks after the last dose (Topham 1980a, 1980b).

Aniline was investigated in a dominant lethal test carried out in Wistar rats in conformity with test guidelines. Groups of 40 fertile male rats were given daily intraperitoneal injections of aniline for five days in doses of 75, 150 or 200 mg/kg body weight. The male rats were then mated 1:1 with female rats of the same strain ten times for one week. The fertility of the rats was not impaired by aniline (Aniline Association 1998).

5.5.2 Developmental toxicity

There is one study with rats available, which was previously described in the 1992 MAK documentation (Price *et al.* 1985). In this study of prenatal and postnatal developmental toxicity, aniline hydrochloride doses of 10, 30 or 100 mg/kg body weight and day were administered by gavage to pregnant rats from days 7 to 20 of gestation ($\geq$ 20 animals/group) or from day 7 of gestation to the end of pregnancy ($\geq$ 12 animals/group). A dose-related increase in spleen weights was observed as a sign of maternal toxicity after doses of 10 mg/kg body weight and day and above (LOAEL for maternal toxicity: 10 mg aniline hydrochloride per kg body weight and day). In addition, the body weight gains of the dams treated with 100 mg/kg body weight were significantly reduced. The NOAEL for teratogenic effects was 100 mg/kg body weight and day, the highest dose tested in this prenatal and postnatal study. The NOAEL given by the authors for foetal effects was 10 mg/kg body weight and day.

Cleft palates, heart and rib anomalies and reduced foetal weights due to maternal methaemoglobinaemia (25%–42% methaemoglobin) induced by subcutaneous aniline hydrochloride doses of 195 mg/kg body weight and day and above were described in rats (Izumi *et al.* 2005; Matsumoto *et al.* 2001a, 2001b, 2002). Since administration of methylene blue reduced the incidence of heart anomalies and cleft palates, the authors attribute the effects to hypoxia induced by methaemoglobin formation (Matsumoto *et al.* 2001a, 2001b). Doses that did not lead to maternal methaemoglobin formation were not investigated.

5.6 Genotoxicity

A recent review analyzed in detail the comprehensive data available for the genotoxicity of aniline and its metabolites (Bomhard and Herbold 2005). Some of the results described below (effects on bacteria, mammalian cells and *in vivo*; Tables 6 and 7) were taken from this publication, unless they were included in the 1992 MAK documentation.

5.6.1 *In vitro*

The incubation of various *Escherichia coli* and *Bacillus subtilis* strains with aniline did not lead to the induction of DNA damage or DNA repair. The SOS chromotest with *Escherichia coli* and the umu test with *Salmonella typhimurium* also yielded negative results (Oda *et al.* 1995; Sakagami *et al.* 1986). Positive results (Brennan and Schiestl 1997; Schiestl 1989; Schiestl *et al.* 1989) were reported for aniline in the test for mitotic gene conversion in yeasts. As described in the 1992 MAK documentation, aniline produced negative results in standard mutagenicity tests with *Salmonella typhimurium*. In numerous studies with and without S9 mix from various species, aniline did not induce gene mutation in the strains TA98, TA100, TA1535, TA1537, TA1538 or *Escherichia coli* WP2 uvrA (Table 6). The incubation of aniline with the co-mutagen norharman led to positive results with *Salmonella typhimurium* TA98 in the presence of S9 mix (Nagao *et al.* 1977), but the formation of the mutagenic reaction product aminophenylnorharman rather than the aniline itself was responsible for this finding (Totsuka *et al.* 1998). As was demonstrated recently, the cytochromes CYP3A4 and CYP1A2 are significantly involved in aminophenylnorharman formation in human liver microsomes (Nishigaki *et al.* 2004).

The data available for the DNA-damaging effect of aniline in indicator tests with mammalian cells are inconsistent. Aniline did not induce DNA repair in primary hepatocytes of rats, mice, hamsters and humans in most UDS (unscheduled DNA synthesis) tests (Butterworth *et al.* 1989; McQueen *et al.* 1981; Williams 1980). A UDS test with rat hepatocytes which used an unusual evaluation method yielded a weakly positive result (BG Chemie 1986a). In most SCE (sister chromatid exchange) tests, aniline led to slightly increased SCE frequencies in various mammalian cell types irrespective of external metabolism (Takehisa *et al.* 1988; Tohda *et al.* 1983; Wilmer *et al.* 1984). In one study, aniline induced SCE only in human whole blood lymphocytes, but not in cultivated, isolated human lymphocytes. Haemoglobin (1000 µg/ml) alone led to a slight increase of the SCE frequency in these isolated lymphocytes (Wilmer *et al.* 1984). In the test for the induction of DNA strand breaks (alkaline elution), aniline did not cause any relevant DNA damage in mouse lymphoma cells or V79 cells (Garberg *et al.* 1988; Swenberg 1981).

Table 6. Genotoxicity of aniline and aniline hydrochloride *in vitro*

Test system		Concentration	Result		References
			– m. a.	+ m. a.	
Bacteria					
DNA damage; DNA repair; rec assay	*Escherichia coli* and *Bacillus subtilis*, various strains	up to 2000 mg/ml (about 20 µM)	–	–	Bomhard and Herbold 2005
SOS chromotest	*E. coli* PQ37	145–9310 µg/ml (about 1.5–100 µM)	–	–	Brams *et al.* 1987
SOS chromotest	*E. coli* PQ37	up to 100 mM	–	–	von der Hude *et al.* 1988
umu test	*Salmonella typhimurium* TA1535, NM2000, NM2009	250–2000 µg/ml (about 2.5–20 µM)	not carried out	–	Oda *et al.* 1995
umu test	*S. typhimurium* TA1535	4000 µg/ml (about 40 µM)	–	–	Sakagami *et al.* 1988
BMT	*S. typhimurium* TA92, TA94, TA97, TA98, TA100, TA102, TA1535, TA1537, TA1538 and *E. coli* WP2 uvrA		–	–	Bomhard and Herbold 2005
Yeasts					
mitotic gene conversion	*Saccharomyces cerevisiae* D3	5–12 mg/ml (about 50–120 µM)	not carried out	+	Brennan and Schiestl 1997
mitotic gene conversion	*S. cerevisiae* D3	1–7 mg/ml (about 10–70 µM)	not carried out	+	Schiestl 1989; Schiestl *et al.* 1989
Mammalian cells					
UDS	primary hepatocytes (rat)	up to 1 mM		–	Butterworth *et al.* 1989
UDS	primary hepatocytes (rat)	0.01 mM, no data on toxicity		–	McQueen *et al.* 1981
UDS	primary hepatocytes (rat)	1 mM		–	Williams 1980
UDS	primary hepatocytes (rat)	up to 5 mM		(+)	BG Chemie 1986a
UDS	primary hepatocytes (mouse; hamster)	0.01 mM, no data on toxicity		–	McQueen *et al.* 1981

Table 6. continued

Test system		Concentration	Result – m. a.	Result + m. a.	References
UDS	primary hepatocytes (human)	up to 1 mM		–	Butterworth *et al.* 1989
SCE	CHO	0.01–1 mM	(+)	not carried out	Takehisa *et al.* 1988
SCE	human lympho-blastoid cells NL3	0.1 mM	not carried out	(+)	Tohda *et al.* 1983
SCE	human lymphocytes	0.01–10 mM	(+)	–	Takehisa and Kanaya 1982
SCE	human lymphocytes	0.05–1.0 mM	–	not carried out	Wilmer *et al.* 1984
SCE	human whole blood	0.05–1.0 mM	(+)	not carried out	Wilmer *et al.* 1984
DNA SB (alkaline elution)	mouse lymphoma cells	1.07–21.5 mM	–	+ (at toxic concen-tration)	Garberg *et al.* 1988
DNA SB (alkaline elution)	V79 cells	0.03–3.0 mM	–	–	Swenberg 1981
HPRT	V79 cells	10–60 mM	–	+ (in 70% survivors)	Fassina *et al.* 1990
HPRT	V79 cells	2–10 mM	not carried out	– (RH)	Fassina *et al.* 1990
$TK^{+/-}$	mouse lymphoma cells	62.5–2400 μg/ml (about 0.6–24 μM)	+	+	McGregor *et al.* 1991
$TK^{+/-}$	mouse lymphoma cells	40–2400 nl/ml (about 0.4–24 μM)	+	+	Mitchell *et al.* 1988
$TK^{+/-}$	mouse lymphoma cells	62.5–2000 nl/ml (about 0.6–20 μM)	+	+	Myhr and Caspary 1988
CA	CHO-K1	444–2664 μg/ml (about 4–26 μM)	+	not carried out	Chung *et al.* 1995, 1996
CA	V79	0.4–4.3 μl/ml (about 4–40 μM)	+	+	BG Chemie 1986b
MNT	SHE	not specified	–	not carried out	Fritzenschaf *et al.* 1993
MNT	CHL/IU	125–2000 μg/ml (about 1.2–20 μM)	–	+	Matsushima *et al.* 1999

+ = positive; (+) = weakly positive; – = negative; ? = questionable
BMT: bacterial mutagenicity test; CA: chromosomal aberration; HPRT: hypoxanthine guanine phosphoribosyl transferase; m. a.: metabolic activation; MNT: micronucleus test; RH: rat hepatocytes; SB: strand breaks; SCE: sister chromatid exchange; SOS: SOS chromotest, $TK^{+/-}$: thymidine kinase assay; UDS: test for DNA repair synthesis

In several studies, high concentrations of aniline (> 1 mg/ml) led to clastogenic effects in the chromosomal aberration test and in the micronucleus test in mammalian cells (Matsushima *et al.* 1999); in two of these studies these effects were produced without metabolic activation (Chung *et al.* 1995, 1996; BG Chemie 1986b). A negative result in the micronucleus test without activation was also reported (Fritzenschaf *et al.* 1993).

Aniline had no clearly mutagenic effects on mammalian cells (HPRT (hypoxanthine guanine phosphoribosyl transferase) test with V79 cells) in the gene mutation test with and without the addition of a metabolic system. A weakly positive result ($p < 0.02$) was observed only in the high concentration range of > 50 mM (> 4.7 mg/ml) with metabolic activation (Fassina *et al.* 1990). Various mouse lymphoma tests with aniline, however, yielded positive results with and without activation (McGregor *et al.* 1991; Mitchell *et al.* 1988; Myhr and Caspary 1988). The effects were induced by concentrations of > 1 mg/ml without S9 mix and about 500 μg/ml and above with activation.

5.6.2 *In vivo*

In vivo tests with aniline (Table 7) indicate the substance has clastogenic potential, but it was detected only at high doses in the LD_{50} range. In the micronucleus test in most studies, the treatment of mice with aniline led to increased micronucleus frequencies in the bone marrow or peripheral blood, irrespective of the route of absorption (intraperitoneal or oral) (see Table 7). Positive results were generally obtained after oral doses in the toxic range of 1000 mg/kg body weight (Westmoreland and Gatehouse 1991) or after two doses of 470 mg/kg body weight (Ress *et al.* 2002) or after one or two intraperitoneal injections of 300 mg/kg body weight, and above (Aniline Association 1989; Ashby *et al.* 1991; Westmoreland and Gatehouse 1991). Negative results were described for the lower oral dose range of 125–250 mg/kg body weight (Harper *et al.* 1984) and 610 mg/kg body weight (BG Chemie 1985). Micronucleus induction was increased 2 to 2.6 times after single intraperitoneal doses of aniline of 50 to 200 mg/kg body weight in only one study (Sicardi *et al.* 1991). In mice given aniline doses of 500, 1000 or 2000 mg/kg diet (about 75, 150 and 300 mg/kg body weight and day) for 90 days, slightly increased micronucleus frequencies in the peripheral blood were found at all dose levels, but the effect was not clearly related to the dose. A three to four-fold dose-dependent increase in the number of polychromatic erythrocytes indicates severe haematopoiesis (Witt *et al.* 2000). Since aniline induced no tumours in mice, unlike in rats, it is assumed that the micronuclei in mice develop indirectly from the aniline-induced erythrotoxicity and the resulting compensatory haematopoiesis in the bone marrow (Ashby *et al.* 1991). Westmoreland and Gatehouse (1991) observed unusually shaped micronuclei and a high interindividual variability of the micronucleus frequencies in their studies and question the validity of the results from the high toxic dose range. The theory that the mechanism of micronucleus formation in mice is non-genotoxic is supported by the observation that increased chromosomal aberration was not observed in the bone marrow of mice after two intraperitoneal injections with aniline of 380 mg/kg body weight, a dose that induces micronuclei (Jones and Fox 2003).

Table 7. Genotoxicity of aniline and aniline hydrochloride *in vivo*

Test system		Exposure	Dose [mg/kg body weight]*	Result	References
Tests for DNA damage					
alkaline elution	SD rat, ♂ spleen	1× intra-peritoneal	**880**	(+)	Russo *et al.* 1981
viscometry	SD rat, ♂ liver	1× intra-peritoneal	210	–	Brambilla *et al.* 1985
comet assay	ddy mouse, ♂ liver, bladder, lung, brain, bone marrow	1× oral	**100**	+	Sasaki *et al.* 1999
comet assay	ddy mouse, ♂ stomach, intestine, kidney	1× oral	100	–	Sasaki *et al.* 1999
comet assay	Wistar rat, ♂ stomach, intestine, liver, kidney, bladder, lung	1× oral	**150**	+	Sekihashi *et al.* 2002
comet assay	Wistar rat, ♂ brain, bone marrow	1× oral	150	–	Sekihashi *et al.* 2002
Tests for somatic mutations					
MNT	ICR mouse, ♂ bone marrow	1× oral	125, 250	–	Harper *et al.* 1984
MNT	CFLP mouse, ♂, ♀ bone marrow	1× oral	610	–	BG Chemie 1985
MNT	CRH mouse, ♂ bone marrow	1× oral	400, 500, **1000**	+	Westmoreland and Gatehouse 1991
MNT	B6C3F$_1$ mouse, ♂ bone marrow	2× oral	12, 23, 47, 120, **470**	+	Ress *et al.* 2002
MNT	Swiss mouse, ♂, ♀ bone marrow	1× intra-peritoneal	5, **50, 100, 200**	+	Sicardi *et al.* 1991
MNT	ICR mouse, ♂, ♀ bone marrow	2× intra-peritoneal	30, 100, **300**	+	Aniline Association 1989
MNT	CBA mouse, ♂ bone marrow	2× intra-peritoneal	100–300	–	Ashby *et al.* 1991
MNT	CBA mouse, ♂ bone marrow	2× intra-peritoneal	238, **380**	+	Ashby *et al.* 1991
MNT	CRH mouse, ♂ bone marrow	1× intra-peritoneal	**380**	+	Westmoreland and Gatehouse 1991

Table 7. continued

Test system		Exposure	Dose [mg/kg body weight]*	Result	References
MNT	$B6C3F_1$ mouse, ♂, ♀ peripheral blood	90-day feeding study	**500, 1000, 2000** mg/kg (about 75, 150, 300 mg/kg body weight)	+	Witt *et al.* 2000
MNT	PVG rat, ♂ bone marrow	1× oral	500	–	Bomhard 2003
MNT	PVG rat, ♂ bone marrow	1× oral	300, **400**, **500**	+	Bomhard 2003
MNT	PVG rat, ♂ bone marrow	1× oral	215, **287, 400, 500**	+	George *et al.* 1990
CA	CBA mouse, ♂ bone marrow	2× intra-peritoneal	220, 300, 380	–	Jones and Fox 2003
CA	PVG rat, ♂ bone marrow	1× oral	300, 400, **500**	+	Bomhard 2003
Germ cell tests					
dominant lethal test	SD rat, ♂	5× intra-peritoneal	75, 150, 200	–	Aniline Association 1998

+ = positive; (+) = weakly positive; – = negative
* figures in bold type show the doses that yielded positive results

An increase in the incidence of micronuclei (Bomhard 2003; George *et al.* 1990) and chromosomal aberrations (Bomhard 2003) was found in the bone marrow of rats after single oral doses of aniline of about 300 mg/kg body weight and above. In the micronucleus tests of Bomhard (2003), positive results were observed only 24 hours, but not 48 hours after the beginning of treatment, whereas George *et al.* (1990) found such effects 24 and 48 hours after treatment with doses of 287 mg/kg body weight and above. George *et al.* (1990) observed in addition to the induction of micronuclei a decrease (24 hours) and subsequent increase (48 hours) in the ratio of polychromatic erythrocytes (PCE) to normochromatic erythrocytes (NCE) in the bone marrow, evidence of compensatory medullary erythropoiesis.

Aniline was not found to cause germ cell mutagenicity in a dominant lethal test carried out in Wistar rats in conformity with test guidelines. Groups of 40 fertile male rats were given daily intraperitoneal injections of aniline in doses of 75, 150 or 200 mg/kg body weight for 5 days. The male rats were then mated 1:1 with female rats of the same strain for 10 weeks. The animals of the high dose group showed signs of toxicity (pallor, ruffled fur, etc.), and two males died (days 4 and 5). The number of dead implantations was slightly increased in this group in the third week. Since this finding was attributed to a subgroup of 7 animals with a reduced number of live implantations and was within the historical control data of the laboratory, the test results are regarded as negative (Aniline Association 1998).

There are also several studies of the induction of DNA strand breaks (Brambilla *et al.* 1985; Cesarone *et al.* 1982; Parodi *et al.* 1982; Russo *et al.* 1981; Sasaki *et al.* 1999; Sekihashi *et al.* 2002) in various organs of mice and rats (see Table 7 and the 1992 MAK documentation). The results cannot be evaluated conclusively since not only positive but also negative results (rat: bone marrow only one negative result) were reported for the bone marrow, liver and kidneys, which were the main organs tested (Table 7). No DNA strand breaks were detected in the rat spleen, the target organ of carcinogenicity, after single intraperitoneal aniline doses of 210 mg/kg body weight (Parodi *et al.* 1982), whereas a weakly positive result was obtained at the very high intraperitoneal dose of 880 mg/kg body weight (corresponding to twice the LD_{50}) (Russo *et al.* 1981). Such investigations of the mouse spleen are not available.

In vivo studies of the binding of aniline to the DNA of various organs did not yield results consistent with the carcinogenicity data. Weak binding of ^{3}H-aniline to DNA was detected by Roberts and Warwick (1966) in the liver, kidneys and spleen of rats. In rats pre-treated with oral doses of unlabelled aniline of 50 mg/kg body weight and day for seven days and then given single intraperitoneal doses of ^{14}C-aniline of 250 mg/kg body weight, McCarthy *et al.* (1985) observed DNA binding with a covalent binding index (CBI) of 14.2 in the kidneys, 3.7 in the spleen and 4.3 in the large intestine. No DNA binding was detected in the liver of rats and in various organs of mice, including the spleen. In all the DNA binding studies the DNA was not purified to constant specific activity; the CBI values can therefore be regarded as maximum values.

In the spleen of rats treated with aniline, oxidative DNA damage was observed parallel to the accumulation of free unchelated iron. Five male rats were given oral doses of aniline of 1 mmol/kg body weight and day for seven days; five further animals were used as vehicle controls. Increases in the total iron content of 200% and in free iron of 375% compared with the levels in the controls were observed in the spleens of the treated animals 24 hours after the last dose. The level of 8-hydroxy-2′-deoxyguanosine (8-OHdG) in the DNA of the spleen of these rats was increased by 83% compared with that in the control animals (Wu *et al.* 2005). The formation of 8-OHdG is considered to be the most frequent DNA damage induced by reactive oxygen species and can lead to mutations through base mismatch (Hartwig and Schlepegrell 1995). No 8-OHdG damage was observed in rats after single aniline doses which did not cause iron accumulation or oxidative stress or lipid peroxidation in the spleen (Wu *et al.* 2005).

Daily treatment of mice with five intraperitoneal aniline doses between 17 and 200 mg/kg body weight did not induce increased incidences of sperm-head abnormalities five weeks after the last administration (Topham 1980a, 1980b).

Genotoxicity of the aniline metabolites

There are comprehensive data for the genotoxicity of aniline metabolites. The main metabolites of aniline in humans and rodents are *p*-aminophenol and *p*-hydroxyacetanilide. Although the quality and quantity of the data for the genotoxicity of these metabolites varies greatly depending on the end point, their mutagenic profile is similar to that of the parent substance (Bomhard and Herbold 2005). This suggests that *p*-aminophenol or *p*-hydroxyacetanilide could be responsible for the clastogenicity of aniline in rats *in vitro* and *in vivo*.

The data available for the genotoxicity of the secondary metabolites phenylhydroxylamine and nitrobenzene, which are considered to be responsible for methaemoglobin formation, are inadequate for a conclusive comparison.

Assessment of genotoxicity

Aniline does not induce gene mutation in bacteria. Results from gene mutation tests in mammalian cells are inconsistent. In the HPRT test, weakly positive results were obtained only at extremely high concentrations. The positive results observed in the $TK^{+/-}$ test cannot be clearly attributed to the end point gene mutation because the colony size was rarely evaluated. Since high concentrations of aniline (> 1 mg/ml) clearly have the potential to induce chromosomal aberrations, the effects observed in the $TK^{+/-}$ test could also be the result of chromosomal aberrations. The available data do not allow any conclusive evaluation.

In animal studies, high aniline doses (about 300 mg/kg body weight and above) led to an increase in the incidence of chromosomal aberrations and micronuclei in the bone marrow of rats. Micronuclei were induced in mice, but not chromosomal aberrations. The results from studies of the induction of DNA adducts and DNA strand breaks are inconsistent and do not correlate with the tumour findings obtained in rats. Overall, the positive results in indicator tests confirm the data from the mutagenicity tests, which indicate *in vitro* and *in vivo* that aniline has mutagenic potential.

Numerous studies investigated whether the clastogenic effects are induced by direct interaction of aniline or its metabolites with DNA or by indirect effects mediated by erythrotoxicity or increased medullary erythropoiesis. The following observations indicate a causal relationship between erythrotoxicity, the induction of iron-mediated oxidative stress, increased erythropoiesis and DNA damage.

The clastogenic effects of aniline were first found at doses at which marked toxic effects on the erythrocytes and haematopoietic system were evident. As described in Section 2, the release of iron in the spleen leads to oxidative stress, resulting in cellular and DNA damage. A relationship was demonstrated between the level of free iron in the spleen of rats treated with aniline and an increase in the incidence of oxidative DNA damage in the spleen (8-OHdG) (Wu *et al.* 2005). Parallel to this, increased compensatory medullary and extramedullary haematopoiesis was induced. As a result of the marked increase in the ratio of PCE to NCE, it could be shown that micronucleus induction in the bone marrow of rats and in the peripheral blood of mice is accompanied by increased erythropoiesis after exposure to aniline (George *et al.* 1990; Witt *et al.* 2000). Stimulation of erythropoiesis, as shown in tests with erythropoietin and prostaglandin E2, increases micronucleus formation by numerous mutagens in the bone marrow of mice (Suzuki *et al.* 1989, 1994). Possible causes of increased micronucleus formation are thought to be the reduced generation time and more rapid maturation (fewer mitotic divisions) of the erythroblasts, leading to decreased DNA repair efficiency (Suzuki *et al.* 1989), and also the increased incidence of spontaneous DNA damage induced by increased cell proliferation (Kirkland *et al.* 1992). The increased iron turnover can also lead to oxidative damage to the bone marrow. Although there are important arguments that the clastogenic effects of aniline are of secondary origin, it cannot be ruled out that very high doses do induce chromosomal aberrations.

The aniline metabolites *p*-aminophenol and *p*-hydroxyacetanilide (pHAA) have a genotoxic profile similar to that of the parent substance (Bomhard and Herbold 2005), but the clastogenic effects on the bone marrow occur even at low doses. This suggests that one or both of the metabolites could be responsible for the clastogenic effects.

pHAA is the active ingredient of a pharmaceutical used worldwide in humans. In a comprehensive review of the genotoxicity of this substance, Bergman *et al.* (1996) showed that the genotoxic effects of pHAA first occur at doses which cause marked liver and bone marrow toxicity and that the threshold for genotoxic effects is not reached at doses used therapeutically. This observation is supported by a number of studies with volunteers. Provided that pHAA is responsible for the mutagenic effect, the same could also be assumed for aniline since the genotoxic profile of pHAA and aniline are similar.

5.7 Carcinogenicity

Carcinogenicity studies with rats and mice are available for aniline hydrochloride. Some earlier studies, in which the purity of the substance was not documented or the study period did not comply with present-day standards, were described in the 1992 MAK documentation. The two carcinogenicity studies relevant for assessment that were included in the 1992 MAK documentation are reviewed here.

Rat

In a long-term feeding study (CIIT 1982; Table 8), F344 rats were given aniline hydrochloride in the diet for 104 weeks. The daily doses were 0, 10, 30 or 100 mg/kg body weight, corresponding to doses of aniline of 7, 22 and 72 mg/kg body weight and day. 130 animals were used per group and sex. The treatment did not affect body weight gain, but mortality was slightly increased in the males of the high dose group, which is probably because of the development of tumours in this group. Haematological examinations revealed decreases in some blood parameters in the males of all dose groups and the high dose group females: haematocrit, haemoglobin concentration and erythrocyte count. In addition, liver weights were increased in the high dose group animals. No splenic tumours occurred in any of the control animals; 34 of the males in the high dose group (26%) developed sarcomas in the spleen and 9 of these died prematurely. Only one splenic tumour was found in the male rats of the middle dose group (a fibrosarcoma) and one in the female rats of the high dose group (a haemangiosarcoma). With regard to non-neoplastic changes, stromal hyperplasia and fibrosis of the splenic red pulp were seen in the males of the high dose group and, to a lesser degree, in the females of this group. These types of hyperplasia can possibly be regarded as pre-neoplastic lesions since they were found to be morphologically similar to the stromal sarcoma cells. Fatty degeneration of the splenic red pulp in the males and chronic capsulitis in both sexes were observed in the high dose groups. Considerable iron deposits and increased extramedullary haematopoiesis were observed in the splenic red pulp of the male rats. Congestion in the spleen of the males and the high dose females, increasing in severity with the dose, indicates disturbed blood circulation in the spleen (see also Section 5.2.2).

Table 8. Splenic tumours and splenic toxicity in F344 rats

Author:	CIIT 1982
Substance:	aniline hydrochloride
Species:	F344 rats, groups of 130 ♂/130 ♀; interim necropsies after 26 weeks (10 ♂/10 ♀), 52 weeks (10 ♂/10 ♀) and 78 weeks (20 ♂/20 ♀)
Administration route:	with the diet
Concentration:	0, 10, 30 and 100 mg/kg body weight and day (aniline doses of about 7, 22 and 72 mg/kg body weight and day)
Duration:	104 weeks
Toxicity:	see "other findings"

		controls	aniline (mg/kg body weight and day)		
		0	7	22	72
evaluated animals	♂	123	129	128	130
	♀	129	129	130	130
premature deaths	♂	16	14	15	26
	♀	20	21	14	13
splenic tumours					
fibrosarcomas	♂	0	0	0	3 (1[a])
	♀	0	0	0	0
stromal sarcomas	♂	0	0	0	21 (1)
	♀	0	0	0	0
capsular sarcomas	♂	0	0	0	1
	♀	0	0	0	0
haemangiosarcomas	♂	0	0	0	6 (4)
	♀	0	0	0	1
osteogenic sarcomas	♂	0	0	0	3 (3)
	♀	0	0	0	0
other findings in week 104 (90 ♂/90 ♀)					
stromal hyperplasia	♂	1	0	0	31
	♀	0	0	0	9
chronic capsulitis	♂	1	1	2	62
	♀	0	4	4	70
fatty degeneration	♂	0	0	0	14
	♀	0	0	0	0

[a] number of animals that died prematurely in brackets

In an earlier long-term study with markedly higher dosages (NCI 1978; Table 9), Fischer 344 rats were given aniline hydrochloride doses of 3000 or 6000 mg/kg diet (corresponding to aniline doses of about 174 and 350 mg/kg body weight and day) for 103 weeks, and were then observed for four to five weeks. Body weight gains were slightly reduced only at the high dose. Both doses induced mesenchymal tumours, mainly in the spleen but, to a lesser degree, also in other organs of the body cavity. In this study, too, the tumour incidences were markedly higher in male rats than in female rats. Haemangiosarcomas were the most frequent malignant splenic tumours, followed by fibrosarcomas. The growth of the fibrosarcomas was very invasive with widespread extension into the body cavity. Therefore, mesenchymal tumours in the spleen and organs of the body cavity were frequently observed in the same animals. The description of the sarcomas in the spleen and body cavity, which were not specified in detail, suggests very weakly differentiated fibrosarcomas. There was a dose-dependent but not significant increase in the incidence of adrenal phaeochromocytomas in the males, but only one malignant tumour was diagnosed in each group. Numerous non-neoplastic lesions were observed in the animals treated with aniline. They included fibrosis of the splenic capsule, fatty degeneration and papillary hyperplasia in the spleen, and haemosiderosis in the renal tubules and liver. Survival, even of the high dose animals, was not reduced compared with the control values. The tumour incidence was markedly lower in the females. The total of fibrosarcomas and sarcomas in the spleen and body cavity was significantly increased in the Cochran-Armitage test, but not in Fisher's exact test (0/24, 1/50 and 7/50). Since these types of tumours are rare, the results were regarded as of importance despite the unclear statistical significance. Data for the blood parameters were not included in this study. The National Cancer Institute concluded from the study data that aniline hydrochloride is carcinogenic for male and female rats.

In a later review (Weinberger *et al.* 1985), the rat spleen tissue from the above carcinogenicity study (NCI 1978) was subjected to another histopathological examination to assess the relationship between the non-neoplastic changes and splenic sarcomas. The study revealed a clear correlation between the incidence and severity of non-neoplastic changes (fibrosis and fatty degeneration) and tumour incidence. In addition, morphological similarity was shown between the fibrosis cells and sarcoma cells. The authors concluded from their results that the splenic lesions caused by aniline, such as haemorrhage, fibrosis and fatty degeneration, are to be regarded as precursors of carcinogenicity.

Another review (Bus and Popp 1987), which evaluated the two carcinogenicity studies with aniline described above (CIIT 1982; NCI 1978) and other studies with structurally related substances, came to a similar conclusion. The authors suspected that the splenic tumours are a secondary response resulting from erythrocyte toxicity rather than a result of the direct effects of aniline or its metabolites on macromolecules of the spleen. However, it must be pointed out that the haemangiosarcomas induced by different monocyclic arylamines are not restricted to the spleen and are regarded as typical of this class of substance (Neumann 1985).

Table 9. Mesenchymal tumours in the spleen, body cavity and other organs

Author:	NCI 1978
Substance:	aniline hydrochloride
Species:	F344 rats; groups of 50 ♂/50 ♀; controls: 25 ♂/25 ♀
Administration route:	with the diet
Concentration:	0, 0.3%, 0.6% (approximate aniline doses of 174 and 350 mg/kg body weight and day)
Duration:	103 weeks
Toxicity:	body weight gains decreased at 350 mg/kg body weight

		controls	aniline (mg/kg body weight and day)	
		0	174	350
tumours in the spleen and capsule:				
tested animals	♂	25	50	46
	♀	23	50	50
sarcomas (not specified)	♂	0	4	2
	♀	0	0	3
fibromas	♂	0	7	6
	♀	0	0	0
fibrosarcomas	♂	0	3	7
	♀	0	0	0
haemangiosarcomas	♂	0	19**	20**
	♀	0	1	2
lipomas	♂	0	0	0
	♀	0	0	1
haemangiomas	♂	0	0	0
	♀	0	0	1
tumours in body cavities and other organs:				
evaluated animals	♂	25	50	48
	♀	24	50	50
fibrosarcomas	♂	0	2	8*
	♀	0	1	3
leiomyosarcomas	♂	0	0	2
	♀	1	0	0
haemangiosarcomas	♂	0	0	1
	♀	0	0	0
sarcomas	♂	0	0	1
	♀	0	0	1
adrenal:				
tested animals	♂	24	50	44
	♀	24	50	48
phaeochromocytomas				
benign/malignant	♂	1/1	5/1	11/1
benign/malignant	♀	1/0	0/0	3/2

*$p < 0.05$; **$p \leq 0.001$

Mouse

No correlation was detected between the ingestion of aniline hydrochloride in the diet and tumour formation in mice. In an NCI study (1978), B6C3F$_1$ mice were given aniline hydrochloride doses of 6000 or 12000 mg/kg feed (corresponding to aniline doses of about 733 or 1510 mg/kg body weight and day for the females and about 737 or 1560 mg/kg body weight and day for the males) for 103 weeks, and then observed for four weeks. No haematological examinations were carried out. No increase in the incidence of tumours or proliferative, non-neoplastic changes was observed compared with that in the controls although the doses used were higher than in the rat study. In the spleen, the target organ of tumourigenicity in rats, one haemangioma occurred in the high dose group (1/49), one haemangiosarcoma in the low dose group (1/49) and one malignant lymphoma in the control group (1/38) of the male mice. Examination of the spleens of the female mice revealed one haemangiosarcoma (1/45) in one control animal and three malignant lymphomas in the low dose group (3/48). Body weight gain was reduced and hair loss was observed in the males. Mortality was not increased compared with the control values.

5.7.1 Assessment of carcinogenicity

Aniline is carcinogenic in rats, but not in mice, although doses used in the long-term feeding study with mice were twice as high as those used in the rat study and, with regard to body weight, even five times as high. This species specificity could be the result of markedly higher erythrocytic methaemoglobin reductase activity and the better capacity in mice for eliminating aniline compared with that in rats.

Aniline was carcinogenic mainly in the spleen of rats; mesenchymal sarcomas and haemangiosarcomas were the principal forms of tumours found. The incidence of splenic tumours was markedly increased in the males, the gradient of the dose-response relationship following a steep course, whereas splenic tumours were only sporadically observed in female rats. Both the range-finding studies of subchronic toxicity and determinations carried out during the course of the long-term studies demonstrated that the methaemoglobin level and erythrocyte turnover were increased under the specific test conditions. Increased extramedullary haematopoiesis can be detected as pigmentation in the spleen and also in the sinusoidal cells of the liver and in the tubular epithelium of the kidneys. In addition, focal and multifocal thickening of the splenic capsule with fibrous tissue, mononuclear cells and granular pigment deposits (capsulitis) were observed, particularly in the group of high dose males most affected. These results pose the question of whether the development of mesenchymal tumours is associated with methaemoglobin formation and its consequences and whether capsulitis and stromal hyperplasia can be considered to be among these consequences.

The organ-specific, species-specific and, to a limited extent, also the sex-specific pattern of the carcinogenicity of aniline, and also numerous mechanistic studies, lend plausibility to the hypothesis that the splenic tumours in rats are caused by indirect mechanisms rather than by a primary genotoxic effect of aniline. This hypothesis is substantiated by the observation that the formation of aniline–DNA adducts in various organs of the rat does not correlate with the carcinogenicity. DNA adduct formation was lowest in the spleen, but marked DNA damage (8-OHdG formation) caused by oxidative stress was observed. The steep gradient of the dose-response relationship for tumour formation, which is untypical of genotoxic carcinogens, also indicates an indirect mechanism. Despite the high tumour incidences after doses of 100 mg/kg body weight and day in the long-term study, no clastogenic effects were observed in rats at this dose or even at doses up to three times as high. In the spleen of rats, high, erythrotoxic aniline doses lead to considerable overloading with cell debris and to the release of iron, resulting in chronic oxidative stress (see Section 2), which, in turn, causes DNA and tissue damage and can contribute to carcinogenicity. In addition, compensatory medullary and extramedullary haematopoiesis is induced by the erythrotoxicity and subsequent anaemia. The resulting increased iron turnover and increased proliferation processes can enhance the occurrence of spontaneous DNA lesions.

The formation of splenic tumours in rats seems to be causally related to the toxic effects of aniline on the erythrocytes and their consequences. The following observations support this assumption:

- The spleen is the only target organ of carcinogenicity for erythrotoxic substances.
- In mice, which possess a high methaemoglobin reductase activity, aniline does not induce tumours.
- The steep gradient of the dose-response relationship for the carcinogenic effects of aniline in rats confirms that carcinogenic effects occur only at high doses.
- Severe erythrotoxicity and prolonged splenic damage is a requirement for tumour formation.
- The incidence and severity of non-neoplastic lesions (fibrosis and fatty degeneration) correlate closely with the tumour incidence.
- There is a striking morphological similarity between the fibrotic cells and sarcoma cells.
- The higher sensitivity of male rats for the postulated iron-dependent mechanism of carcinogenicity is in agreement with the results for ferric nitrilotriacetate, which is also tumourigenic mainly in male rats (Deguchi *et al.* 1995).

The same mechanism is assumed for the tumours observed at even higher doses in the abdominal and thoracic cavities but not specified in detail, unless they are metastases.

6 Manifesto (MAK value/classification)

In animal studies, aniline led to splenic tumours in rats but not in mice. Numerous mechanistic studies suggest an indirect mechanism for the tumourigenicity. The splenic tumours in rats are causally related to the toxic effects of aniline on the erythrocytes and their consequences. An increase in tumour incidence is therefore not expected if damage to the erythrocytes is avoided. Since there are fundamental physiological differences between rats and humans and disorders observed in humans that accompany methaemoglobinaemia or haemolysis, aniline is not expected to cause such effects in human spleens.

There are comprehensive data for the genotoxicity of aniline. The available data indicate a genotoxic potential that is evident *in vitro* and *in vivo*, but mainly at high, erythrotoxic doses. All the available findings provide evidence that indirect mechanisms mediated via erythrotoxicity are responsible for the clastogenic effects. Since the genotoxic effects of aniline are found only at doses which lead to marked damage to the haematopoietic system in toxicity studies, it can be assumed that the genotoxicity is of subordinate relevance if the MAK value is observed at the workplace. Aniline is therefore classified in Carcinogen Category 4.

As shown in the 1992 MAK documentation, assuming an airborne concentration at the MAK value established in 1983 of 2 ml/m^3 (about 8 mg/m^3) and absorption of aniline via the respiratory tract of 90% and via the skin of about 25%, 72 mg aniline is absorbed through the respiratory tract (90% of 80 mg), and about 20 mg aniline is absorbed through the skin over an 8-hour working day (inhaled volume of 10 m^3). This results in a total of 92 mg aniline for an 8-hour working day or about 11 mg per hour. According to a study by Jenkins *et al.* (1972), a single oral dose of 45 mg is the lower limit at which healthy persons show no symptoms induced by methaemoglobinaemia. Taking into account the half-life of aniline of 3.5 hours for humans (1992 MAK documentation) and that the onset of aniline is considerably slower after inhalation than after ingestion, the MAK value of 2 ml/m^3 thus protects healthy workers from relevant methaemoglobinaemia if skin contact with liquid aniline is ruled out. Experience from occupational medicine has shown that no health effects occur if the MAK value or BAT value is observed. The existing MAK value of 2 ml/m^3 has therefore been retained.

Classification in Peak Limitation Category II with an excursion factor of 2 has been retained since systemic effects are decisive for deriving the MAK value ("Aniline", Supplement 2002, this volume).

As there is relevant dermal absorption from the liquid and gaseous phases ("Aniline", Volume 6, present series; Pauluhn 2005), designation with an "H" has been retained.

It is difficult to interpret the available findings from patch tests with aniline since there are no data for previous exposure and relatively high test concentrations were used in some cases. Nor can the results of a study with volunteers, in which contact sensitization was determined, be assessed conclusively. Since several clinical studies have demonstrated possible cross-reactions to aniline in patients with existing sensitization to (disubstituted) aromatic amino compounds, skin sensitization in humans seems to be plausible even if the available findings could not explicitly demonstrate contact sensitization induced solely by aniline. Results from animal studies suggest a slight contact sensitizing potential. Since skin contact with aniline can induce an allergic reaction at least in persons with existing sensitization to some disubstituted aromatic amines, aniline is designated with an "Sh". There are no studies available for respiratory tract sensitization. Aniline is therefore not designated with an "Sa".

In a relevant prenatal and postnatal developmental toxicity study with oral administration in rats (Price *et al.* 1985), maternal toxicity was evident even at the lowest dose of 10 mg/kg body weight and day. The NOAEL for developmental toxicity is

10 mg/kg body weight and day (corresponding to about 18 ml/m^3 assuming an inhaled volume of 10 m^3 and 70 kg body weight). The margin between this and the MAK value of 2 ml/m^3 is thus sufficiently large, and after renewed assessment of the data aniline has been classified in Pregnancy Risk Group C.

Since no germ cell mutagenicity was found in a dominant lethal test, aniline is not classified in any category for germ cell mutagenicity.

7 References

Albrecht W, Neumann H-G (1985) Biomonitoring of aniline and nitrobenzene. *Arch Toxicol 57*: 1–5

Angelini G, Rantuccio F, Meneghini CL (1975) Contact dermatitis in patients with leg ulcers. *Contact Dermatitis 1*: 81–87

Aniline Association (1989) *Mouse bone marrow micronucleus assay of Aniline*. Haskell Lab. Report No. Du Pont HLR 263-89. Aniline Association, Washington, DC, USA, unpublished

Aniline Association (1998) *Aniline: Dominant lethal study in the rat*, Report No. CTL/P/5404. Central Toxicology Laboratory, Alderly Park Macclesfield, Chesire, UK, Aniline Association, Washington, DC, USA, unpublished

Anonymous (2001) Aniline – Acute exposure guideline levels. *Inhal Toxicol 13, Suppl*: 7–42

Ashby J, Vlachos DA, Tinwell H (1991) Activity of aniline in the mouse bone marrow micronucleus assay. *Mutat Res 263*: 115–117

BASF AG (1970) *Bericht über die Prüfung der Hämoglobin (Methämoglobin)-bildenden Wirkung von Anilin p.a. (MERCK) an der Katze* (Report about the investigation of the haemoglobin (methaemoglobin)-producing effects of aniline p.a. (MERCK) in the cat) (German). BASF AG, 16.04.1970/k, unpublished

BASF AG (1971) *Bericht über die Prüfung der hämoglobinbildenden Wirkung von Anilin bei oraler Gabe an der Katze* (Report about the investigation of the haemoglobin-producing effects of oral doses of aniline in the cat) (German). BASF AG, 11.03.1971, unpublished

Basketter DA, Scholes EW (1992) Comparison of the local lymph node assay with the guinea-pig maximization test for the detection of a range of contact allergens. *Food Chem Toxicol 30*: 65–69

Basketter DA, Smith Pease CK, Patlewicz GY (2003) Contact allergy: the local lymph node assay for the prediction of hazard and risk. *Clin Exp Dermatol 28*: 218–221

Bayer AG (1984) *Anilin. Untersuchungen zur akuten oralen Toxizität an der Katze. Einfluß auf Met-Hämoglobingehalt und Zahl der Heinz-Innenkörper im peripheren Blut* (Aniline. Investigations of the acute oral toxicity in the cat. Influence on methaemoglobin level and number of Heinz bodies in peripheral blood) (German). Bayer AG, Wuppertal, 23.01.1984, unpublished

Bergman K, Müller L, Weberg Teigen S (1996) The genotoxicity and carcinogenicity of paracetamol: a regulatory (re)view. *Mutat Res 349*: 263–288

BG Chemie (Berufsgenossenschaft der chemischen Industrie) (1985) *Mouse micronucleus test on aniline*. Toxicol Laboratories Ltd, Lebury, UK, Report No. 54/8505, October 1985, BG Chemie, Heidelberg, unpublished

BG Chemie (1986a) *Aniline – Test for the induction of DNA repair in rat hepatocyte primary cultures*. Gesellschaft für Strahlen- und Umweltforschung, D-8042 Neuherberg, 28.07.1986, BG Chemie, Heidelberg, unpublished

BG Chemie (1986b) *Chromosome aberrations in cells of Chinese hamster cell line V79*. Laboratorium für Mutagenitätsprüfung (LMP) TH Darmstadt, 03.04.1986, BG Chemie, Heidelberg, unpublished

Bio-Fax (1969) *Aniline. Acute oral LD_{50} male albino rats. Acute eye irritation albino rabbits. Primary skin irritation albino rabbits. Acute dermal LD_{50} albino rabbits. Acute inhalation LC_{50} t=1 hr. Male albino rats. Subacute feeding (28 days) male albino rats*. Industrial Bio-Test Laboratories, Northbrook, IL, USA

Birner G, Neumann H-G (1988) Biomonitoring of aromatic amines II: Hemoglobin binding of some monocyclic aromatic amines. *Arch Toxicol 62*: 110–115

Bomhard EM (2003) High-dose clastogenic activity of aniline in the rat bone marrow and its relationship to the carcinogenicity in the spleen of rats. *Arch Toxicol 77*: 291–297

Bomhard EM, Herbold BA (2005) Genotoxic activities of aniline and its metabolites and their relationship to the carcinogenicity of aniline in the spleen of rats. *Crit Rev Toxicol 35*: 783–835

Bonnevie P (1939) *Aetiologie und Pathogenese der Ekzemkrankheiten* (Etiology and pathogenesis of dermatitis) (German), JA Barth, Leipzig, 217

Bradshaw TP, McMillan DC, Crouch RK, Jollow DJ (1995) Identification of free radicals produced in rat erythrocytes exposed to hemolytic concentrations of phenylhydroxylamine. *Free Radical Biol Med 18*: 279–285

Brambilla G, Carlo P, Finollo R, Ledda A (1985) Viscometric detection of liver DNA fragmentation in rats treated with ten aromatic amines. Discrepancies with results provided by the alkaline elution technique. *Carcinogenesis 6*: 1285–1288

Brams A, Buchet JP, Crutzen-Fayt MC, de Meester C, Lauwerys R, Léonard A (1987) A comparative study, with 40 chemicals, of the efficiency of the Salmonella assay and the SOS chromotest (kit procedure). *Toxicol Lett 38*: 123–133

Brennan RJ, Schiestl RH (1997) Aniline and its metabolites generate free radicals in yeast. *Mutagenesis 12*: 215–220

Bus JS, Popp JA (1987) Perspectives on the mechanism of action of the splenic toxicity of aniline and structurally-related compounds. *Food Chem Toxicol 8*: 619–626

Butterworth BE, Smith-Oliver T, Earle L, Loury DJ, White RD, Doolittle DJ, Working PK, Cattley RC, Jirtle R, Michalopoulos G, Strom S (1989) Use of primary cultures of human hepatocytes in toxicology studies. *Cancer Res 49*: 1075–1084

Calas E, Castelain PY, Piriou A (1978) Épidémiologie des dermatoses de contact à Marseille (Epidemiology of contact dermatitis in Marseilles) (French). *Ann Dermatol Venereol (Paris) 105*: 345–347

Carpenter CP, Smyth HF, Pozzani UC (1949) The assay of acute vapor toxicity, and the grading and interpretation of results on 96 chemical compounds. *J Ind Hyg Toxicol 31*: 343–346

Cesarone CF, Bolognesi C, Santi L (1982) Evaluation of damage to DNA after *in vivo* exposure to different classes of chemicals. *Arch Toxicol 5*: 355–359

Chung K-T, Murdock CA, Stevens SE Jr, Li Y-S, Wei C-I, Huang T-S, Chou M-W (1995) Mutagenicity and toxicity studies of *p*-phenylenediamine and its derivatives. *Toxicol Lett 81*: 23–32

Chung K-T, Murdock CA, Zhou Y, Stevens SE Jr, Li Y-S, Wei C-I, Fernando SY, Chou M-W (1996) Effects of the nitro-group on the mutagenicity and toxicity of some benzamines. *Environ Mol Mutagen 27*: 67–74

Ciccoli L, Ferrali M, Rossi V, Signorini C, Alessandrini C, Comporti M (1999) Hemolytic drugs aniline and dapsone induce iron release in erythrocytes and increase the free iron pool in spleen and liver. *Toxicol Lett 110*: 57–66

CIIT (Chemical Industry Institute of Technology) (1977) *Four week pilot study in rats, aniline hydrochloride, final report, 26.05.1977*. Hazleton Laboratories America, Vienna, VA, USA, CIIT, Research Triangle Park, NC, USA

CIIT (1982) *104-week chronic toxicity study in rats. Aniline hydrochloride*. Final report, Project No. 2010-101. Hazleton Laboratories America, Vienna, VA, USA, CIIT, Research Triangle Park, NC, USA

Cronin MTD, Basketter DA (1994) Multivariate QSAR analysis of a skin sensitization database. *SAR QSAR Environ Res 2*: 159–179

Czarnecki N (1977) Zur Klinik und Pathogenese des Friseurekzems (Clinical studies and pathogenesis of hairdressers' dermatitis) (German). *Z Hautkr 52*: 1–10

Deguchi J, Miyamoto M, Okada S (1995) Sex hormone-dependent renal cell carcinogenesis induced by ferric nitrilotriacetate in Wistar rats. *Jpn J Cancer Res 86*: 1068–1071

Dickinson DA, Forman HJ (2002) Cellular glutathione and thiols metabolism. *Biochem Pharmacol 64*: 1019–1026

Düngemann H, Borelli S (1966) Untersuchungen zur Gruppenallergie bei aromatischen Amino-Verbindungen (Studies of group allergies of aromatic amino compounds) (German). *Berufsdermatosen 14*: 281–295

Eberhartinger C (1984) Beobachtungen zur Häufigkeit von Kontaktallergien (Comments on the incidence of contact allergy) (German). *Z Hautkr 59*: 1283–1289

Ebner H, Lindemayr H (1977) Ulcus cruris und allergisches Kontaktekzem (Leg ulcer and allergic eczematous contact dermatitis) (German). *Wien Klin Wochenschr 89*: 184–188

ECB (European Chemicals Bureau) (2004) *Aniline*. European Union Risk Assessment Report, 1st priority list, volume 50, Office for Official Publications of the European Communities, Luxemburg, Luxemburg

Enders F (1986) *Häufigkeiten positiver Epikutantestreaktionen an 12993 Patienten der Dermatologischen Klinik und Poliklinik der Universität München von 1977–1983* (Frequency of patch test reactions in 12993 patients of the Dermatological Clinic and Outpatient Department of the University of Munich from 1977–1983) (German). Dissertation, Ludwig-Maximilians-Universität, München

Ethyl Corporation (1980) *Acute oral toxicity in Albino rats*, Project No. 12085. Bio-Research Laboratories, Ethyl Corporation, Baton Rouge, LA, USA, EPA-OTS 86-870001696, NTIS, Springfield, VA, USA

Fassina G, Abbodandolo A, Mariani L, Taningher M, Parodi S (1990) Mutagenicity in V79 cells does not correlate with carcinogenicity in small rodents for 12 aromatic amines. *J Toxicol Environ Health 29*: 109–130

Fedorowicz A, Singh H, Soderholm S, Demchuk E (2005) Structure-activity models for contact sensitization. *Chem Res Toxicol 18*: 954–969

Flandin C, Rabeau H, Ukrainczyk M (1936) Intolérances à certains anesthésiques et à l'aniline. Réactions de groupe (Intolerance to certain anaesthetics and to aniline. Group reactions) (French). *Bull Soc Fr Dermatol Syphiligr 43*: 1638–1640

Foussereau J, Escande JP, Lantz JP, Grosshans E, Wick P (1973) Sensitisation to ortho-aminoazotoluene. *Trans St John's Hosp Dermatol Soc 59*: 251–260

Freedman MH, Saunders EF (1981) Hematopoiesis in the human spleen. *Am J Hematol 11*: 271–275

Fritzenschaf H, Kohlpoth M, Rusche B, Schiffmann D (1993) Testing of known carcinogens and noncarcinogens in the Syrian hamster embryo (SHE) micronucleus test *in vitro*; correlations with *in vivo* micronucleus formation and cell transformation. *Mutat Res 319*: 47–53

Garberg P, Åkerblom E-L, Bolcsfoldi G (1988) Evaluation of a genotoxicity test measuring DNA-strand breaks in mouse lymphoma cells by alkaline unwinding and hydroxyapatite elution. *Mutat Res 203*: 155–176

George E, Andrews M, Westmoreland C (1990) Effects of azobenzene and aniline in the rodent bone marrow micronucleus test. *Carcinogenesis 11*: 1551–1555

Gerberick GF, Ryan CA, Kern PS, Schlatter H, Dearman RJ, Kimber I, Patlewicz GY, Basketter DA (2005) Compilation of historical local lymph node data for evaluation of skin sensitization alternative methods. *Dermatitis 16*: 157–202

Goodwin BFJ, Crevel RWR, Johnson AW (1981) A comparison of three guinea-pig sensitization procedures for the detection of 19 reported human contact sensitizers. *Contact Dermatitis 7*: 248–258

Grossman SJ, Simson J, Jollow DJ (1992) Dapsone-induced hemolytic anemia: effect of *N*-hydroxy dapsone on the sulfhydryl status and membrane proteins of rat erythrocytes. *Toxicol Appl Pharmacol 117*: 208–217

Haneke KE, Tice RR, Carson BL, Margolin BH, Stokes WS (2001) ICCVAM evaluation of the murine local lymph node assay. III. Data analyses completed by the National Toxicology Program Interagency Center for the Evaluation of Alternative Toxicological Methods. *Regul Toxicol Pharmacol 34*: 274–286

Harper BL, Ramanujam VMS, Gad-El-Karim MM, Legator MS (1984) The influence of simple aromatics on benzene clastogenicity. *Mutat Res 128*: 105–114

Harrison JH Jr, Jollow DJ (1987) Contribution of aniline metabolites to aniline-induced methemoglobinemia. *Mol Pharmacol 32*: 432–431

Hartwig A, Schlepegrell R (1995) Induction of oxidative DNA damage by ferric iron in mammalian cells. *Carcinogenesis 16*: 3009–3013

Harvey JW (1989) Erythrocyte metabolism. in: Kaneko JJ (Ed.) *Clinical biochemistry of domestic animals*, Academic Press, San Diego, CA, USA, 185–234

Henschler D, Lehnert G (Eds) (1986) Biologische Arbeitsstoff-Toleranz-Werte (BAT-Werte) und Expositionsäquivalente für krebserzeugende Arbeitsstoffe (EKA), 3. Lieferung, *Anilin*, VCH-Verlagsgesellschaft, Weinheim (in English translation in *Biological Exposure Values*, Volume 2 of the series now called *BAT Value Documentations*, available from the publisher)

von der Hude W, Behm C, Gürtler R, Basler A (1988) Evaluation of the SOS chromotest. *Mutat Res 203*: 81–94

IVDK (Informationsverbund Dermatologischer Kliniken) (2005) Datenbankauszug (database excerpt) (German), IVDK, Göttingen

Izumi Y, Matsumoto K, Fujimoto N, Ooshima Y (2005) Axial skeletal abnormalities in aniline hydrochloride-treated rats. *Congenit Anom (Kyoto) 45*: A48–A49

Jarolim P, Lahav M, Liu S-C, Palek J (1990) Effect of hemoglobin oxidation products on the stability of red cell membrane skeletons and the associations of skeletal proteins: correlation with a release of hemin. *Blood 76*: 2125–2131

Jenkins FP, Robinson JA, Gellatly JBM, Salmond GWA (1972) The no-effect dose of aniline in human subjects and a comparison of aniline toxicity in man and the rat. *Food Cosmet Toxicol 10*: 671–679

Jirásek L, Schwank R, Voborová A (1966) Skupinová prˇecitlivelost u azobarviv (Cross hypersensitivity to azodyes) (Czech). *Cesk Dermatol 41*: 381–393

Jollow DJ, McMillan DC (2001) Oxidative stress, glucose-6-phosphate dehydrogenase and the red cell. *Adv Exp Med Biol 500*: 595–606

Jones E, Fox V (2003) Lack of clastogenic activity of aniline hydrochloride in the mouse bone marrow. *Mutagenesis 18*: 283–286

van Joost T, Heule F, de Boer J (1987) Sensitization to methylenedianiline and para-structures. *Contact Dermatitis 16*: 246–248

Khan MF, Kaphalia BS, Boor PJ, Ansari GAS (1993) Subchronic toxicity of aniline hydrochloride in rats. *Arch Environ Contam Toxicol 24*: 368–374

Khan MF, Kaphalia BS, Ansari GAS (1995a) Erythrocyte-aniline interaction leads to their accumulation and iron deposition in rat spleen. *J Toxicol Environ Health 44*: 415–421

Khan MF, Boor PJ, Kaphalia BS, Alcock NW, Ansari GAS (1995b) Hematopoietic toxicity of linoleic acid anilide: Importance of aniline. *Fundam Appl Toxicol 25*: 224–232

Khan MF, Boor PJ, Gu Y, Alcock NW, Ansari GAS (1997) Oxidative stress in the splenotoxicity of aniline. *Fundam Appl Toxicol 35*: 22–30

Khan MF, Green SM, Ansari GAS, Boor PJ (1998) Phenylhydroxylamine: role in aniline-associated splenic oxidative stress and induction of subendocardial necrosis. *Toxicol Sci 42*: 64–71

Khan MF, Wu X, Alcock NM, Boor PJ, Ansari GAS (1999a) Iron exacerbates aniline-associated splenic toxicity. *J Toxicol Environ Health A 57*: 173–184

Khan MF, Wu X, Boor PJ, Ansari GAS (1999b) Oxidative modification of lipids and proteins in aniline-induced splenic toxicity. *Toxicol Sci 48*: 134–140

Khan MF, Wu X, Ansari GAS (2000a) Contribution of nitrosobenzene to splenic toxicity of aniline. *J Toxicol Environ Health A 60*: 263–273

Khan MF, Wu X, Ansari GS (2000b) Induction of transforming growth factor-beta 1 in splenocytes of aniline-treated rats. *Toxicologist 54*: 255

Khan MF, Wu X, Ansari GAS, Boor PJ (2003a) Malondialdehyde-protein adducts in the spleens of aniline-treated rats: immunochemical detection and localization. *J Toxicol Environ Health A 66*: 93–102

Khan MF, Wu X, Boor PJ, Ansari GS (2003b) Nitrotyrosine and splenic toxicity of aniline. *Toxicologist 72*: 78

Khan MF, Wu X, Kaphalia BS, Boor PJ, Ansari GAS (2003c) Nitrotyrosine formation in splenic toxicity of aniline. *Toxicology 194*: 95–102

Kiese M (1974) *Methemoglobinemia – A comprehensive treatise*, CRC Press, Cleveland, OH, USA, 100

Kim YC, Carlson GP (1986) The effect of an unusual workshift on chemical toxicity. II. Studies on the exposure of rats to aniline. *Fundam Appl Toxicol 7*: 144–152

Kirkland DJ, Dresp JH, Marshall RR, Baumeister M, Gerloff C, Gocke E (1992) Normal chromosomal aberration frequencies in peripheral lymphocytes of healthy human volunteers exposed to a maximum daily dose of paracetamol in a double blind trial. *Mutat Res 279*: 181–194

Kleniewska D (1975) Studies on hypersensitivity to "para group". *Berufsdermatosen 23*: 31–36

Kligman AM (1966) The identification of contact allergens by human assay. III. The maximization test: a procedure for screening and rating contact sensitizers. *J Invest Dermatol 12*: 393–409

Kondrashov VA (1969) [The toxic influence of the vapours of chloroanilines and aniline on the organism upon exposure via the intact skin] (Russian). *Gig Tr Prof Zabol 13*: 29–32

Lewalter J, Korallus U (1985) Blood protein conjugates and acetylation of aromatic amines. *Int Arch Occup Environ Health 56*: 179–196

Lewalter J, Leng G, Sturm S (2002) Bedeutung der Proteinaddukt-Befunde in der Risikobewertung des Anilinumgangs (Importance of protein adduct findings in risk assessment after contact with aniline) (German). in: Nowak D, Praml G (Eds) *Verhandlungen der DGAUM, Dokumentationsband*, Rindt-Druck, Fulda, 486–487

Li S, Fedorowicz A, Singh H, Soderholm SC (2005) Application of the random forest method in studies of local lymph node assay based skin sensitization data. *J Chem Inf Model 45*: 952–964

Malten KE (1969) Results of an additional standard series patch testing series in 1000 patients. *Contact Dermatitis Newslett 6*: 139

Malten KE, Kuiper JP, van der Staak WBJM (1973) Contact allergic investigations in 100 patients with ulcus cruris. *Dermatologica 147*: 241–254

Matsumoto K, Matsumoto S, Fukuta K, Ooshima Y (2001a) Cardiovascular malformations associated with maternal hypoxia due to methemoglobinemia in aniline hydrochloride-treated rats. *Congenit Anom (Kyoto) 41*: 118–123

Matsumoto K, Seki N, Fukuta K, Ooshima Y (2001b) Induction of cleft palate in aniline hydrochloride-treated rats: possible effect of maternal methemoglobinemic hypoxia. *Congenit Anom (Kyoto) 41*: 112–117

Matsumoto K, Seki N, Ooshima Y (2002) Investigation of the mechanism of aniline hydrochloride inducible cleft palate in rat fetuses. *Congenit Anom (Kyoto) 42*: 274–275

Matsushima T, Hayashi M, Matsuoka A, Ishidate M Jr, Miura KF, Shimizu H, Suzuki Y, Morimoto K, Ogura H, Mure K, Koshi K, Sofuni T (1999) Validation study of the *in vitro* micronucleus test in a Chinese hamster lung cell line (CHL/IU). *Mutagenesis 14*: 569–580

Mawatari S, Murakami K (2004) Different types of glutathionylation of hemoglobin can exist in intact erythrocytes. *Arch Biochem Biophys 421*: 108–114

Mayer RL (1928) Die Überempfindlichkeit gegen Körper von Chinonstruktur (Oversensitivity to compounds with quinone structure) (German). *Arch Dermatol Syph 156*: 331–354

Mayer RL (1929) Die Hautüberempfindlichkeit gegen Körper von Chinonstruktur. III. Mitteilung. Ursol- und Pellidolüberempfindlichkeit, anderer Mechanismus in gewissen Fällen nebst Bemerkungen über die Natur des reizenden Stoffes in den gefärbten Pelzen (Oversensitivity of the skin to compounds with quinone structure. III. Communication. Oversensitivity to Ursol and Pellidol, other mechanisms in certain cases and comments on the nature of the irritant substance in dyed furs) (German). *Arch Dermatol Syph 158*: 266–274

Mayer RL (1954) Group-sensitization to compounds of quinone structure and its biochemical basis; role of these substances in cancer. *Prog Allergy 4*: 79–172

McCarthy DJ, Waud WR, Struck RF, Hill DL (1985) Disposition and metabolism of aniline in Fischer 344 rats and C57BL/6×C3HF1 mice. *Cancer Res 45*: 174–180

McGregor DB, Brown AG, Howgate S, McBride D, Riach C, Caspary WJ (1991) Responses of the L5178Y mouse lymphoma cell forward mutation assay. V: 27 coded chemicals. *Environ Mol Mutagen 17*: 196–219

McQueen CA, Maslansky CJ, Crescenzi SB, Williams GM (1981) The genotoxicity of 4,4′-methylenebis-2-chloroaniline in rat, mouse, and hamster hepatocytes. *Toxicol Appl Pharmacol 58*: 231–235

Mellert W, Deckardt K, Gembardt C, Zwirner-Baier I, Jäckh R, van Ravenzwaay B (2004) Aniline: early indicators of toxicity in male rats and their relevance to spleen carcinogenicity. *Hum Exp Toxicol 23*: 379–389

Meltzer L, Baer RL (1949) Sensitization to monoglycerol para-aminobenzoate. *J Invest Dermatol 12*: 31–39

Meneghini CL, Rantuccio F, Riboldi A (1963) Klinisch-allergologische Beobachtungen bei beruflich ekzematösen Kontakt-Dermatosen (Clinical allergological observations on occupational eczematous contact dermatitis) (German). *Kontakt-Dermatosen 11*: 280–293

Meneghini CL, Rantuccio F, Riboldi A, Hofmann MF (1967) Beobachtungen über das Persistieren der experimentellen ekzematösen Kontaktsensibilisierung auf einige chemische Substanzen beim Menschen (Studies of the persistence in man of experimental eczematous contact sensitization to certain chemical substances) (German). *Berufsdermatosen 15*: 103–111

Mierzecki H, Miklaszewska M (1958) Die Aktivitätsveränderungen der Katalase im Laufe der Sensibilisierung und Vergiftung der Haut durch industrielle chemische Verbindungen (Activity changes of catalase in the course of sensitization and poisoning of the skin by industrial chemical compounds) (German). *Arch Gewerbepathol Gewerbehyg 16*: 387–395

Mitchell AD, Rudd CJ, Caspary WJ (1988) Evaluation of the L5178Y mouse lymphoma cell mutagenesis assay: intralaboratory results for sixty-three coded chemicals tested at SRI International. *Environ Mol Mutagen 12, Suppl 13*: 37–101

Moriearty PL, Pereira C, Guimaraes NA (1978) Contact dermatitis in Salvador, Brazil. *Contact Dermatitis 4*: 185–189

Myhr BC, Caspary WJ (1988) Evaluation of the L5178Y mouse lymphoma cell mutagenesis assay: intralaboratory results for sixty-three coded chemicals tested at Litton Bionetics, Inc. *Environ Mol Mutagen 12, Suppl 13*: 103–194

Nagao M, Yahagi T, Honda M, Sieno Y, Matsushima T, Sugimura T (1977) Demonstration of mutagenicity of aniline and *o*-toluidine by norharman. *Proc Jpn Acad Ser B Phys Biol Sci 53*: 34–37

NCI (National Cancer Institute) (1978) *Bioassay of aniline hydrochloride for possible carcinogenicity*. CAS No. 142-04-1, Technical Report Series No. 130 (NTIS PB-287539), National Cancer Institute, Bethesda, MD, USA

Neumann H-G (1985) Die Rolle der Promotion bei der Einwirkung krebserzeugender Stoffe (The role of promotion in the effects of carcinogenic substances) (German). in: Appel KE, Hildebrand AG (Eds) *Tumorpromotoren. Erkennung, Wirkungsmechanismen und Bedeutung*, bga Schriften 6/85, Medizin Verlag, München, 98–107

Nishigaki R, Totsuka Y, Takamura-Enya T, Sugimura T, Wakabayashi K (2004) Identification of cytochrome P-450s involved in the formation of APNH from norharman with aniline. *Mutat Res 562*: 19–25

Oda Y, Yamazaki H, Watanabe M, Nohmi T, Shimada T (1995) Development of high sensitive umu test system: rapid detection of genotoxicity of promutagenic aromatic amines by *Salmonella typhimurium* strain NM2009 possessing high *O*-acetyltransferase activity. *Mutat Res 334*: 145–156

Pambor M (1971) Ein *p*-Phenetidin-haltiges Antioxydans als berufliches Ekzematogen bei der Herstellung von Spezialfuttermischungen (A *p*-phenetidine-containing antioxidant as an occupational eczematogenic agent in the manufacturing of special feed mixtures) (German). *Berufsdermatosen 19*: 285–291

Papanikolaou G, Pantopoulos K (2005) Iron metabolism and toxicity. *Toxicol Appl Pharmacol 202*: 199–211

Parodi S, Pala M, Russo P, Zunino A, Balbi C, Albini A, Valerio F, Cimberle MR, Santi L (1982) DNA damage in liver, kidney, bone marrow, and spleen of rats and mice treated with commercial and purified aniline as determined by alkaline elution assay and sister chromatid exchange induction. *Cancer Res 42*: 2277–2283

Paschoud JM (1963) Quelques cas d'eczéma de contact avec sensibilisation de groupe (Some cases of contact eczema with group sensitization) (French). *Dermatologica 127*: 349–364

Pauluhn J (2002) Aniline-induced methemoglobinemia in dogs: pitfalls of route-to route extrapolations. *Inhal Toxicol 14*: 959–973

Pauluhn J (2004) Subacute inhalation toxicity of aniline in rats: analysis of time-dependence and concentration-dependence of hematotoxic and splenic effects. *Toxicol Sci 81*: 198–215

Pauluhn J (2005) Concentration-dependence of aniline-induced methemoglobinemia in dogs: a derivation of an acute reference concentration. *Toxicology 214*: 140–150

Pauluhn J, Mohr U (2001) Inhalation toxicity of 4-ethoxyaniline (*p*-phenetidine): critical analysis of results of subacute inhalation exposure studies in rats. *Inhal Toxicol 13*: 993–1013

Pepelko WE (1970) Effects of hypoxia and hypercapnea, singly and combined, on growing rats. *J Appl Physiol 28*: 646–651

Piotrowski J (1972) *Pracov Lék 24*: 94–97

Price CJ, Tyl RW, Marks TA, Paschke LL, Ledoux TA, Reel JR (1985) Teratologic and postnatal evaluation of aniline hydrochloride in the Fischer 344 rat. *Toxicol Appl Pharmacol 77*: 465–478

Ress NB, Witt KL, Xu J, Haseman JK, Bucher JR (2002) Micronucleus induction in mice exposed to diazoaminobenzene or its metabolites, benzene and aniline: implications for diazoaminobenzene carcinogenicity. *Mutat Res 521*: 201–208

Roberts JJ, Warwick GP (1966) The covalent binding of metabolites of dimethylaminoazobenzene, β-naphthylamine and aniline to nucleic acids *in vivo*. *Int J Cancer 1*: 179–196

Rockwood GA, Armstrong KR, Baskin SI (2003) Species comparison of methemoglobin reductase. *Exp Biol Med 228*: 79–83

Roudabush RL, Terhaar CJ, Fassett DW, Dziuba SP (1965) Comparative acute effects of some chemicals on the skin of rabbits and guinea pigs. *Toxicol Appl Pharmacol 7*: 559–563

Rudzki E, Grzywa Z (1975) Contact sensitivity in atopic dermatitis. *Contact Dermatitis 1*: 285–287

Rudzki E, Kleniewska D (1970) The epidemiology of contact dermatitis in Poland. *Br J Dermatol 83*: 543–545

Rudzki E, Kleniewska D (1971) Cross reactions between parabens and the para group. *Contact Dermatitis Newslett 9*: 199

Rudzki E, Rebandel P, Zawadzka A (1995) Sensitivity to diaminodiphenylmethane. *Contact Dermatitis 32*: 303

Russo P, Vecchio D, Balbi C, Parodi S, Santi L (1981) Danno al DNA, dopo trattamento "*in vivo*" con alcune ammine aromatiche: 2,4-diamminotoluene, 4-amminobenzene (anilina) e paradimethylamminobenzene (giallo burro) (DNA damage after treatment "*in vivo*" with several aromatic amines: 2,4-diaminotoluene, 4-aminobenzene (aniline) and paradimethylaminobenzene (butter yellow)) (Italian). *Boll Soc Ital Biol Sper 57*: 131–137

Sakagami Y, Yokoyama H, Ose Y, Sato T (1986) Screening test for carcinogenicity of chlorhexidine digluconate and its metabolites (Japanese). *Eisei Kagaku 32*: 171–175

Sakagami Y, Yamazaki H, Ogasawara N, Yokoyama H, Ose Y, Sato T (1988) The evaluation of genotoxic activities of disinfectants and their metabolites by umu test. *Mutat Res 209*: 155–160

Sasaki YF, Fujikawa K, Ishida K, Kawamura N, Nishikawa Y, Ohta S, Satoh M, Madarame H, Ueno S, Susa N, Matsusaka N, Tsuda S (1999) The alkaline single cell gel electrophoresis assay with mouse multiple organs: results with 30 aromatic amines evaluated by the IARC and U. S. NTP. *Mutat Res 440*: 1–18

Scarpa C, Ferrea E (1966) Group variation in reactivity to common contact allergens. *Arch Dermatol 94*: 589–591

Schiestl RH (1989) Nonmutagenic carcinogens induce intrachromosomal recombination in yeast. *Nature 337*: 285–288

Schiestl RH, Gietz RD, Mehta RD, Hastings PJ (1989) Carcinogens induce intrachromosomal recombination in yeast. *Carcinogenesis 10*: 1445–1455

Schultheiss E (1959) *Gummi und Ekzem* (Rubber and dermatitis) (German), Editio Cantor, Aulendorf

Schulz KH (1962a) Allergien gegenüber aromatischen Amino- und Nitro-Verbindungen (Allergies to aromatic amino and nitro compounds) (German). *Berufsdermatosen 10*: 69–91

Schulz KH (1962b) *Chemische Struktur und allergene Wirkung – Unter besonderer Berücksichtigung von Kontaktallergenen* (Chemical structure and allergenic effects – in particular of contact allergens) (German), Editio Cantor, Aulendorf

Schwarz A, Gottmann-Lückerath J (1982) Allergenhäufigkeit bei Kontaktallergien in der Universitäts-Hautklinik Köln (1970–1971 und 1976–1979) (The incidence of contact allergens at the Department of Dermatology of the University of Cologne (1970–1971 and 1976–1979)) (German). *Z Hautkr 57*: 951–960

Seifert MF, Marks SC Jr (1985) The regulation of hemopoiesis in the spleen. *Experientia 41*: 192–199

Sekihashi K, Yamamoto A, Matsumura Y, Ueno S, Watanabe-Akanuma M, Kassie F, Knasmüller S, Tsuda S, Sasaki YF (2002) Comparative investigation of multiple organs of mice and rats in the Comet assay. *Mutat Res 517*: 53–75

Sicardi SM, Martiarena JL, Iglesias MT (1991) Mutagenic and analgesic activities of aniline derivatives. *J Pharm Sci 80*: 761–764

Singh H, Purnell ET (2005a) Aniline derivative-induced methemoglobin in rats. *J Environ Pathol Toxicol Oncol 24*: 57–65

Singh H, Purnell ET (2005b) Hemolytic potential of structurally related aniline halogenated hydroxylamines. *J Environ Pathol Toxicol Oncol 24*: 67–76

Smith RP (1986) Toxic responses of the blood. in: Amdur MO, Doull J, Klaassen CD (Eds) *Casarett and Doull's Toxicology*, Pergamon Press, New York, USA, 257–281

Srivastava S, Alhomida AS, Siddiqi NJ, Puri SK, Pandey VC (2002) Methemoglobin reductase activity and *in vitro* sensitivity towards oxidant induced methemoglobinemia in Swiss mice and Beagle dogs erythrocytes. *Mol Cell Biochem 232*: 81–85

Stone K, Ksebati MB, Marnett LJ (1990) Investigation of the adducts formed by reaction of malon-dialdehyde with adenosine. *Chem Res Toxicol 3*: 33–38

Suzuki Y, Nagae Y, Ishikawa T, Watanabe Y, Nagashima T, Matsukubo K, Shimizu H (1989) Effect of erythropoietin on the micronucleus test. *Environ Mol Mutagen 13*: 314–318

Suzuki Y, Shimizu H, Ishikawa T, Sakaba H, Fukumoto M, Okonogi H, Kadokura M (1994) Effects of prostaglandin E2 on the micronucleus formation in mouse bone marrow cells by various mutagens. *Mutat Res 311*: 287–293

Swenberg JA (1981) Utilization of the alkaline elution assay as a short-term test for chemical carcinogens. in: Stich HF, San RHC (Eds) *Short-term tests for chemical carcinogens*, Springer, New York, 48–58

Sziza M, Podhragyai L (1957) Toxikologische Untersuchung einiger in der ungarischen Industrie zur Anwendung gelangender aromatischer Amidoverbindungen (Toxicological study of certain aromatic amido compounds used in the Hungarian industry) (German). *Arch Gewerbepathol Gewerbehyg 15*: 447–456

Takehisa S, Kanaya N (1982) SCE induction in human lymphocytes by combined treatment with aniline and norharman. *Mutat Res 101*: 165–172

Takehisa S, Kanaya N, Rieger R (1988) Promutagen activation by *Vicia faba*: an assay based on the induction of sister-chromatid exchanges in Chinese hamster ovary cells. *Mutat Res 197*: 195–205

Tohda H, Tada M, Sugawara R, Oikawa A (1983) Actions of amino-β-carbolines on induction of sister-chromatid exchanges. *Mutat Res 116*: 137–147

Topham JC (1980a) The detection of carcinogen-induced sperm head abnormalities in mice. *Mutat Res 69*: 149–155

Topham JC (1980b) Do induced sperm-head abnormalities in mice specifically identify mammalian mutagens rather than carcinogens? *Mutat Res 74*: 379–387

Totsuka Y, Hada N, Matsumoto K-I, Kawahara N, Murakami Y, Yokohama Y, Sugimura T, Waka-bayashi K (1998) Structural determination of a mutagenic aminophenylnorharman produced by the co-mutagen norharman and aniline. *Carcinogenesis 19*: 1995–2000

Wang J, Kannan S, Li H, Khan MF (2005) Cytokine gene expression and activation of NF-χB in aniline-induced splenic toxicity. *Toxicol Appl Pharmacol 203*: 36–44

Weinberger MA, Albert RH, Montgomery ST (1985) Splenotoxicity associated with splenic sarcomas in rats fed high doses of D&C red No. 9 or aniline hydrochloride. *J Natl Cancer Inst 75*: 681–687

Westmoreland C, Gatehouse DG (1991) Effects of aniline hydrochloride in the mouse bone marrow micronucleus test after oral administration. *Carcinogenesis 12*: 1057–1059

Wilhelm J, Herget J (1999) Hypoxia induces free radical damage to rat erythrocytes and spleen: analysis of the fluorescent end-products of lipid peroxidation. *Int J Biochem Cell Biol 31*: 671–681

Williams GM (1980) DNA repair and mutagenesis in liver cultures as indicators in chemical carcinogen screening. in: Mishra N, Dunkel V, Mehlman M (Eds) *Advances in modern environmental toxicology*, Volume 1, Mammalian cell transformation by chemical carcinogens, Senate Press, Princeton Junction, NJ, USA, 273–296

Wilmer JL, Erexson GL, Kligerman AD (1984) The effect of erythrocytes and hemoglobin on sister chromatid exchange induction in cultured human lymphocytes exposed to aniline HCl. *Basic Life Sci 293*: 561–567

Witt KL, Knapton A, Wehr CM, Hook GJ, Mirsalis J, Shelby MD, MacGregor JT (2000) Micro-nucleated erythrocyte frequency in peripheral blood of B6C3F1 mice from short-term, prechronic, and chronic studies of the NTP carcinogenesis bioassay program. *Environ Mol Mutagen 36*: 163–194

Wu X, Kannan S, Ramanujam VM-S, Khan MF (2005) Iron release and oxidative DNA damage in splenic toxicity of aniline. *J Toxicol Environ Health 68*: 657–666

Wulferink M, González J, Goebel C, Gleichmann E (2001) T cells ignore aniline, a prohapten, but respond to its reactive metabolites generated by phagocytes: possible implications for the pathogenesis of toxic oil syndrome. *Chem Res Toxicol 14*: 389–397

Zwirner-Baier I, Deckart K, Jäckh R, Neumann H-G (2003) Biomonitoring of aromatic amines VI: determination of hemoglobin adducts after feeding aniline hydrochloride in the diet of rats for 4 weeks. *Arch Toxicol 77*: 672–677

Zündel W (1936) Erfahrungen mit Hautfunktionsprüfungen an 2000 Patienten (Experiences with skin function testing in 2000 patients) (German). *Arch Dermatol Syph 173*: 435–472

completed 29.03.2006

2-Butoxyethanol[1] (Ethylene glycol monobutyl ether)

Supplement 2007

MAK value (2006)	**10 ml/m³ (ppm) ≙ 49 mg/m³**
Peak limitation (2006)	**Category I, excursion factor 2**
Absorption through the skin (1980)	**H**
Sensitization	**–**
Carcinogenicity (2006)	**Category 4**
Prenatal toxicity (1985)	**Pregnancy Risk Group C**
Germ cell mutagenicity	**–**
BAT value (1995, 2008)	**100 mg butoxyacetic acid/l urine 200 mg total butoxyacetic acid/l urine**
Synonyms	*n*-butoxyethanol *O*-butyl ethylene glycol butyl glycol EGBE ethylene glycol *n*-butyl ether glycol butyl ether monobutyl glycol ether 3-oxa-1-hepatanol
Chemical name (CAS)	2-butoxyethanol
CAS number	111-76-2

Since the MAK documentation from 1983 and 1986 ("2-Butoxyethanol", Volume 6, present series), two other studies have been published which necessitate a re-evaluation of 2-butoxyethanol.

[1] MAK value for the sum concentration of 2-butoxyethanol and 2-butoxyethylacetate in air

The MAK-Collection Part I: MAK Value Documentations, Vol. 26. DFG, Deutsche Forschungsgemeinschaft

ISBN: 978-3-527-32306-7

1 Toxic Effects and Mode of Action

Increases were observed in the incidence of benign and malignant phaeochromocytomas in the adrenal cortex of female rats, in the incidence of liver cell carcinomas and haemangiosarcomas in the liver of male mice and in the incidence of squamous cell papillomas of the forestomach (combined with a concentration-dependent increase in ulcers and epithelial hyperplasia) of female mice after inhalation exposure to 2-butoxyethanol for two years. The available data from genotoxicity studies *in vitro* and *in vivo* indicate that neither 2-butoxyethanol nor its metabolite 2-butoxyacetic acid have genotoxic effects. The data from *in vitro* studies with 2-butoxyacetaldehyde indicate the substance has mutagenic potential, but are not conclusive.

After inhalation exposure, oral administration or dermal application, 2-butoxyethanol is readily absorbed, rapidly distributed in tissue, metabolized, mainly by alcohol dehydrogenase and aldehyde dehydrogenase, to form 2-butoxyacetaldehyde and 2-butoxyacetic acid, and eliminated with the urine.

Haemolysis of the erythrocytes is a species-specific toxic effect, caused by the metabolite 2-butoxyacetic acid. It is especially pronounced in the rat, mouse and rabbit. Changes in haematological parameters occur in female rodents after inhalation exposure to 2-butoxyethanol concentrations of 31 ml/m^3 for 14 weeks. Other target organs are the liver, kidneys, spleen and the central nervous system (CNS).

At 2-butoxyethanol concentrations of 100 ml/m^3 and above, eye, nose and throat irritation, nausea and headaches occur in humans. No clinical signs of haemolysis were observed after exposure to 2-butoxyethanol concentrations of 195 ml/m^3 for four to eight hours.

There are no clinical reports of sensitizing effects of 2-butoxyethanol. Nor did a maximization test in guinea pigs indicate that the substance has contact-sensitizing effects.

Reproductive toxicity occurs in the rat, mouse and rabbit only at doses that also produce general toxicity.

2 Mechanism of Action

2.1 Haemolytic effect

The characteristic effect of 2-butoxyethanol in various species is its haemolytic effect; this is caused by its metabolite 2-butoxyacetic acid. Signs of 2-butoxyethanol-induced haemolysis are swelling of the erythrocytes, morphological changes and decreasing deformability (IARC 2006). Data from *in vivo* and *in vitro* studies show that rats react more sensitively than other species. Changes in haematological parameters occurred in rats after long-term exposure to 2-butoxyethanol concentrations of 31 ml/m^3 and above.

Comparative investigations with human and rat erythrocytes showed human erythrocytes to be markedly less sensitive. Thus, only minimal swelling or haemolysis of the erythrocytes, but no decrease in the blood ATP concentration, was observed after incubation of human blood with 2-butoxyacetic acid concentrations of up to 4 mmol/l. 2-Butoxyacetic acid concentrations of 0.5 to 2.0 mmol/l caused complete haemolysis, however, in rat erythrocytes. No haemolysis was found in blood samples collected from 97 persons of various ages and states of health exposed to 2-butoxyacetic acid concentrations of 10 mmol/l for four hours. A slight decrease in deformability and increasing osmotic fragility were described in human erythrocytes at 2-butoxyacetic acid concentrations of 7.5 to 10 mmol/l, while in rats these effects were seen at concentrations as low as 0.05 mmol/l (ECETOC 2005).

2.2 Carcinogenesis

As the published data do not indicate that 2-butoxyethanol causes genotoxic effects, the following mechanisms have been suggested for its carcinogenicity.

Liver neoplasms in male mice

The liver neoplasms in male mice do not seem to be caused by 2-butoxyethanol directly, but indirectly via its haemolytic effects. 2-Butoxyethanol acts haemolytically in rodents, producing iron deposits in the liver. On the one hand, iron can form reactive oxygen species via Fenton or Haber-Weiss reactions, which can result in tumour formation via lipid peroxidation or oxidative DNA damage. On the other hand, excess iron can activate Kupffer's cells, and thus release reactive oxygen species and other biologically active molecules such as cytokines, which can also contribute to carcinogenesis (Boatman *et al.* 2004; Siesky *et al.* 2002).

An increase in oxidative DNA damage (8-hydroxydeoxyguanosine) and lipid peroxidation (malondialdehyde) in the liver and in DNA synthesis in hepatocytes and endothelial cells was observed in male mice after oral administration of 2-butoxyethanol (see Section 5.2.2). In addition, it was shown in rat and mouse hepatocytes that neither 2-butoxyethanol nor 2-butoxyacetic acid causes direct oxidative DNA damage or lipid peroxidation. Iron(II) sulfate, on the other hand, produced a significant increase in 8-hydroxydeoxyguanosine and malondialdehyde. Mouse hepatocytes were found to react more sensitively than rat hepatocytes (Park *et al.* 2002a). Neither 2-butoxyethanol nor 2-butoxyacetic acid induced cell transformation in SHE cells (a cell line derived from Syrian hamster embryo). Iron(II) sulfate, however, caused a significant increase in the frequency of transformations. After incubation of the SHE cells with iron(II) sulfate and additionally with an antioxidant such as vitamin E or (–)-epigallocatechin-3-gallate, significantly fewer transformations were observed than after incubation of the cells with iron(II) sulfate alone. The amount of oxidatively damaged DNA (8-hydroxydeoxyguanosine) and the number of DNA strand breaks detected in the comet assay after treating the cells with iron(II) sulfate and additionally with an antioxidant such as vitamin E or (–)-epigallocatechin-3-gallate were significantly lower than after treatment of the cells with iron(II) sulfate alone (Park *et al.* 2002b).

Although 2-butoxyethanol causes haemolysis in rats and mice, an increase in oxidative damage in the liver and in DNA synthesis in the hepatocytes and endothelial cells (see Section 5.2.2) was observed only in male B6C3F$_1$ mice, but not in male F344 rats after oral administration of 2-butoxyethanol. The reason for this difference in sensitivity is considered to be the lower antioxidative capacity in the liver of male mice compared with that in rats and female mice (Boatman *et al.* 2004). The level of vitamin E was approximately 2.5 times higher in the liver of untreated male F344 rats than in that of untreated male B6C3F$_1$ mice. After oral administration of 2-butoxyethanol, the level of vitamin E in the liver decreased both in mice and in rats. Nevertheless, the lowest level of vitamin E (reduced by exposure to 2-butoxyethanol) in the liver of male F344 rats was still higher than that found in untreated male B6C3F$_1$ mice (Siesky *et al.* 2002).

That excess iron could play a part in carcinogenesis is seen also in the fact that hepatocellular carcinomas are produced in about 30 % of haemochromatosis patients with a pathological excess of iron. Other types of cancer, such as oesophageal tumours, likewise correlate with hereditary haemochromatosis (Papanikolaou and Pantopoulos 2005).

130 NTP studies with B6C3F$_1$ mice were evaluated for a possible relationship between haemangiosarcomas in the liver and excess iron. The three substances (2-butoxyethanol, *p*-nitroaniline and *p*-chloroaniline), for which a relatively high incidence of Kupffer's cell pigmentation with haemosiderin was found in both sexes, had also resulted in an incidence—albeit relatively low—of substance-related haemangiosarcomas in male mice. Other substances had induced either a low incidence of Kupffer's cell pigmentation and no substance-related haemangiosarcomas or substance-related haemangiosarcomas, but no Kupffer's cell pigmentation (Nyska *et al.* 2004).

Squamous cell papillomas, ulcers and epithelial hyperplasia in the forestomach of female mice

In the NTP (2000) publication, it was suggested that 2-butoxyethanol or its metabolites induced the squamous cell papillomas observed in the forestomach of female mice via chronic irritation of the forestomach and the resultant regenerative hyperplasia. This mechanism of action is discussed extensively in EPA (2005). It can be summarized in the following steps:

1. Deposition in the forestomach of 2-butoxyethanol or its metabolites directly or via systemic distribution after exposure. This was shown in studies with rats and mice after inhalation, intravenous and intraperitoneal administration, subcutaneous injection and oral exposure (see Section 3.1 and 5.2) (EPA 2005).
2. Retention of 2-butoxyethanol or 2-butoxyacetic acid in the forestomach tissue and contents (feed) so that, for example, 24 hours after intraperitoneal or oral administration of 2-butoxyethanol, high (no other details) concentrations of 2-butoxyethanol and 2-butoxyacetic acid were determined in the forestomach, but were not detectable in other tissues or blood after 30 minutes (see Section 3.1).

3. Metabolic conversion of 2-butoxyethanol to 2-butoxyacetic acid via 2-butoxyacetaldehyde.
4. Irritation of the target cells, leading to hyperplasia and ulcers (see Sections 5.2.1, 5.2.2 and 5.2.4).
5. Persistent damage and degeneration, resulting in high cell proliferation and a high turnover.
6. High cell proliferation and a high turnover, leading to the clonal expansion of spontaneously initiated forestomach cells. The fact that exposure to 2-butoxyethanol does not increase the mutation pattern and mutation frequency of the H-*ras* gene in forestomach neoplasms also speaks for this non-genotoxic mechanism (see Section 5.6.2) (EPA 2005).

In male mice, no significant increase was observed in the incidence of squamous cell papillomas or carcinomas in the forestomach compared with that in the concurrent controls; however, the marked, concentration-dependent increase in the incidence of epithelial hyperplasia in the forestomach (see Section 5.2.1) and the incidence of squamous cell papillomas or carcinomas in the forestomach outside the range of variation of the historical controls (NTP 2000) could indicate the early stages of carcinogenesis and would lend support to the postulated effect mechanism.

Investigations of the alcohol dehydrogenase or aldehyde dehydrogenase in the stomach of mice and rats (Green *et al.* 2002) showed that the ratio of V_{max} to K_M for the conversion of 2-butoxyethanol to aldehyde is considerably smaller than the V_{max} to K_M ratio for the oxidation of the aldehyde to the acid (EPA 2005). Accumulation of the aldehyde in the stomach is, therefore, unlikely. Accordingly, very low aldehyde concentrations (maximum: about 33 µmol/l in female animals and about 19 µmol/l in males) were detected in the forestomach of mice after oral 2-butoxyethanol doses of 600 mg/kg body weight (Deisinger and Boatman 2004). The high local concentrations of metabolite to be expected in this region with the dehydrogenase concentration in the stratified squamous epithelium of the forestomach could explain the higher sensitivity of the forestomach compared with the glandular stomach, in which the enzymes are distributed more evenly (Green *et al.* 2002). Whether the effects in the forestomach can be explained by the higher maximum conversion rate of alcohol dehydrogenase in mice compared with that in rats was also discussed (Green *et al.* 2002). However, as the alcohol dehydrogenase in mice has a higher K_M value than that in rats and consequently the conversion to aldehyde in the stomach in the relevant concentration range of < 4.5 mM 2-butoxyethanol takes place more rapidly in rats than in mice (EPA 2005), the occurrence of effects in the forestomach only in mice cannot be explained by this. The mice, however, were exposed to a 2-butoxyethanol concentration two times higher than that in the rats, which certainly would have resulted in a higher dose in the target organ (EPA 2005).

In a drinking water study in mice exposed for 2 and 13 weeks, no signs of irritation in the forestomach were found after doses of up to 1400 mg/kg body weight and day. It was suggested that continuous administration with the drinking water, unlike a bolus dose, e.g. after gavage, produced an insufficiently high and consequently non-irritative dose in the forestomach. In addition, the systemic availability and thus also the local dose could be reduced as a result of the first pass effect after oral administration (EPA 2005).

On the other hand, in another study, clearly dose-dependent irritation of the forestomach was demonstrated after oral administration (gavage) of the undiluted substance to mice for four days (see Section 5.2.2) (Poet *et al.* 2003).

It was also discussed whether 2-butoxyacetaldehyde contributes to tumour formation by means of (possibly) weak genotoxic activity: however, structural comparisons between aldehydes indicate that longer-chain aldehydes such as 2-butoxyacetaldehyde interact with DNA to a lesser extent than short-chain aldehydes such as acetaldehyde. In addition, corresponding investigations with 2-butoxyethanol yielded no results which would indicate a genotoxic metabolite (EPA 2005).

3 Toxicokinetics and Metabolism

3.1 Absorption, distribution, elimination

2-Butoxyethanol is readily absorbed after inhalation and oral and dermal exposure and is rapidly distributed in the tissues, metabolized and eliminated (mainly with the urine, but some is exhaled in the form of CO_2) (NTP 2000).

Male volunteers exposed to 2-butoxyethanol concentrations of 20 ml/m^3 for 2 hours while performing light physical exercise absorbed on average 10.1 µmol/min (1.2 mg/min) or 57 % of the inhaled 2-butoxyethanol. Within 1 to 2 hours, a maximum 2-butoxyethanol concentration (plateau) of 7.4 µmol/l (0.9 mg/l) was attained in the capillary blood. When inhalation took place through the mouth only, a steady state appeared to be reached in the capillary blood of male volunteers at approximately 3 µmol/l during the second hour after exposure to a 2-butoxyethanol concentration of 50 ml/m^3 for 2 hours. In all, 1.3 mmol 2-butoxyethanol (11 µmol/min) was absorbed during the 2-hour mouth-only inhalation exposure (ATSDR 1998). The elimination half-time for 2-butoxyethanol in the blood was 40 minutes. Of the total amount of 2-butoxyethanol absorbed, less than 0.03 % was eliminated with the urine in the form of 2-butoxyethanol and 17 % to 55 % in the form of 2-butoxyacetic acid (NTP 2000).

The influence of room temperature, humidity and clothing on the dermal absorption of 2-butoxyethanol from the vapour phase (2 hours, 50 ml/m^3) was investigated in 4 volunteers. The amount of 2-butoxyethanol absorbed was determined in the 34-hour urine. At 25°C and 40 % relative humidity, around 11 % of the absorbed dose of 2-butoxyethanol was absorbed dermally; the volunteers were wearing shorts and T-shirts. With increasing temperature and humidity, dermal absorption increased and reached a mean value of 39 % at 30°C, 60 % relative humidity and additional clothing (the volunteers wore an overall as in the industrial scenario) (Jones *et al.* 2003).

A study in which the dermal absorption of 2-butoxyethanol was estimated via capillary blood samples from the fingertips yielded a very high percentage for dermal absorption. Therefore, in another investigation, one arm each of 6 volunteers was exposed to a 2-butoxyethanol concentration of 50 ml/m^3 for 2 hours. In the capillary blood obtained from the finger of the exposed arm, a 2-butoxyethanol concentration 1500 times higher than that found in the venous blood of the arm not exposed was obtained. It was thus assumed that the capillary blood reflected a local accumulation and not a general systemic concentration (ACGIH 2002; NTP 2000).

A 40 cm^2 area of skin on the forearms of 6 male volunteers was occlusively exposed to a 50 % or 90 % aqueous solution or undiluted 2-butoxyethanol for 4 hours; dermal absorption of the substance from the aqueous solutions was higher than that from undiluted 2-butoxyethanol. From the total amount of 2-butoxyacetic acid (conjugated and free 2-butoxyacetic acid) eliminated with the 24-hour urine, dermal fluxes of 1.34 ± 0.49, 0.92 ± 0.60 and 0.26 ± 0.17 mg/cm^2 and hour were calculated for 50 %, 90 % and undiluted 2-butoxyethanol. It was thus demonstrated that the dermal flux of aqueous dilutions of 2-butoxyethanol is higher than that of the undiluted substance (for dilutions of 50 % and 90 % by factors of about 5 and 3.5) (Jakasa *et al.* 2004).

Using the model calculations of Fiserova-Bergerova *et al.* (1990), Guy and Potts (1993) and Wilschut *et al.* (1995), fluxes of 10.81, 1.34 and 1.73 mg/cm^2 and hour are obtained for the undiluted substance.

The elimination half-time of 2-butoxyethanol in the blood after inhalation exposure (2-butoxyethanol concentrations of 20 and 50 ml/m^3 for 2 hours) was about one hour in humans; in F344 rats and B6C3F$_1$ mice (no other details) it was less than 10 and 5 minutes, respectively (IARC 2006).

Using data from 48 workers, the elimination half-times for non-conjugated and total (free and conjugated) 2-butoxyacetic acid was estimated to be about 6 hours. This means that both metabolites are eliminated at the same speed (ECETOC 2005).

Studies with mice indicate that, after inhalation exposure, 2-butoxyethanol or its metabolites reach the forestomach (and are retained there) not only directly (for example, by licking the fur, the cage walls or swallowing the material condensed in the nose and throat area), but also via systemic distribution: thus, lesions in the forestomach were observed also after subcutaneous and intraperitoneal administration of 2-butoxyethanol (see Section 5.2.4).

Female B6C3F$_1$ mice were given intravenous injections of 2-butoxy[1-^{14}C]ethanol of 10 mg/kg body weight. Even 48 hours after administration, radioactivity was detected autoradiographically in the buccal cavity and the oesophagus, as well as in the mucosa of the forestomach and glandular stomach. Also the secretory glands in the head region, Harderian gland and salivary gland, contained high concentrations of radioactive material (Green *et al.* 2002).

Toxicokinetic studies showed that 2-butoxyethanol and, to a lesser extent, 2-butoxyacetic acid were eliminated more slowly from the forestomach tissue of mice than from the blood or other tissues after oral administration (2-butoxyethanol doses of 250 mg/kg body weight) or intraperitoneal injection (2-butoxyethanol doses of 50 or 250 mg/kg body weight). After intraperitoneal doses of 50 mg/kg body weight, the forestomach was the only tissue with a detectable amount of 2-butoxyethanol after 24 hours, whereas in the blood, the 2-butoxyethanol level was below the detection limit within an hour, independent of the dose or route of administration. Up to 3 hours after intraperitoneal injection of 250 mg 2-butoxyethanol, the level of 2-butoxyacetic acid was highest in the blood, kidneys and liver; afterwards, the concentration in the forestomach was higher than in other tissues. The presence of 2-butoxyethanol and 2-butoxyacetic acid was detected in the saliva after oral administration and intraperitoneal injection (Poet *et al.* 2003).

Several PBPK models were developed for 2-butoxyethanol, the results of which were confirmed in part by *in vivo* data. The models show that, as a result of the relatively rapid degradation of 2-butoxyethanol in humans, accumulation of the substance is not to be expected, and that the 2-butoxyacetic acid concentration in blood is lower in humans than in rats (IARC 2006).

3.2 Metabolism

As shown in Figure 1, 2-butoxyethanol can be degraded via oxidation, dealkylation or conjugation: 2-butoxyethanol is metabolized mainly by alcohol dehydrogenase and aldehyde dehydrogenase to form 2-butoxyacetaldehyde and 2-butoxyacetic acid. 2-Butoxyacetic acid can be degraded to CO_2 or conjugated to *N*-butoxyacetylglycine or to *N*-butoxyacetylglutamine, which has been demonstrated only in human urine samples. In addition, 2-butoxyethanol can be dealkylated by CYP2E1 to form ethylene glycol and butyraldehyde, especially after high doses. Ethylene glycol is further metabolized via oxalic acid to CO_2, and butyraldehyde to butyric acid. The 2-butoxyethanol glucuronide or 2-butoxyethanol sulfate formed by direct conjugation has to date been found only in the rat (ATSDR 1998).

In rats exposed to 2-butoxyethanol concentrations of 4.3 to 438 ml/m^3 via inhalation, there was a concentration-dependent increase in the ratio of butoxyethanol glucuronide to 2-butoxyacetic acid and a decrease in the amount of 2-butoxyethanol exhaled as CO_2 (NTP 2000).

The absorption of 2-butoxyethanol and its metabolism to 2-butoxyacetic acid was linear in rats up to 2-butoxyethanol concentrations of 400 ml/m^3 (IARC 2006).

The differences in metabolism found after different routes of administration are attributed to the different internal doses (NTP 2000). As major metabolite, 2-butoxyacetic acid was eliminated in the urine with 60 % to 75 % of the absorbed 2-butoxyethanol both after inhalation and after dermal or oral (via gavage and drinking water) exposure in the male rat (NTP 2000). In humans, about 17 % to 55 % of the inhaled amount was detected in the urine in the form of unconjugated 2-butoxyacetic acid (IARC 2006).

The *N*-butoxyacetylglutamine formed in the human body was not found in rodents (ECETOC 2005). According to the International Agency for Research on Cancer (IARC 2006), the extent of glutamine conjugation varies considerably both intraindividually (practically 0 % to 100 %) and interindividually. *O*-dealkylation to ethylene glycol is possible in rodents, but in humans there is no reliable evidence of this (IARC 2006).

Approximately 80 % of the metabolites were eliminated with the urine, about 20 % exhaled as CO_2. Female and older animals eliminated 2-butoxyacetic acid more slowly than male and younger animals (ECETOC 2005).

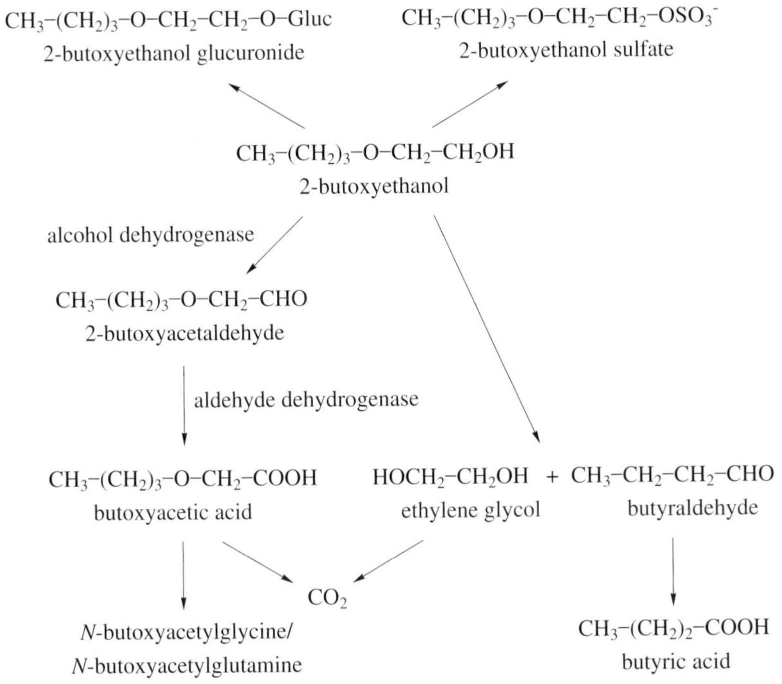

Figure 1. Metabolism of 2-butoxyethanol (ATSDR 1998)

4 Effects in humans

4.1 Single exposures

4.1.1 Inhalation

In male and female volunteers exposed to 2-butoxyethanol for 4 to 8 hours, irritative effects in the eyes, nose and throat and a disturbed sense of taste, headaches and nausea were observed after 2-butoxyethanol concentrations of 100 ml/m^3, but no clinical signs of haemolysis before concentrations above 195 ml/m^3 (WHO 1998).

No clinical signs of adverse effects were found nor were subjective complaints reported in 7 male volunteers exposed to 2-butoxyethanol concentrations of 20 ml/m^3 for 2 hours while performing light physical exercise (ATSDR 1998).

In another study, 4 male volunteers were initially exposed to 2-butoxyethanol concentrations of 50 ml/m^3 inhaled through the mouth for two hours and, after an interval of one hour, dermally to 50 ml/m^3 for a further two hours. No toxic effects (no other details) were observed (NTP 2000).

Seven workers exposed to 2-butoxyethanol for 0.5 to 4 hours suffered during the exposure from irritation of the eyes and respiratory tract, shortness of breath, nausea and weakness. These symptoms continued for three days. In view of the symptoms, the exposure concentration was estimated to be 100 to 300 ml/m^3 (IARC 2006). Irritation of the eyes and the respiratory tract, a dry cough and headaches occurred repeatedly later on (ECETOC 2005).

4.1.2 Ingestion

In a number of case reports, the main symptoms described after the ingestion of 2-butoxyethanol concentrations of about 25 to 60 g via cleaning agents were haemoglobinuria, erythropoenia, hypotension, metabolic acidosis, shock and haematuria (WHO 1998), and also CNS depression and coma (Gualtieri *et al.* 2003).

4.2 Repeated exposures

The erythrocyte count, haemoglobin, haematocrit, MCV (mean corpuscular volume), MCH (mean corpuscular haemoglobin), MCHC (mean corpuscular haemoglobin concentration), haptoglobin, reticulocytes and osmotic pressure in the blood, and the concentrations of free 2-butoxyacetic acid, retinol-binding proteins and creatinine in the urine were investigated in 31 male workers. The workers had been exposed to 2-butoxyethanol concentrations (geometric mean) of 0.6 ml/m^3 for 1 to 6 years (20 workers to an average of 0.75 ml/m^3 and 11 to an average of 0.46 ml/m^3). They were also exposed to 2-butanone (concentration not given), which has not been found to cause haematological effects in animal studies. Twenty-one unexposed workers from the same plant, who were matched as regards age, sex and smoking habits, served as controls; the air at their workplace was, however, not analyzed. There was a significant correlation ($r = 0.55$; $p = 0.0012$) between the concentration of free 2-butoxyacetic acid in the urine after the end of the shift and the 2-butoxyethanol concentration in the workplace air. In addition, the haematocrit values were significantly ($p = 0.03$) lower, and the MCHC values significantly ($p = 0.02$) higher in the exposed workers compared with those in the control group. The haematocrit and MCHC values were, however, within the reference ranges (ATSDR 1998).

4.3 Local effects on skin and mucous membranes

Volunteers who had immersed two or four fingers in undiluted 2-butoxyethanol for two hours suffered no adverse effects. The skin had more wrinkles and seemed less elastic. These mild effects on the skin reached their maximum two to four hours after exposure, and had disappeared within one to two days (ATSDR 1998).

4.4 Allergenic effects

In a skin sensitization study, 201 volunteers were exposed nine times for 24 hours to 0.2 ml 10 % aqueous 2-butoxyethanol via occlusive patches on the back. At the reading 72 hours after the challenge treatment with 2-butoxyethanol, 12 volunteers produced a slight erythematous reaction, and one person a marked reaction, which the authors did not regard as signs of sensitization. At the reading after 48 hours, slight erythema was found in seven volunteers. Already during the induction treatment, slight or marked erythematous reactions were found with increasing frequency, and after the eighth and ninth challenge treatments in about a quarter of the volunteers. Thirteen other volunteers terminated the study prematurely for reasons not related to the treatment (Greenspan *et al.* 1995). Therefore, there are no positive clinical findings of skin sensitization, and no data for the sensitizing effects of the substance on the respiratory tract.

4.5 Reproductive toxicity

There are no data available for the effects of 2-butoxyethanol on reproduction in humans.

4.6 Genotoxicity

The frequency of micronuclei or SCE (sister chromatid exchange) in workers exposed to 2-ethoxyethanol and 2-butoxyethanol was not found to be increased. The concentrations of 2-ethoxyethanol and 2-butoxyethanol in the workplace air were in the range of below 0.1 to 15.2 ml/m^3 and below 0.1 to 1.4 ml/m^3, respectively. After the end of the shift, concentrations of ethoxyacetic acid of 53.8 mg/l and of 2-butoxyacetic acid of 16.4 mg/l were detected in the urine (no other details; ATSDR 1998).

4.7 Carcinogenicity

A case control study of the relationship between acute myeloid leukaemia or myelodysplasia and exposure to a group of glycol ethers, including 2-butoxyethanol, showed no increase in the relative risk. As 2-butoxyethanol was only one of a number of components, the IARC (2006) considers the exposure index to be relatively unspecific and the relevance of this study for 2-butoxyethanol to be limited (IARC 2006).

5 Animal Experiments and *in vitro* Studies

5.1 Acute toxicity

In rodents, short-term exposure to 2-butoxyethanol can produce haemolytic effects, damage to or congestion of the kidneys, liver, lungs and spleen, and result in death from coma or respiratory arrest (OECD 1997). The haemolytic effects are especially pronounced in rats, mice and rabbits, but less so in primates, dogs and guinea pigs (1983 MAK documentation, "2-Butoxyethanol", Volume 6, present series).

5.1.1 Inhalation

The 4-hour LC_{50} values for inhalation exposure to 2-butoxyethanol in male and female rats were 486 and 450 ml/m^3, respectively, and the 7-hour value in mice was 700 ml/m^3. Exposure of rats to 2-butoxyethanol concentrations of 62 ml/m^3 for 4 hours produced increased fragility of the erythrocytes (NTP 2000).

5.1.2 Ingestion

Oral LD_{50} values given for 2-butoxyethanol were 530 to 3000 mg/kg body weight in rats, about 1230 mg/kg body weight in mice, about 350 mg/kg body weight in rabbits and 1200 mg/kg body weight in guinea pigs (NTP 2000).

Changes in haematological parameters, which are characteristic of haemolytic anaemia, occurred in rats after single oral 2-butoxyethanol doses of 125 mg/kg body weight. In older rats (no other details), haemoglobinuria was described after single doses of 32 mg/kg body weight (IARC 2006). Older and female rats reacted more sensitively than younger and male animals. Congested or haemorrhagic lungs, mottled liver, congested kidneys and haemoglobinuria were observed (ECETOC 2005).

In the blood of male rats given 2-butoxyethanol doses of 0, 150, 250 or 500 mg/kg body weight, an early (after one and two hours) dose and time-dependent increase in the haematocrit, PCV (packed cell volume) and MCV was detected. The haematocrit and PCV then dropped below the values in the controls. This indicates that 2-butoxyethanol-induced haemolysis is preceded by massive swelling, presumably of the erythrocyte membrane (ECETOC 2005).

5.1.3 Dermal absorption

The dermal LD_{50} for 2-butoxyethanol was above 2000 mg/kg body weight in rats (ECETOC 2005), between 72 and 638 mg/kg body weight in rabbits, and between 205 and 4800 mg/kg body weight in guinea pigs (NTP 2000).

5.1.4 Intraperitoneal and intravenous injection

In an earlier study with intraperitoneal injection of 2-butoxyethanol, the LD_{50} for rats was 550 mg/kg body weight. After intravenous injection, the LD_{50} was 300 to 500 mg/kg body weight for rats, 1130 mg/kg body weight for mice and 250 to 500 mg/kg body weight for rabbits (ECETOC 2005).

5.2 Subacute, subchronic and chronic toxicity

5.2.1 Inhalation

Studies of the effects of 2-butoxyethanol after repeated inhalation exposure published since the 1983 documentation, and in which also concentrations below 100 ml/m^3 were used, are summarized in Table 1.

In male and female F344 rats exposed to 2-butoxyethanol concentrations of 5, 25 or 77 ml/m^3 for 13 weeks, the erythrocyte count and MCHC were decreased at 77 ml/m^3. The NOAEC (no observed adverse effect concentration) for haematological effects was 25 ml/m^3 in both sexes (ATSDR 1998; Table 1).

In a 14-week inhalation study with 10 male and 10 female F344/N rats and $B6C3F_1$ mice, the LOAEC (lowest observed adverse effect concentration) with regard to changes in haematological parameters in the female rodents was 31 ml/m^3; male animals were less sensitive (NTP 2000; Table 1).

Table 1. Effects of 2-butoxyethanol after repeated inhalation (includes studies since the 1983 documentation with exposures below 100 ml/m³)

Species, strain, number of animals	Exposure	Findings	References
rat, F344, ♂/♀, no other details	**9 days**, 0, 20, 86, 245 ml/m³, 6 hours/day, 5 days/week	**20 ml/m³: NOAEC for haematological parameters** **86 ml/m³ and above**: haematological findings **245 ml/m³**: haematuria; effects reversible during 14-day recovery period	ATSDR 1998
rat, F344, ♂/♀, no other details	**13 weeks**, 5, 25, 77 ml/m³, 6 hours/day, 5 days/week	**25 ml/m³: NOAEC for haemolytic effects** **77 ml/m³**: ♀/♂: erythrocyte count decreased, no unusual findings in lungs, heart; ♀: haemoglobin decreased, mean corpuscular haemoglobin concentration decreased	ATSDR 1998
rat, F344/N, groups of 10 ♂/10 ♀	**14 weeks**, 0, 31, 62.5, 125, 250, 500 ml/m³, 6 hours/day, 5 days/week	**31 ml/m³: LOAEC**: ♂: erythrocyte count decreased **31 ml/m³ and above**: ♀: anaemia (haematocrit decreased, erythrocyte count decreased, haemoglobin decreased) **62.5 ml/m³ and above**: ♀: changes in further haematological parameters, Kupffer's cell pigmentation, bone marrow hyperplasia, relative heart weights increased, relative kidney weights increased **125 ml/m³ and above**: ♂: anaemia (haematocrit decreased, erythrocyte count decreased, haemoglobin decreased), Kupffer's cell pigmentation; ♀: relative liver weights increased, proliferation of haematopoietic cells in the spleen, pigmentation of renal tubules; ♂/♀: breathing difficulties, lethargy, eye and nose irritation, haematuria **250 ml/m³ and above**: ♂: relative kidney weights increased, relative liver weights increased, bone marrow hyperplasia, proliferation of haematopoietic cells in the spleen, pigmentation of renal tubules **500 ml/m³**: ♀: mortality 5/10, body weights decreased, relative thymus weights decreased	NTP 2000
rat, F344/N, groups of 50 ♂/50 ♀	**2 years**, 0, 31, 62.5, 125 ml/m³, 6 hours/day, 5 days/week	**31 ml/m³: LOAEC for haemolytic effects and irritant effects in the nose** **31 ml/m³ and above**: ♀: anaemia (haematocrit decreased, erythrocyte count decreased, haemoglobin decreased); ♂: MCV increased; ♂/♀: concentration-dependent increase in hyaline degeneration of olfactory epithelium (significant after 62.5 ml/m³ and above) **62.5 ml/m³ and above**: ♂: fibrosis of the spleen, anaemia; ♂/♀: Kupffer's cell pigmentation **125 ml/m³**: ♀: body weights decreased, bone marrow cellularity increased; phaeochromocytomas increased (see Table 6); ♂/♀: M/E (myeloid/erythroid) ratio decreased by 15 % to 35 %; no substance-related clinical findings **see also Table 2 and Section 5.7.2**	NTP 2000

Table 1. continued

Species, strain, number of animals	Exposure	Findings	References
mouse, B6C3F$_1$, groups of 10 ♂/10 ♀	**14 weeks,** 0, 31, 62.5, 125, 250, 500 ml/m^3, 6 hours/day, 5 days/week	**31 ml/m^3**: ♀: **LOAEC for haemolytic effects** **31 ml/m^3 and above**: ♀: anaemia (haemoglobin decreased, erythrocyte count decreased) **62.6 ml/m^3 and above**: ♂: **NOAEC** **125 ml/m^3 and above**: ♂: anaemia (haemoglobin decreased, erythrocyte count decreased), body weights decreased, haemosiderin pigmentation in the spleen; ♀: epithelial hyperplasia in the forestomach **250 ml/m^3 and above**: ♂: proliferation of haematopoietic cells in the spleen, relative liver weights increased; ♀: haemosiderin pigmentation in the spleen, Kupffer's cell pigmentation, forestomach inflammation **500 ml/m^3 and above**: ♂: Kupffer's cell pigmentation; ♀: proliferation of haematopoietic cells in the spleen, relative heart weights increased, relative kidney weights increased, relative liver weights increased; ♂/♀: pigmentation of renal tubules, breathing difficulties, lethargy, haematuria, mortality 6/10	NTP 2000
mouse, B6C3F$_1$, groups of 50 ♂/50 ♀	**2 years,** 0, 62.5, 125, 250 ml/m^3, 6 hours/day, 5 days/week	**62.5 ml/m^3**: **LOAEC**: ♀: hyaline degeneration of the olfactory and respiratory epithelium **62.5 ml/m^3 and above**: ♀: anaemia (haematocrit decreased, erythrocyte count decreased, haemoglobin decreased), thrombocytes increased, Kupffer's cell pigmentation, forestomach ulcer; ♂: thrombocytes increased, haemosiderin pigmentation in the spleen; ♂/♀: body weights decreased, epithelial hyperplasia of the forestomach **125 ml/m^3**: ♀: hyaline degeneration of the respiratory epithelium; ♂: M/E ratio decreased, kidneys (glomerulosclerosis, hydronephrosis), ulcers in bladder and forestomach **125 ml/m^3 and above**: ♀: haemosiderin pigmentation in the spleen; ♂: anaemia, Kupffer's cell pigmentation, proliferation of haematopoietic cells in the spleen, bone marrow hyperplasia, preputial inflammation and ulcers, cystitis **250 ml/m^3**: ♀: proliferation of haematopoietic cells in the spleen; ♂: inflammation of the kidneys, preputial glands and prostate; ♂/♀: tumours in liver and forestomach increased (see Table 6) **see also Table 3 and Section 5.7.2**	NTP 2000

Groups of 50 male and 50 female F344/N rats were exposed to 2-butoxyethanol concentrations of 0, 31.2, 62.5 or 125 ml/m^3 for 104 weeks (5 days per week, 6 hours per day) (neoplasms, see Section 5.7.2; toxic effects, see Tables 1 and 2). In addition, 9 male and 9 female F344/N rats were exposed to 2-butoxyethanol concentrations of 0, 62.5 or 125 ml/m^3 for 3, 6 or 12 months, and 9 male and 9 female F344/N rats to 2-butoxyethanol concentrations of 31.2 ml/m^3 for 3 or 6 months for haematology and bone marrow investigations. Local effects in the respiratory tract were observed at the lowest concentration of 31.2 ml/m^3 and above: thus, the incidence of hyaline degeneration of the olfactory epithelium in male and female rats (see Table 2) increased from the lowest concentration upwards in a concentration-dependent manner, and was found to be significant at 62.5 ml/m^3 and above. The severity of the damage was not changed, however, by the level of exposure. The hyaline degeneration, which also occurred in control animals, was considered to be a common age-related change. The increased incidence in exposed rats could possibly be viewed as an adaptive/protective change. As an expression of the haemolytic effects of 2-butoxyethanol, mild macrocytic, normochromic, regenerative anaemia was diagnosed in female rats after only 3 months exposure to concentrations of 31.2 ml/m^3, while in male rats this was not observed until concentrations of 62.5 ml/m^3 and exposure for 12 months. After concentrations of 62.5 ml/m^3 and above, the incidence of Kupffer's cell pigmentation in the liver of male and female rats increased in a concentration-dependent manner. In female animals of the 125 ml/m^3 group, bone marrow cellularity was increased, the mean body weight was lower compared with that of the control animals and the incidence of phaeochromocytomas was increased (see Table 6) (NTP 2000). The LOAEC for the irritative effects histologically detectable in the nose of rats was thus 31.2 ml/m^3. A NOAEC could not be derived. For this reason, a benchmark dose was calculated using the EPA programme. The Gamma and the Weibull models provided the best adaptation. For a 5 % increase in incidence, a benchmark concentration of about 15 ml/m^3 was obtained.

Table 2. Incidences of histological findings in rats after exposure for 2 years to 2-butoxyethanol (NTP 2000)

Findings		Exposure concentration (ml/m^3)			
		0	31.2	62.5	125
nose:					
hyaline degeneration of the olfactory epithelium	♂	13/48 (27 %)	21/49 (43 %)	23/49 (47 %)*	40/50 (80 %)**
	♀	13/50 (26 %)	18/48 (38 %)	28/50 (56 %)**	40/49 (82 %)**
liver:					
pigmentation of Kupffer's cells	♂	23/50 (46 %)	30/50 (60 %)	34/50 (68 %)	42/50 (84 %)
	♀	15/50 (30 %)	19/50 (38 %)	36/50 (72 %)	47/50 (94 %)
adrenal cortex:					
hyperplasia	♀	11/50 (22 %) 1.9[a]	11/50 (22 %) 2.3[a]	8/49 (16 %) 2.1[a]	17/49 (35 %) 2.5[a]

[a] average degree of severity: from 1 (minimal) to 4 (marked)
* $p \leq 0.05$; ** $p \leq 0.01$ (significance given only for effects on the nose)

Groups of 50 male and 50 female B6C3F$_1$ mice were exposed to 2-butoxyethanol concentrations of 0, 62.5, 125 or 250 ml/m^3 for 104 weeks (5 days per week, 6 hours per day) (neoplasms, see Section 5.7.2; toxic effects, see Tables 1 and 3). In addition, groups of 10 male and 10 female B6C3F$_1$ mice were exposed to 2-butoxyethanol concentrations of 0, 62.5, 125 or 250 ml/m^3 for 3, 6 or 12 months for the haematology and bone marrow investigations. Local effects on the respiratory tract were found in female mice in the form of increased hyaline degeneration of the olfactory epithelium (at 62 ml/m^3) and respiratory epithelium of the nose (at 62 ml/m^3 and 125 ml/m^3; see Table 3). The severity of the damage was not changed, however, by the level of exposure. In the female animals, the incidence of tumours (see Section 5.7.2), epithelial hyperplasia and ulcers of the forestomach increased significantly and in a concentration-dependent manner. In the male mice, a significant concentration-dependent increase in the incidence of epithelial hyperplasia of the forestomach was observed. Forestomach ulcers were significantly increased only in the medium dose group. The haemolytic effects of 2-butoxyethanol were manifest in the female animals in the form of mild macrocytic, normochromic, regenerative anaemia after concentrations of 62.5 ml/m^3 and above; in male mice, this was observed at 125 ml/m^3 and above. As a result of the haemolytic effects, the incidence of haemosiderin pigmentation in Kupffer's cells in the male animals of the 125 ml/m^3 and 250 ml/m^3 groups and in the exposed female animals was significantly increased. The incidence of haematopoietic cell proliferation and haemosiderin pigmentation in the spleen was increased in male and female mice, and the incidence of bone marrow hyperplasia was increased in male animals. An increased number of haemangiosarcomas and hepatocellular carcinomas were observed in the male mice of the 250 ml/m^3 group (see Table 6, Section 5.7.2). A NOAEC cannot be derived from this study (NTP 2000).

Table 3. Incidences of histological findings in mice after exposure for 2 years to 2-butoxyethanol (NTP 2000)

Findings		Exposure concentration (ml/m^3)			
		0	62.5	125	250
nose:					
hyaline degeneration of the olfactory epithelium	♀	6/50 (12 %)	14/50 (28 %)*	11/49 (22 %)	12/50 (24 %)
hyaline degeneration of the respiratory epithelium	♀	17/50 (34 %)	35/50 (70 %)**	26/49 (53 %)*	23/50 (46 %)
liver:					
pigmentation of Kupffer's cells	♂	0/50 (0 %)	0/50 (0 %)	8/49 (16 %)	30/49 (61 %)
	♀	0/50 (0 %)	5/50 (10 %)	25/49 (51 %)	44/50 (88 %)
forestomach:					
ulcer	♂	1/50 (2 %)	2/50 (4 %)	9/49 (18 %)**	3/48 (6 %)
	♀	1/50 (2 %)	7/50 (14 %)*	13/49 (27 %)**	22/50 (44 %)**
epithelial hyperplasia	♂	1/50 (2 %)	7/50 (14 %)*	16/49 (33 %)**	21/48 (44 %)**
	♀	6/50 (12 %)	27/50 (54 %)**	42/49 (86 %)**	44/50 (88 %)**

Table 3. continued

Findings		Exposure concentration (ml/m³)			
		0	62.5	125	250
spleen:					
haematopoietic cell proliferation	♂	12/50 (24 %)	11/50 (22 %)	26/48 (54 %)	42/49 (86 %)
	♀	24/50 (48 %)	29/50 (58 %)	32/49 (32 %)	35/50 (70 %)
haemosiderin pigmentation	♂	0/50 (0 %)	6/50 (12 %)	45/48 (94 %)	44/49 (90 %)
	♀	39/50 (39 %)	44/50 (88 %)	46/49 (94 %)	48/50 (96 %)
bone marrow:					
hyperplasia	♂	0/50 (0 %)	1/50 (2 %)	9/49 (18 %)	5/50 (10 %)

* $p \le 0.05$; ** $p \le 0.01$ (significance given only for effects in nose and forestomach)

5.2.2 Ingestion

Studies of the effects of 2-butoxyethanol after repeated ingestion are presented in Table 4.

Table 4. Effects of 2-butoxyethanol after oral administration

Species, strain, number of animals	Exposure	Findings	References
rat, Sprague-Dawley, ♀, no other details	**21 days**, 0, 204, 444 mg/kg body weight and day, drinking water	**204 mg/kg body weight**: natural killer cells decreased **204 mg/kg body weight and above**: body weights decreased **444 mg/kg body weight**: no unusual findings: thymus, liver and kidneys	OECD 1997
rat, Sprague-Dawley, ♂, no other details	**21 days**, 0, 180, 506 mg/kg body weight and day, drinking water	**180 mg/kg body weight**: natural killer cells decreased **506 mg/kg body weight**: body weights decreased, no unusual findings: thymus, liver, kidneys, testes	OECD 1997
rat, COBS CD(SD)BR, groups of 10 ♂	**6 weeks**, 0, 222, 443, 885 mg/kg body weight and day, 5 days/week, gavage	**222 mg/kg body weight and above**: erythrocytes decreased, haemoglobin decreased, relative liver weights increased, haemoglobinuria **443 mg/kg body weight and above**: relative kidney, heart, brain, spleen weights increased **885 mg/kg body weight**: body weight gains decreased; no unusual findings: testes, thymus, leukocytes, bone marrow; clinical finding (dose not specified): lethargy	OECD 1997

Table 4. continued

Species, strain, number of animals	Exposure	Findings	References
rat, F344, groups of 20 ♂	**7, 14, 28 or 90 days**, 0, 225, 450 mg/kg body weight and day, gavage	**225 mg/kg body weight and above**: vitamin E level in the liver decreased (after 7 and 90 days); haematocrit decreased, relative spleen weights increased (after 7 days); Kupffer's cell pigmentation (after 14 days) **450 mg/kg body weight**: Kupffer's cell pigmentation (after 7 days); no increase in oxidative damage	Siesky *et al.* 2002
rat, F344, groups of 10 ♀	**13 weeks**, 0, 82, 151, 304, 363, 470 mg/kg body weight and day, drinking water	**82 mg/kg body weight**: **LOAEL** **82 mg/kg body weight and above**: anaemia **304 mg/kg body weight and above**: bone marrow cellularity increased, leukocytosis **363 mg/kg body weight and above**: body weights decreased **470 mg/kg body weight**: no unusual findings in the stomach; complete necropsy and histopathology performed, positive findings in the liver, spleen, bone marrow (no other details)	EPA 1999, OECD 1997
rat, F344, groups of 10 ♂	**13 weeks**, 0, 69, 129, 281, 367, 452 mg/kg body weight and day, drinking water	**69 mg/kg body weight**: **NOAEL** **129 mg/kg body weight and above**: haemo-globin decreased **281 mg/kg body weight and above**: anaemia, reticulocytes increased, leukocytes increased **367 mg/kg body weight and above**: body weight decreased, thrombocytopaenia, bone marrow cellularity increased; complete necropsy and histopathology performed, effects in liver, spleen, bone marrow (no other details)	EPA 1999, OECD 1997
mouse, $B6C3F_1$, groups of 16 ♂/16♀ (8/sex fore-stomach/8/sex liver)	**2 days**, 0, 400, 800, 1200 mg/kg body weight and day, then because of mortality: **2 days**, 0, 200, 400, 600 mg/kg body weight and day, gavage	**400/200 mg/kg body weight and above**: haematocrit decreased, epithelial hyperplasia and inflammation of the forestomach (degree of severity increased)	Poet *et al.* 2003

Table 4. continued

Species, strain, number of animals	Exposure	Findings	References
mouse, $B6C3F_1$, groups of 5 ♀	**10 days**, 0, 50, 150, 500 mg/kg body weight and day, in polyethylene glycol, gavage	**150 mg/kg body weight**: NOAEL **500 mg/kg body weight**: in 2/5: minimal hyperkeratosis in the forestomach; no adverse clinical and no gross pathological findings, no unusual findings in glandular stomach	Green *et al.* 2002
mouse, $B6C3F_1$, groups of 5 ♀	**10 days**, **2-butoxyacetic acid** concentrations of 0, 50, 150, 500 mg/kg body weight and day, in polyethylene glycol, gavage	**50 mg/kg body weight**: NOAEL **150 mg/kg body weight**: in 3/5 minimal hyperkeratosis in the forestomach **500 mg/kg body weight**: slight hyper-keratosis in the forestomach in 3/5 and moderate in 1/5; no clinical and no gross pathological findings, no unusual findings in glandular stomach	Green *et al.* 2002
mouse, $B6C3F_1$, groups of 60 ♂	**7, 14, 28 or 90 days**, 0, 225, 450, 900 mg/kg body weight and day, gavage	**225 mg/kg body weight and above**: vitamin E level in the liver decreased (after 7 and 90 days), DNA synthesis in liver endothelial cells increased (after 7 and 14 days), DNA synthesis in the hepatocytes increased (90 days) **450 mg/kg body weight and above**: relative liver and spleen weights increased, haematocrit decreased, Kupffer's cell pigmentation (after 7 days) **450 mg/kg body weight and above**: oxidative damage increased (after 7 and 90 days)	Siesky *et al.* 2002
mouse, $B6C3F_1$, groups of 10 ♂/10 ♀	**13 weeks**, ♀: 0, 185, 370, 676, 861, 1306 or ♂: 118, 223, 553, 676, 694 mg/kg body weight and day, drinking water	**370 or 223 mg/kg body weight**: **NOAEL** **676 or 553 mg/kg body weight and above**: body weight gains decreased; no gross pathological or microscopic findings	EPA 1999, OECD 1997

In order to clarify whether the formation of the liver neoplasms observed in male mice in the carcinogenicity study of the NTP (2000) is caused by oxidative damage resulting from the deposition of haemolytic iron in the liver, 2-butoxyethanol concentrations of 0, 225, 450 or 900 mg/kg body weight were administered orally to male $B6C3F_1$ mice, and 0, 225 or 450 mg/kg body weight to male F344 rats once a day, five times a week, for 7, 14, 28 or 90 days. A dose-dependent increase in haemolysis (determined via the decrease in the haematocrit value and the increase in relative spleen weights) was observed in rats and mice. An increasing number of iron-stained Kupffer's cells occurred in mice of the 450 and 900 mg/kg groups and in rats of the 225 and 450 mg/kg groups. The oxidative damage (8-hydroxydeoxyguanosine and malondialdehyde) was increased in the liver of mice from the 450 mg/kg group after 7 and 90 days treatment, but not in the rat liver. The vitamin E levels were reduced by exposure to 2-butoxyethanol in both mice and rat liver; the initial value in rats was approximately 2.5 times higher than that in mice. Induction of DNA synthesis was observed in the hepatocytes of the mice treated for 90 days and in the endothelial cells of the liver after treatment for 7 and 14 days in all dose groups (Siesky *et al.* 2002). The biphasic course is explained by the fact that younger erythrocytes seem to be more resistant to the haemolytic effects of the substance than older ones. Thus, rats from which blood had been taken several days prior to the single oral 2-butoxyethanol dose (no other details) reacted less sensitively than rats not pretreated (IARC 2006). Also the tolerance in rats against 2-butoxyethanol-induced haemolysis reported after long-term exposure (ECETOC 2005) is possibly explained by the selective removal of the older erythrocytes (Boatman *et al.* 2004). No change in DNA synthesis was observed in the treated rats. No apparent differences were observed in apoptosis and mitosis in the liver of treated and untreated mice or rats. From these results it may be assumed that the increased DNA synthesis, which occurred selectively in the mouse liver and especially in the endothelial cells—a target site for the carcinogenic effects of 2-butoxyethanol—is caused by oxidative damage resulting from exposure to 2-butoxyethanol (Siesky *et al.* 2002).

After oral administration of 2-butoxyethanol doses of 0, 400, 800 or 1200 mg/kg body weight for 2 days followed by gavage doses of 0, 200, 400 or 600 mg/kg body weight and day for another 2 days, a dose-dependent increase in forestomach lesions (inflammation and epithelial hyperplasia) was observed in the groups of 8 male and 8 female $B6C3F_1$ mice (Poet *et al.* 2003).

Hyperkeratosis developed in the forestomach of female $B6C3F_1$ mice given single oral 2-butoxyethanol or 2-butoxyacetic acid doses of 0, 50, 150 or 500 mg/kg body weight daily for 10 days (see Table 4). 2-Butoxyacetic acid, with a NOAEL of 50 mg/kg body weight and day, was found to be markedly more effective as regards forestomach damage than 2-butoxyethanol, with a NOAEL of 150 mg/kg body weight and day (Green *et al.* 2002).

The NOAEL for haemolytic effects in a 13-week drinking water study was 69 mg/kg body weight and day for male rats. In female rats, the lowest dose tested of 82 mg/kg body weight and day was already a LOAEL and resulted in anaemia (EPA 1999; OECD 1997).

5.2.3 Dermal absorption

In a 13-week study with groups of 10 male and 10 female New Zealand White rabbits with dermal (no other details) application of 2-butoxyethanol doses of 0, 10, 50 or 150 mg/kg body weight and day for 6 hours a day, on 5 days a week, neither haematological nor clinicochemical or histopathological changes were observed up to the highest dose of 150 mg/kg body weight and day (OECD 1997).

5.2.4 Intraperitoneal and subcutaneous injection

After intraperitoneal or subcutaneous injection of 2-butoxyethanol doses of 400 mg/kg body weight and day for 5 days, minimal forestomach lesions (epithelial hyperplasia and inflammation) were observed in 1/6 and 2/6 mice, respectively. Damage to the forestomach was not found in the control animals (Poet *et al.* 2003).

5.3 Local effects on skin and mucous membranes

5.3.1 Skin

In rabbits, the 24-hour occlusive application (Draize Test) of 0.5 ml undiluted 2-butoxyethanol produced severe skin irritation; 4-hour occlusive application was irritative (ATSDR 1998). Moderate skin irritation occurred in a 24-hour percutaneous toxicity study, and slight irritation after 4-hour, non-occlusive exposure (no other details) (ECETOC 2005).

5.3.2 Eyes

In the rabbit eye, 2-butoxyethanol caused conjunctival irritation and moderate to severe, transient corneal injury (ACGIH 2002). In a study performed in accordance with OECD Test Guideline 405, the average scores given for irritation (24–72 hours, 3 animals) were 0.9 on a scale with a maximum of 4 for corneal opacity, 0.6 on a scale with a maximum of 2 for iritis, 2.6 on a scale with a maximum of 3 for conjunctival redness and 1.8 on a scale with a maximum of 4 for chemosis. The irritation was reversible within 21 days (ECETOC 2005).

5.4 Allergenic effects

2-Butoxyethanol was not found to have skin-sensitizing effects in guinea pigs in a maximization test according to Magnusson and Kligman. After provocation with 1 % 2-butoxyethanol (no other details), none of the 20 animals produced a reaction. The concentrations and vehicles used for intradermal and epicutaneous induction treatment are not documented (Zissu 1995).

5.5 Reproductive toxicity

5.5.1 Fertility

5.5.1.1 Male fertility

Specific effects on the reproductive organs were not detected in the large number of short, medium and long-term studies with 2-butoxyethanol (Fastier *et al.* 2005).

In inhalation studies, changes in testicular weights were not observed in Alpk/Ap rats 15 days after exposure to 2-butoxyethanol concentrations of 800 ml/m^3 for three hours, or in F344 rats after exposure to concentrations of 867 ml/m^3 for four hours, or to 77 ml/m^3 for 13 weeks (5 days a week, 6 hours a day) (ATSDR 1998). Also the inhalation of 2-butoxyethanol concentrations of 125 and 250 ml/m^3 for two years (see Section 5.2.1) did not have any effect on the male reproductive organs in F344 rats and B6C3F$_1$ mice (NTP 2000). In short-term studies with Sprague-Dawley, Crl:COBS CD(SD)BR and F344 rats and JCL-ICR mice, no effects were found on testicular weights and histopathology after oral 2-butoxyethanol doses of up to 506, 885, 1000 and 2000 mg/kg body weight and day, respectively (IARC 2006). No histopathological changes were found in the seminal vesicles, testes and epididymides of B6C3F$_1$ mice after 2-butoxyethanol concentrations of up to 627 mg/kg body weight and day (drinking water) for 13 weeks (ATSDR 1998).

Effects on fertility parameters were found only at doses that also produced haematological effects (IARC 2006). In a 13-week drinking water study in F344 rats, sperm concentrations were significantly decreased after doses of 281 mg/kg body weight and day and above compared with the values for the controls. No treatment-related effects on sperm parameters were found up to 2-butoxyethanol doses of 627 mg/kg body weight and day in B6C3F$_1$ mice, however (ATSDR 1998).

5.5.1.2 Female fertility

The inhalation of 2-butoxyethanol concentrations of 125 and 250 ml/m^3 (see Section 5.2.1) for two years did not lead to effects on the female reproductive organs in F344 rats and B6C3F$_1$ mice, respectively (NTP 2000).

In rodents, effects on fertility occurred only at doses that also resulted in general toxicity. In F344 rats, the oestrous cycle was altered after 2-butoxyethanol doses of 363 mg/kg body weight and day and above administered in the drinking water for 13 weeks (ATSDR 1998).

The NOAEL for fertility was found to be 700 mg/kg body weight and day in Swiss mice in a 2-generation study with continuous mating (see Section 5.5.2). In dams given 2-butoxyethanol doses of 1300 mg/kg body weight and day in the drinking water for 21 weeks and mated with control animals, the number of litters was 58 % lower and the number of pups per litter 66 % lower than that in untreated pairs (Heindel *et al.* 1990).

5.5.2 Developmental toxicity

Studies of developmental toxicity in rats, mice and rabbits with oral and inhalation exposure to 2-butoxyethanol showed that damage to embryos does not occur until maternally toxic concentrations are reached. Increased mortality, the resorption of foetuses or embryos and delayed ossification were observed in particular (Fastier *et al.* 2005; IARC 2006), but there were no malformations.

The NOAEC for developmental toxicity after inhalation of 2-butoxyethanol was 50 ml/m^3 in rats and 100 ml/m^3 in rabbits (Tyl *et al.* 1984); the NOAEL after oral doses was 100 mg/kg body weight and day in rats (NTP 1989) and 650 mg/kg body weight and day in mice (Wier *et al.* 1987). The individual studies are listed in Table 5.

Table 5. Studies of developmental toxicity after the administration of 2-butoxyethanol

Species, strain, number of animals	Exposure	Findings	References
Inhalation			
rat, F344, groups of about 30 ♀	**GD 6–15**, 0, 25, 50, 100, 200 ml/m^3, 6 hours/day, examined on GD 21	**50 ml/m^3: NOAEC** **100 ml/m^3 and above**: <u>dams</u>: haematological effects; <u>foetuses</u>: delayed ossification **200 ml/m^3 and above**: <u>foetuses</u>: number of non-viable implantations increased, resorptions increased, number of live foetuses per litter decreased	Tyl *et al.* 1984
rat, Sprague-Dawley, 34, 19, 18 ♀	**GD 7–15**, 0, 150, 200 ml/m^3, 7 hours/day, examined on GD 20	**200 ml/m^3**: <u>dams</u>: haematuria on GD 7; <u>foetuses</u>: no toxic effects on development	Nelson *et al.* 1984
rabbit, New Zealand White, groups of 24 ♀	**GD 6–18**, 0, 25, 50, 100, 200 ml/m^3, 6 hours/day, examined on GD 29	**100 ml/m^3: NOAEC** **200 ml/m^3**: <u>dams</u>: mortality increased (not significant), body weights decreased (GD 15), weights of pregnant uterus decreased; <u>foetuses</u>: abortions increased (not significant), number of live foetuses per litter decreased, number of total implantations decreased	Tyl *et al.* 1984
Oral administration			
rat, F344, groups of 28–35 ♀	**GD 9–11**, 0, 30, 100, 200 mg/kg body weight, examined on GD 12, 20	**30 mg/kg body weight**: <u>dams</u>: **NOAEL** **100 mg/kg body weight and above:** <u>dams</u>: haematological effects, body weights decreased; <u>foetuses</u>: **NOAEL** **200 mg/kg body weight and above:** <u>foetuses</u>: mortality increased	NTP 1989

Table 5. continued

Species, strain, number of animals	Exposure	Findings	References
mouse, CD-1, groups of 6 ♀	**GD 8–14,** 0, 350, 650, 1000, 1500, 2000 mg/kg body weight, examined on GD 18 or PND 22	**350 mg/kg body weight:** dams: **NOAEL** **650 mg/kg body weight and above**: dams: characteristic green/reddish brown discoloration of bedding; offspring: **NOAEL** **1000 mg/kg body weight and above**: foetuses: resorptions increased **up to 1000 mg/kg body weight**: juvenile animals: no effects on survival or growth up to postnatal day 22 **1500 mg/kg body weight and above**: dams: lethargy, abnormal respiration, mortality increased **2000 mg/kg body weight**: dams: mortality 6/6	Wier *et al.* 1987
mouse, CD-1, no other details	**GD 7–14,** 0, 1180 mg/kg body weight, no other details	**1180 mg/kg body weight:** dams: mortality increased, number of viable litters decreased; juvenile animals: no effect on weight or survival (postnatal)	IARC 2006
mouse, CD-1, groups of 20 ♂/20 ♀ (controls 40 ♂/40 ♀)	**"continuous breeding",** 0, 700, 1300, 2000 mg/kg body weight	**700 mg/kg body weight and above:** newborn animals (F_1): body weights decreased **1300 mg/kg body weight and above:** dams: body weights decreased, drinking water consumption decreased, kidney weights increased, mortality increased; juvenile animals: number of litters per pair decreased, litter size decreased	Heindel *et al.* 1990
Dermal administration			
rat, Sprague-Dawley, no other details	**GD 7–16,** 0, 1920, 5600 mg/kg body weight, dermal, no other details	**1920 mg/kg**: NOAEL **5600 mg/kg body weight**: dams: lethal dose	IARC 2006

GD: gestation day; PND: postnatal day

5.6 Genotoxicity

The data for the genotoxicity of 2-butoxyethanol published up to 2004 are summarized and evaluated in the reports by the IARC (2006) and the ECETOC (2005) and in the publication by Fastier *et al.* (2005). The genotoxicity data published up to 1997 were summarized in addition in Elliott and Ashby (1997).

The available data for 2-butoxyethanol indicate that neither the substance itself (WHO 1998; EPA 1999; IARC 2006; NTP 2000) nor its metabolite 2-butoxyacetic acid has genotoxic effects. In an *in vitro* study with the metabolite 2-butoxyacetaldehyde, genotoxic effects were found for various genetic parameters (Elias *et al.* 1996). No *in vivo* data are available (IARC 2006). Structural comparison of aldehydes suggests, however, that aldehydes with longer chains, such as 2-butoxyacetaldehyde, interact with DNA to a lesser extent than short-chained aldehydes, such as acetaldehyde. In addition, corresponding investigations with 2-butoxyethanol did not yield evidence of a genotoxic metabolite (EPA 2005). There are, nevertheless, insufficient data to be able to assess the genotoxic potential of 2-butoxyacetaldehyde conclusively.

5.6.1 *In vitro*

5.6.1.1 2-Butoxyethanol

Bacterial mutagenicity tests with *Salmonella typhimurium* strains TA97, TA98, TA100, TA1535 and TA1537 and *Escherichia coli* WP2*uvra* yielded no evidence of mutagenic effects up to 2-butoxyethanol concentrations of 10 mg/plate either in the presence or absence of a metabolic activation system (EPA 1999; Fastier *et al.* 2005; IARC 2006; NTP 2000). The positive result described in one study with *Salmonella typhimurium* strain TA97a could not be reproduced in another study (EPA 1999; IARC 2006) and was possibly caused by an impurity in the test material (EPA 1999).

As 2-butoxyethanol in concentrations of 8.5 mmol/l and above increased the induction of chromosomal aberrations by methylmethane sulfonate in V79 cells, but did not have any clastogenic effects itself, it was assumed that 2-butoxyethanol inhibits the DNA repair process. This assumption was supported by the observation that in SHE cells treated with methylmethane sulfonate the cellular concentration of poly(ADP-ribose) increased, but subsequent exposure to 2-butoxyethanol inhibited the synthesis of poly(ADP-ribose). 2-Butoxyethanol alone did not influence the concentration of poly(ADP-ribose) in SHE cells (IARC 2006).

In CHO cells (a cell line derived from Chinese hamster ovary), 2-butoxyethanol induced a delay in the cell cycle (EPA 1999; NTP 2000), but did not induce sister chromatid exchange (SCE) after concentrations of up to 5 mg/ml in the presence of a metabolic activation system and after concentrations of up to 3.5 mg/ml in the absence of a metabolic activation system (Elliott and Ashby 1997; NTP 2000). An increased incidence of SCE was observed in V79 cells (less than double the control value in a delayed cell cycle) and in human lymphocytes. The evaluation is not described in detail in either study, however, so that indirect effects cannot be excluded (Elliott and Ashby 1997; IARC 2006). For this reason, these studies are not included in this evaluation.

In the investigations of one working group it is reported that 2-butoxyethanol (up to 10 mmol/l) did not induce an increase in DNA damage (determined by means of comet assay) in mouse endothelial cells (SVEC4-10) after treatment for 2, 4 and 24 hours (Corthals *et al.* 2005, 2006; Klaunig and Kamendulis 2005; Reed *et al.* 2003).

A UDS test with primary rat hepatocytes and 2-butoxyethanol concentrations of up to 0.1 % in the culture medium produced a statistically significant increase in DNA repair synthesis only at the two lowest doses. Because a dose–response relationship was not apparent, the results were considered by the authors to be questionable (Elliott and Ashby 1997).

In the presence and absence of a metabolic activation system, chromosomal aberration tests yielded negative results with CHO cells after 2-butoxyethanol concentrations of up to 5000 µg/ml, and with V79 cells (no other details) and human lymphocytes without metabolic activation after concentrations of up to 3000 µg/ml (Elliott and Ashby 1997; IARC 2006).

After high 2-butoxyethanol concentrations, the weak induction of micronuclei (after concentrations of 8.5 mmol/l and above), aneuploidy (after about 17 mmol/l and above) and gene mutations at the HPRT locus (at 8.5 mmol/l) were reported in one study with V79 cells without metabolic activation (Elias *et al.* 1996).

2-Butoxyethanol was not found to be mutagenic in the HPRT gene mutation test with CHO cells after concentrations of up to 1 % (v/v) either in the presence or absence of a metabolic activation system (Elliott and Ashby 1997; Fastier *et al.* 2005).

5.6.1.2 The metabolite 2-butoxyacetaldehyde

The metabolite 2-butoxyacetaldehyde did not cause mutagenic effects in the *Salmonella typhimurium* strains TA97a, TA98, TA100, TA102 after exposure to concentrations of up to 7 mg/plate either in the presence or absence of a metabolic activation system (Elliott and Ashby 1997; IARC 2006). In a study with V79 cells (Elias *et al.* 1996), 2-butoxyacetaldehyde induced SCE (after concentrations of about 0.2 mmol/l and above). In an unpublished study, 2-butoxyacetaldehyde concentrations of 0.5 mmol/l led to double the number of SCEs in human lymphocytes compared with that in the controls. At this concentration, 2-butoxyacetaldehyde caused a 50 % reduction in cell count and cell viability (no other details; EPA 2005).

In mouse endothelial cells (SVEC4-10), treatment with 2-butoxyacetaldehyde (up to 1 mmol/l) for 2, 4 or 24 hours did not produce an increase in DNA damage (determined by comet assay) (Corthals *et al.* 2005, 2006; Klaunig and Kamendulis 2005).

In a study with V79 cells (Elias *et al.* 1996), 2-butoxyacetaldehyde induced chromosomal aberrations (after concentrations of about 0.1 mmol/l and above), micronuclei (after 0.04 mmol/l and above) and aneuploidy (after about 0.3 mmol/l and above) (Elliott and Ashby 1997; IARC 2006).

2-Butoxyacetaldehyde was not found to be mutagenic in a xanthine –guanine phosphoribosyl transferase (*gpt*) gene mutation test with CHO-AS52 cells after concentrations of up to 0.2 % (v/v) (Chiewchanwit and Au 1995).

5.6.1.3 The metabolite 2-butoxyacetic acid

The metabolite 2-butoxyacetic acid did not cause mutagenic effects in *Salmonella typhimurium* strains TA97a, TA98, TA100, TA102 after concentrations of up to 1 mg/plate either in the presence or absence of a metabolic activation system (Elliott and Ashby 1997; IARC 2006). In mouse endothelial cells (SVEC4-10), treatment with 2-butoxyacetic acid (up to 10 mmol/l) for 2, 4 and 24 hours did not induce an increase in DNA damage (determined by comet assay) (Corthals *et al.* 2005, 2006; Klaunig and Kamendulis 2005; Reed *et al.* 2003).

2-Butoxyacetic acid did not cause genotoxic effects in V79 cells in an SCE test (up to 0.8 mmol/l) and a chromosomal aberration test (no details), but did lead to a slight increase in micronuclei (after 5 mmol/l and above) (Elias *et al.* 1996).

5.6.2 *In vivo*

5.6.2.1 2-Butoxyethanol

In Sprague-Dawley rats given single gavage doses of 120 mg/kg body weight, no increase in DNA adducts in the brain, liver, kidneys, testes or spleen was detected using ^{32}P-postlabelling 24 hours after the administration of 2-butoxyethanol (IARC 2006). Tests for DNA methylation in the brain, liver, kidneys, testes and spleen of Sprague-Dawley rats given single oral 2-butoxyethanol doses of 120 mg/kg body weight, and in the brain, bone marrow, testes and spleen of transgenic FVB/N mice bearing the v-H-*ras* oncogene, which had been treated with about 120 mg/kg body weight and day by means of a subcutaneously implanted minipump for 2 weeks, did not reveal any substance-related effects (IARC 2006).

Groups of male B6C3F$_1$ mice and male F344 rats were given single oral doses of 2-butoxyethanol of 0, 225, 450 or 900 mg/kg body weight and 0, 225 or 450 mg/kg body weight, respectively, for 7, 14, 28 or 90 days (see Section 5.2.2). The levels of 8-hydroxydeoxyguanosine and malondialdehyde, resulting from oxidative damage, were significantly increased in the liver of the mice of the 450 and 900 mg/kg groups after 7 and 90 days treatment, but not in the liver of rats. In the mice of all dose groups, the induction of DNA synthesis was found to be significant in the hepatocytes of the animals treated for 90 days and in the endothelial cells of the liver after 7 and 14 days treatment. No significant increases in oxidative damage or DNA synthesis developed in the treated rats. No apparent differences were found between treated and untreated mice and rats as regards apoptosis and mitosis in the liver (Siesky *et al.* 2002).

To clarify the 2-butoxyethanol-induced formation of tumours in the forestomach of mice, the mutation spectrum of the H-*ras* oncogene in forestomach tumours of mice exposed to 2-butoxyethanol for two years (see Section 5.7.2) was compared with that of B6C3F$_1$ mice not exposed (control animals in the 2-butoxyethanol study and other NTP studies): mutations in codon 61 of the H-*ras* gene were found in 57 % (8/14) of the 2-butoxyethanol-induced forestomach tumours and in 45 % (5/11) of the spontaneous tumours. Comparison of the mutation spectra and incidence of mutations revealed no significant difference between the spontaneous and the 2-butoxyethanol-induced tumours. Mutations in exon 1 of the H-*ras* gene or in exons 1 and 2 of the K-*ras* gene were not found in the forestomach neoplasms of animals exposed to 2-butoxyethanol (NTP 2000). These results indicate that 2-butoxyethanol does not increase the mutation frequency of the H-*ras* gene (IARC 2006).

In F344 rats and B6C3F$_1$ mice given intraperitoneal injections of 2-butoxyethanol three times at 24-hour intervals (in doses of up to 450 and 550 mg/kg body weight, respectively), the number of micronuclei in the polychromatic erythrocytes of the bone marrow was not significantly increased (NTP 2000). Also in CD-1 mice, single intraperitoneal doses of 2-butoxyethanol of up to 800 mg/kg body weight 24 hours after treatment or of up to 600 mg/kg body weight 24, 48 or 72 hours after treatment did not lead to the induction of micronuclei in the polychromatic erythrocytes of the bone marrow (Elias *et al.* 1996).

5.6.2.2 The metabolite 2-butoxyacetaldehyde

There are no *in vivo* genotoxicity studies with 2-butoxyacetaldehyde available.

5.6.2.3 The metabolite 2-butoxyacetic acid

2-Butoxyacetic acid did not induce micronuclei in the polychromatic erythrocytes of the bone marrow of CD-1 mice 24 hours after single intraperitoneal doses of 200 mg/kg body weight (Elias *et al.* 1996).

5.7 Carcinogenicity

5.7.1 Short-term tests

SHE cells were incubated with 2-butoxyethanol concentrations of 0.5 to 20 mmol/l, 2-butoxyacetic acid concentrations of 0.5 to 20 mmol/l or iron(II) sulfate concentrations of 0.5 to 75 µg/ml at 37°C for seven days. Whereas 2-butoxyethanol and 2-butoxyacetic acid did not induce cell transformation, treatment with iron(II) sulfate in concentrations of 2.5 and 5.0 µg/ml significantly increased the frequency of transformations. Iron(II) sulfate concentrations of 10 µg/ml and above had cytotoxic effects. After incubation of the cells with iron(II) sulfate and with an additional antioxidant such as vitamin E or (–)-epigallocatechin-3-gallate, significantly fewer transformations were observed than after incubation of the cells with iron(II) sulfate alone. In addition, the level of oxidative DNA damage (8-hydroxydeoxyguanosine) and DNA strand breaks (determined via comet assay) after treatment of the cells for 24 hours with iron(II) sulfate and an antioxidant such as vitamin E or (–)-epigallocatechin-3-gallate was significantly lower than after treatment of the cells with iron(II) sulfate alone. In view of this, the authors suggest that not only the cell transformations can be explained by oxidative stress, but also the hepatocarcinogenic effects in male mice resulting from oxidative stress caused by haemolysis and the associated accumulation of iron in the liver (Park *et al.* 2002b).

Transgenic FVB/N mice bearing the v-Ha-*ras* oncogene were given 2-butoxyethanol doses of 120 mg/kg body weight and day for 14 days. No increase in tumour formation was found during the 120-day recovery period (Elliott and Ashby 1997).

5.7.2 Long-term tests

In a two-year carcinogenicity study, groups of 50 F344/N rats and 50 B6C3F$_1$ mice inhaled 2-butoxyethanol concentrations of 0, 31.2, 62.5 or 125 ml/m^3 and 0, 62.5, 125 or 250 ml/m^3, respectively (5 days a week for 6 hours a day; see Section 5.2.1 and Table 6).

Table 6. Studies of the carcinogenicity of 2-butoxyethanol

Author:	NTP (2000)
Substance:	2-butoxyethanol (purity > 99 %); peroxide content 105 or 1000 ppm
Species:	**rat,** F344/N, groups of 50 ♂, 50 ♀
Administration route:	inhalation
Concentration:	0, 31.2, 62.5 or 125 ml/m^3
Duration:	104 weeks exposure, 5 days a week, 6 hours a day
Toxicity:	see Section 5.2.1
Tumours:	

		Exposure concentration (ml/m^3)			
		0	31.2	62.5	125
Survivors:	♂	19/50	11/50	21/50	24/50
	♀	29/50	27/50	23/50	21/50
Adrenal cortex:					
hyperplasia	♀	11/50 (22 %) 1.9[a]	11/50 (22 %) 2.3[a]	8/49 (16 %) 2.1[a]	17/49 (35 %) 2.5[a]
benign phaeochromo-cytomas	♀	3/50 (6 %)	4/50 (8 %)	1/49 (2 %)	7/49 (14 %)[b]
benign or malignant phaeochromocytomas	♀	3/50 (6 %)	4/50 (8 %)	1/49 (2 %)	8/49 (16 %)[c]

[a] average degree of severity: from 1 (minimal) to 4 (marked)
[b] historical controls: 47/889; 5.3 % ± 3.9 %; 0 %–13 %
[c] historical controls (benign, complex or malignant phaeochromocytomas): 57/889; 6.4 % ± 3.5 %; 2 %–13 %

Author:	NTP (2000)
Substance:	2-butoxyethanol (purity > 99 %); peroxide content 105 or 1000 ppm
Species:	**mouse,** B6C3F$_1$, groups of 50 ♂, 50 ♀
Administration route:	inhalation
Concentration:	0, 62.5, 125 or 250 ml/m^3
Duration:	104 weeks exposure, 5 days a week, 6 hours a day
Toxicity:	see Section 5.2.1
Tumours:	

Table 6. continued

		Exposure concentration (ml/m³)			
		0	62.5	125	250
Survivors:	♂	39/50	39/50	27/50	26/50
	♀	29/50	31/50	33/50	36/50
Liver:					
haemangiosarcomas	♂	0/50 (0 %)	1/50 (2 %)	2/49 (4 %)	4/49 (8 %)*[a]
liver cell carcinomas	♂	10/50 (20 %)	11/50 (22 %)	16/49 (33 %)	21/49 (43 %)**[b]
Forestomach:					
squamous cell papillomas/carcinomas	♂	1/50 (2 %)	1/50 (2 %)	2/49 (4 %)	2/48 (4 %)[c]
squamous cell papillomas	♀	0/50 (0 %)	1/50 (2 %)	2/50 (4 %)	5/50 (10 %)*[d]
squamous cell papillomas/carcinomas	♀	0/50 (0 %)	1/50 (2 %)	2/50 (4 %)	6/50 (12 %)*[e]

* $p \leq 0.05$; ** $p \leq 0.01$ (Poly-3 test)
[a] historical controls: 14/968; 1.5 % ± 1.5 %; 0 %–4 %
[b] historical controls: 247/968; 25.7 % ± 10.4 %; 11 %–48 %
[c] historical controls: 5/970; 0.5 % ± 0.9 %; 0 %–2 %
[d] historical controls: 7/973; 0.7 % ± 1.0 %; 0 %–2 %
[e] historical controls: 9/973; 0.9 % ± 1.1 %; 0 %–3 %

Survival of the rats and the body weights of the male rats were not affected. After week 17, the body weight gains of the female rats exposed to 125 ml/m³ were reduced by 5 % to 10 % compared with those in the control animals. In this group, the incidence of phaeochromocytomas in the adrenal medulla was outside the range of variation of the historical controls, but was not significantly increased compared with that in the concurrent control group. One of the phaeochromocytomas in the 125 ml/m³ group was malignant, and another, while benign, was bilateral. The incidence of adrenal medullary hyperplasia was not significantly increased compared with that in the controls. A relationship between this occurrence and the inhalation of 2-butoxyethanol seemed questionable to the authors.

The mean body weight of the exposed male mice during the last 6 months was lower than that of the control animals. In the animals of the 250 ml/m³ group, the incidences of hepatocellular carcinomas and haemangiosarcomas of the liver were significantly increased compared with those in the concurrent controls. The incidence of liver haemangiosarcomas was outside the range of variation of the historical controls (Battelle Pacific Northwest Laboratories), while that of hepatocellular carcinomas was within the historical control range.

The exposed female animals gained markedly less weight than did the controls. In the animals of the 250 ml/m³ group, the incidence of squamous cell papillomas of the forestomach was significantly increased. The papillomas in the high dose group first occurred after 582 days, in the lower dose groups after 731 days. In addition, a significant, concentration-dependent increase in ulcers and epithelial hyperplasia of the forestomach was observed mainly in the female mice (see Section 5.2.1). In the male mice of the 250 ml/m^3 group, the incidence of squamous cell papillomas or carcinomas of the forestomach was outside the range of variation of the historical controls. There was no evidence of infection with *Helicobacter hepaticus* (which can result in an increase in liver neoplasms) in the subsequent investigation of tissue samples from 14 animals. Two of the four mice of the 250 ml/m^3 group with haemangiosarcomas of the liver also had haemangiosarcomas in the bone marrow and the heart, or in the bone marrow and the spleen. It could not be determined which were the primary tumours and which the metastases. The mouse with the liver haemangiosarcoma in the 62.5 ml/m^3 group also had haemangiosarcomas in the bone marrow and in the spleen. Of the four male mice in the 250 ml/m^3 group with liver haemangiosarcomas, only 3 had haemosiderin pigmentation in the liver. 2-Butoxyethanol led also to a significant, concentration-dependent increase in the incidences of haematopoietic cell proliferation and haemosiderin deposits in the spleen, but not to an increase in neoplasms in this organ, such as could be expected in the case of an association.

The evaluation by the NTP cited "some evidence" of carcinogenicity in male and female mice, "no evidence" in male rats and "equivocal evidence" in female rats (NTP 2000).

6 Manifesto (MAK value, classification)

The inhalation of 2-butoxyethanol for two years induced an increased number of benign and malignant phaeochromocytomas in the adrenal medulla of female rats, an increased number of hepatocellular carcinomas and haemangiosarcomas in the liver of male mice, and squamous cell papillomas of the forestomach associated with a concentration-dependent increase in ulcers and epithelial hyperplasia in the forestomach of female mice.

Although the incidence of phaeochromocytomas in the adrenal medulla of female rats of the high dose group was outside the range of variation of the historical controls, it was not significantly increased compared with that in the concurrent control group. Nor is a dose-dependent increase apparent. In addition, a non-genotoxic mechanism related to the haemolytic effects seems possible, so that, at present, 2-butoxyethanol-induced phaeochromocytomas do not seem to be of any relevance for humans. It should be noted, however, that the Commission is presently evaluating phaeochromocytomas.

The liver tumours in the male mice are presumably a result of the haemolytic effects of 2-butoxyethanol and subsequent oxidative stress. As human erythrocytes are considerably more resistant to the haemolytic effects of the substance, particularly as the level of vitamin E in the human liver is about 100 times higher than that in mice, it seems unlikely that liver tumours are induced by 2-butoxyethanol in humans (IARC 2006).

The forestomach tumours in female mice are probably caused by chronic irritation and regenerative hyperplasia. Humans do not possess any organ of comparable anatomy and physiology. Although there are histological similarities between the lower section of the human oesophagus and the forestomach of mice, the oesophagus is not used for the storage of food as is the forestomach of mice. It can thus be assumed that the exposure concentration necessary to produce corresponding irritation in humans, if at all attainable, is markedly above that for mice (Boatman *et al.* 2004; EPA 2005). In view of the anatomical, and particularly the physiological differences between the mouse and humans, the forestomach tumours are, therefore, of limited relevance for humans. Data from *in vitro* and *in vivo* studies of genotoxicity indicate that neither 2-butoxyethanol nor 2-butoxyacetic acid are genotoxic. However, a possible genotoxic potential which affects tumour formation cannot be excluded at present for 2-butoxyacetaldehyde. The available data do not permit a conclusive evaluation of the genotoxicity of 2-butoxyethanol. However, as the tumours can probably be attributed to toxic proliferative changes, genotoxic effects do not appear to be of foremost importance and doses that do not lead to carcinogenic effects can be derived, 2-butoxyethanol is classified in Carcinogen Category 4.

Human experience of the irritative effects of 2-butoxyethanol on the nose yielded only one NOAEC and one LOAEC: no adverse effects were observed after exposure to 20 ml/m^3 for 2 hours, and the lowest observed adverse effect concentration was 100 ml/m^3 after exposure for 4 to 8 hours. As these data are not sufficiently reliable, the MAK value is derived from the results of animal experiments. Relevant for the evaluation are the haemolytic and the irritative effects: the NOAEC for haemolytic effects in a 13-week study with rats, which are considerably more susceptible than humans, was 25 ml/m^3. A LOAEC of 31.2 ml/m^3 was obtained in the 2-year NTP inhalation study (2000) for the histological irritation in the nose of rats. Although the observed effects (hyaline degeneration) were considered to be more of an adaptive or protective nature (NTP 2000), they could represent initial degenerative changes. As only one LOAEC was obtained from the 2-year NTP study (2000), a benchmark calculation was carried out. This yielded a benchmark concentration of about 15 ml/m^3 for a 5 % increase in incidence. As the rat, unlike humans, breathes through the nose only, the 2-year inhalation study produced merely an increase in the incidence but not in the severity of the irritative effects, and it can also be assumed that humans do not react more sensitively than rats, the MAK value has been lowered to 10 ml/m^3.

As irritation is the critical effect, 2-butoxyethanol is classified in Peak Limitation Category I. As no sensory irritation was observed after the exposure of 7 volunteers to 2-butoxyethanol concentrations of 20 ml/m^3 for two hours, an excursion factor of 2 has been set.

Studies of developmental toxicity in rats, mice and rabbits with oral or inhalation exposure to 2-butoxyethanol found there to be embryotoxic, but not teratogenic effects, only after the maternally toxic concentration of 100 ml/m^3 and above. The previous classification in Pregnancy Risk Group C has therefore been retained.

Current studies have shown that 2-butoxyethanol readily penetrates the skin and thus confirm the previous evaluation. On the basis of the dermal flux determined by Jakasa *et al.* (2004), it can be calculated that 2680 mg of a 50 % aqueous 2-butoxyethanol solution is absorbed within one hour by 2000 cm^2 of exposed skin. After exposure (respiratory volume: 10 m^3) to a 2-butoxyethanol concentration in the air of 10 ml/m^3 for 8 hours, 450 mg of the substance was absorbed by inhalation. The contribution of dermal exposure to systemic toxicity is, therefore, relevant. Designation of 2-butoxyethanol with an “H” has been retained.

There are no clinical findings of skin or respiratory sensitization. Also an animal experiment, albeit incompletely documented, yielded no evidence of contact-sensitizing effects. The substance is thus neither designated with “Sa” nor with “Sh”.

The available data for genotoxicity do not justify classification of the substance in one of the categories for germ cell mutagens.

7 References

ACGIH (American Conference of Governmental Industrial Hygienists) (2002) 2-Butoxyethanol. in: *Documentation of TLVs and BEIs*, ACGIH, Cincinnati, OH, USA

ATSDR (Agency for Toxic Substances and Disease Registry) (1998) *Toxicological profile for 2-butoxyethanol and 2-butoxyethanol acetate.* U.S. Department of Health and Human Services, Public Health Service, Agency for Toxic Substances and Disease Registry, Atlanta, GA, USA

Boatman RJ, Corley RA, Green T, Klaunig JE, Udden MM (2004) Review of studies concerning the tumorigenicity of 2-butoxyethanol in B6C3F1 mice and its relevance for human risk assessment. *J Toxicol Environ Health B 7*: 385–398

Chiewchanwit T, Au WW (1995) Mutagenicity and cytotoxicity of 2-butoxyethanol and its metabolite, 2-butoxyacetaldehyde, in Chinese hamster ovary (CHO-AS52) cells. *Mutat Res 334*: 341–346

Corthals SM, Kamendulis LM, Klaunig JE (2005) Mechanisms of 2-butoxyethanol carcinogenesis (Abstract). *Toxicologist 84 (1–5)*: 153

Corthals SM, Kamendulis LM, Klaunig JE (2006) Mechanisms of 2-butoxyethanol-induced hemangiosarcomas. *Toxicol Sci 92*: 378–386

Deisinger PJ, Boatman RJ (2004) *In vivo* metabolism and kinetics of ethylene glycol monobutyl ether and its metabolites, 2-butoxyacetaldehyde and 2-butoxyacetic acid, as measured in blood, liver and forestomach of mice. *Xenobiotica 34*: 675–685

ECETOC (European Centre for Ecotoxicology and Toxicology of Chemicals) (2005) *The toxicology of glycol ethers and its relevance to man*, fourth edition. Technical Report No. 95, ECETOC, Brussels, Belgium

Elias Z, Danière MC, Marande AM, Poirot O, Terzetti F, Schneider O (1996) Genotoxic and/or epigenetic effects of some glycol ethers: results of different short-term tests. *Occup Hyg 2*: 187–212

Elliott BM, Ashby J (1997) Review of the genotoxicity of 2-butoxyethanol. *Mutat Res 387*: 89–96

EPA (Environmental Protection Agency) (1999) *Ethylene glycol monobutyl ether.* Washington, DC, USA

EPA (2005) *An evaluation of the human carcinogenic potential of ethylene glycol butyl ether.* EPA 600/R-04/123 (final) Washington, DC, USA

Fastier A, Herve-Bazin B, McGregor DB (2005) INRS activities on risk assessment of glycol ethers. *Toxicol Lett 156*: 59–76

Fiserova-Bergerova V, Pierce JT, Droz PO (1990) Dermal absorption potential of industrial chemicals: criteria for skin notation. *Am J Ind Med 17*: 617–635

Green T, Toghill A, Lee R, Moore R, Foster J (2002) The development of forestomach tumours in the mouse following exposure to 2-butoxyethanol by inhalation: studies on the mode of action and relevance to humans. *Toxicology 180*: 257–273

Greenspan AH, Reardon RC, Gingell R, Rosica KA (1995) Human repeated insult patch test of 2-butoxyethanol. *Contact Dermatitis 33*: 59–60

Gualtieri JF, DeBoer L, Harris CR, Corley R (2003) Repeated ingestion of 2-butoxyethanol: case report and literature review. *J Toxicol Clin Toxicol 41*: 57–62

Guy RH, Potts RO (1993) Penetration of industrial chemicals across the skin: a predictive model. *Am J Ind Med 23*: 711–719

Heindel JJ, Gulati DK, Russell VS, Reel JR, Lawton AD, Lamb JC (1990) Assessment of ethylene glycol monobutyl and monophenyl ether reproductive toxicity using a continuous breeding protocol in Swiss CD-1 mice. *Fundam Appl Toxicol 15*: 683–696

IARC (International Agency for Research on Cancer) (2006) *2-Butoxyethanol*. IARC monographs on the evaluation of the carcinogenic risk of chemicals to humans, Volume 88, IARC, Lyon, France

Jakasa I, Mohammadi N, Krüse J, Kezic S (2004) Percutaneous absorption of neat and aqueous solutions of 2-butoxyethanol in volunteers. *Int Arch Occup Environ Health 77*: 79–84

Jones K, Cocker J, Dodd LJ, Fraser I (2003) Factors affecting the extent of dermal absorption of solvent vapours: a human volunteer study. *Ann Occup Hyg 47*: 145–150

Klaunig JE, Kamendulis LM (2005) Mode of action of butoxyethanol-induced mouse liver hemangiosarcomas and hepatocellular carcinomas. *Toxicol Lett 156*: 107–115

Nelson BK, Setzer JV, Brightwell WS, Mathinos PR, Kuczuk MH, Weaver TE, Goad PT (1984) Comparative inhalation teratogenicity of four glycol ether solvents and an amino derivative in rats. *Environ Health Perspect 57*: 261–271

NTP (National Toxicology Program) (1989) *Teratologic evaluation of ethylene glycol monobutyl-ether (CAS No. 111-76-2) administered to Fischer-344 rats on either gestational days 9 through 11 or days 11 through 13* (NTIS No. PB89-165849). NTP National Institute of Environmental Health Sciences, Research Triangle Park, NC, USA

NTP (2000) *Toxicology and carcinogenesis studies of 2-butoxyethanol (CAS No. 111-76-2) in F344/N rats and B6C3F1 mice (inhalation studies)*. NTP Technical Report Series No. 484, U. S. Department of Health and Human Services, National Institutes of Health, Research Triangle Park, NC, USA

Nyska A, Haseman JK, Kohen R, Maronpot RR (2004) Association of liver hemangiosarcoma and secondary iron overload in B6C3F1 mice – the National Toxicology Program experience. *Toxicol Pathol 32*: 222–228

OECD (Organization for Economic Co-operation and Development) (1997) *SIDS Initial Assessment Report (SIAR) 2-butoxyethanol [CAS 111-76-2]*. OECD, Paris

Papanikolaou G, Pantopoulos K (2005) Iron metabolism and toxicity. *Toxicol Appl Pharmacol 202*: 199–211

Park J, Kamendulis LM, Klaunig JE (2002a) Effects of 2-butoxyethanol on hepatic oxidative damage. *Toxicol Lett 126*: 19–29

Park J, Kamendulis LM, Klaunig JE (2002b) Mechanisms of 2-butoxyethanol carcinogenicity: studies on Syrian Hamster Embryo (SHE) cell transformation. *Toxicol Sci 68*: 43–50

Poet TS, Soelberg JJ, Weitz KK, Mast TJ, Miller RA, Thrall BD, Corley RA (2003) Mode of action and pharmacokinetic studies of 2-butoxyethanol in the mouse with an emphasis on forestomach dosimetry. *Toxicol Sci 71*: 176–189

Reed JM, Kamendulis LM, Klaunig JE (2003) Examination of DNA damage in endothelial cells following treatment with 2-butoxyethanol using the single cell gel electrophoresis (comet) assay. *Toxicol Sci 72 (S-1)*: 206

Siesky AM, Kamendulis LM, Klaunig JE (2002) Hepatic effects of 2-butoxyethanol in rodents. *Toxicol Sci 70*: 252–260

Tyl RW, Millicovsky G, Dodd DE, Pritts IM, France KA, Fisher LC (1984) Teratologic evaluation of ethylene glycol monobutyl ether in Fischer-344 rats and New Zealand white rabbits following inhalation exposure. *Environ Health Perspect 57*: 47–68

WHO (World Health Organization) (1998) *2-Butoxyethanol.* IPCS – Concise international chemical assessment document Nr. 10, WHO, Geneva

Wier PJ, Lewis SC, Traul KA (1987) A comparison of developmental toxicity evident at term to postnatal growth and survival using ethylene glycol monoethyl ether, ethylene glycol monobutyl ether and ethanol. *Teratogen Carcinogen Mutagen 7*: 55–64

Wilschut A, ten Berge WF, Robinson PJ, McKone TE (1995) Estimating skin permeation. The validation of five mathematical skin permeation models. *Chemosphere 30*: 1275–1296

Zissu D (1995) Experimental study of cutaneous tolerance to glycol ethers. *Contact Dermatitis 32*: 74–77

completed 02.12.2005

Chloroform

MAK value (1999)	**0.5 ml/m³ (ppm) ≙ 2.5 mg/m³**
Peak limitation (2001)	**Category II, excursion factor 2**
Absorption through the skin (1999)	**H**
Sensitization	**–**
Carcinogenicity (1999)	**Category 4**
Prenatal toxicity (1999)	**Pregnancy Risk Group C**
Germ cell mutagenicity	**–**
BAT value	**–**
Synonyms	trichloromethane
Chemical name (CAS)	trichloromethane
CAS number	67-66-3
Structural formula	$CHCl_3$
Molecular weight	119.38
Melting point	–63.5°C
Boiling point	61.7°C
Density at 20°C	1.48 g/cm³
Vapour pressure at 20°C	211 hPa
1 ml/m³ (ppm) ≙ 4.962 mg/m³	**1 mg/m³ ≙ 0.202 ml/m³ (ppm)**

In 1999 chloroform was classified in Carcinogen Category 4. According to the criteria valid at the time, classification of the substance in a category for germ cell mutagenicity was not necessary. As, in the meantime, new criteria for the classification of substances in categories for germ cell mutagens have been drawn up (see "Germ cell mutagens", Volume 17, present series), this supplement to Volume 14 of the present series evaluates the data for the genotoxicity of chloroform in the light of the newly defined classification criteria.

The MAK-Collection Part I: MAK Value Documentations, Vol. 26. DFG, Deutsche Forschungsgemeinschaft

ISBN: 978-3-527-32306-7

Genotoxicity

On the basis of the documentation from 1999 (“Chloroform”, Volume 14), the data for the genotoxicity of chloroform *in vitro* and *in vivo* are summarized and their relevance for classification in one of the categories for germ cell mutagens evaluated.

In vitro

As a result of the high volatility of chloroform, testing in a closed system is recommended for *in vitro* investigations. The conditions under which incubation was carried out are therefore always given in the description of the studies available.

Gene mutation tests

In various bacterial test systems chloroform was not found to be mutagenic in either the presence or absence of a metabolic activation system, even in a closed system. Most of the studies with yeasts and other fungi yielded negative results (see Volume 14). In a study with *Saccharomyces cerevisiae* D7 for mitotic recombination and gene conversion with concentrations between 21 and 54 mM in a closed system, the results were positive only at clearly cytotoxic concentrations. This strain of yeast contains an endogenous cytochrome P450-dependent monooxygenase system (Callen *et al.* 1980).

Also in mammalian cells mutagenic effects were caused only by high concentrations. In the HPRT (hypoxanthine guanine phosphoribosyl transferase) gene mutation test with V79 cells, three experiments were carried out under identical conditions with chloroform concentrations between 100 and 1500 µg/ml in a closed system. In two of the three experiments there was a slight increase in the number of mutants by a factor of 1.8 to 4.6 after concentrations of 1000 µg/ml and more (about 8.4 mM) in the presence of a metabolic activation system (S9 from Aroclor 1254-induced rat liver). Without the addition of a metabolic activation system the test yielded negative results. Cytotoxic effects were not observed; at the highest tested concentration precipitation was observed in a cytotoxicity test previously carried out (Hoechst AG 1987). Another HPRT gene mutation test with V79 cells, which is of limited validity as a result of the inadequate description of the method and presentation of the results, yielded negative results. The concentrations used were between 1 % and 2.5 % (about 124 and 310 mM). Incubation took place after gassing the cells in closed bottles for one hour; the test was carried out only without a metabolic activation system and no positive control was included (Sturrock 1977). The $TK^{+/-}$ mutation test with L5178Y mouse lymphoma cells in an open system yielded negative results in the absence of a metabolic activation system in the tested concentration range between 0.39 and 1.5 µl/ml (about 4.9–8.5 mM) (Caspary *et al.* 1988; Mitchell *et al.* 1988). In the presence of a metabolic activation system (S9 from Aroclor 1254-induced rat liver), weak positive, statistically significant effects were obtained in the cytotoxic range of 0.025 µl/ml and above (= about 0.31 mM). The concentration range tested was between 0.007 and 0.06 µl/ml. The frequency of mutants was increased three-fold relative to that in the control (Mitchell *et al.* 1988).

Indicator tests

In human lymphocytes, induction of sister chromatid exchange (SCE) was not detected after treatment of the cells with chloroform concentrations of up to 400 μg/ml without a metabolic activation system (about 3.4 mM) (no details of the incubation conditions; Kirkland *et al.* 1981). After chloroform concentrations of 10 mM and more, an increase in SCE count per cell by a factor of 1.2 to 1.8 was observed (no details of the incubation conditions; Morimoto and Koizumi 1983). In K_3D cells (permanent leukaemia cell line, Long-Evans rats) the SCE frequency increased significantly at 1 mM in the presence of a metabolic activation system (S9 from PCB-induced rat liver) to 10.0 SCEs per cell; in the absence of a metabolic activation system the results (7.4 SCEs) were, according to the authors, negative (no details of the incubation conditions; Fujie *et al.* 1993). The results of the negative control with a metabolic activation system are missing in the publication. Chloroform concentrations of 0.1 to 10 mM caused a statistically significant increase in the frequency of SCE in SHE cells (a cell line derived from Syrian hamster embryo) by a factor of 1.1 to 1.6 relative to that in the control (Suzuki 1987). The gassing of CHO cells (a cell line derived from Chinese hamster ovary) with chloroform concentrations of 7000 ml/m^3 and S9 from Aroclor 1254-induced rat liver for one hour did not cause an increase in the frequency of SCE (White *et al.* 1979). The data available indicate that SCE is induced only at high concentrations in the presence of a metabolic activation system. Numerous UDS tests with different cell types (rat and mouse hepatocytes, human lymphocytes, SHE) yielded negative results even in the presence of an activation system (see "Chloroform", Volume 14); these tests were either carried out in an open system or details of incubation are missing. A test for the induction of DNA strand breaks using the alkaline elution method yielded negative results with rat hepatocytes in the tested concentration range of 0.03 to 3 mM. The highest concentration of 3 mM was not yet cytotoxic; the test was carried out in a closed system (Sina *et al.* 1983). In freshly isolated hepatocytes from $B6C3F_1$ mice and F344 rats, chloroform concentrations of 0.1 to 5 mM led to the time and concentration-dependent induction of DNA strand breaks, which were detected by means of pulsed field gel electrophoresis. The concentrations used did not produce any cytotoxic effects (no other details; Ammann and Kedderis 1997). The incubation of calf thymus DNA with 1 mM ^{14}C-chloroform and liver microsomes from rats treated with phenobarbital led to covalent DNA binding (0.46 nmol bound/mg DNA/h) (no details of incubation conditions; DiRenzo *et al.* 1982).

Cytogenetic tests

A chromosomal aberration test with human lymphocytes yielded negative results in the presence of a metabolic activation system (S9 Aroclor 1254-induced rat liver) up to the maximum chloroform concentration tested of 400 μg/ml (about 3.4 mM) (Kirkland *et al.* 1981). No details were given about the cytotoxicity of the substance or how incubation was carried out.

Overall, the *in vitro* genotoxicity tests with chloroform presented an inconsistent picture. Genotoxic effects were observed, even in those studies in which incubation was carried out in a closed system, as a rule only at high, cytotoxic concentrations.

In vivo

Somatic cells

Gene mutation tests

Chloroform did not induce mutations or mitotic recombination in *Drosophila* larvae. The eye mosaic test yielded negative results after inhalation exposure to concentrations between 2000 and 16000 ml/m^3. The highest concentration was lethal, concentrations as low as 8000 ml/m^3 caused a reduced spot frequency (mosaic light spots) relative to that in the controls (Vogel and Nivard 1993).

Groups of 10 female LacI transgenic B6C3F$_1$ mice were exposed by inhalation to chloroform concentrations of 0, 10, 30 or 90 ml/m^3, for 6 hours a day, on 7 days a week for 10, 30, 90 or 180 days. A gene mutation test with the DNA isolated from the liver of these animals yielded negative results. After concentrations of 30 ml/m^3 and more, histopathological changes and the induction of regenerative cell proliferation in the liver were observed in transgenic and non-transgenic animals tested in parallel. In the DNA isolated from the livers of these animals, no gene mutations were detected in the LacI target gene. The authors draw attention, however, to the two weaknesses of the test system: its insensitivity, i.e. weak mutagens cannot be detected, and its suitability only for detecting gene mutations (Butterworth *et al.* 1998).

A host-mediated assay with mice with the *Salmonella* strains TA1535 and TA1537 is described as yielding weak positive results (San Agustin and Lim-Sylianco 1978). There are no data for the route of administration or the doses used.

Indicator tests

In the bone marrow cells of mice given oral doses of chloroform of 25 to 200 mg/kg body weight for four days, doses of 50 mg/kg body weight and above caused an increase in SCE count per cell, which at the highest dose was of the factor 1.5 (Morimoto and Koizumi 1983). There are no data for toxicity available. Also after inhalation exposure of mice to chloroform concentrations of 300 ml/m^3 for 3 and 6 hours, the number of SCEs per cell was increased (no other details; Iijima *et al.* 1982).

UDS tests in hepatocytes from rats (after chloroform doses of 0, 40, 400 mg/kg body weight) and mice (after doses of 0, 238, 477 mg/kg body weight) yielded negative results even with hepatotoxic doses (Larson *et al.* 1994; Mirsalis *et al.* 1982). In the hepatocytes of male B6C3F$_1$ mice, chloroform doses of 200 and 500 mg/kg body weight induced an increase in DNA synthesis in the S-phase, an indirect measure of hepatocellular proliferation (Mirsalis *et al.* 1989).

Alkaline elution revealed that DNA strand breaks were not induced in the kidneys of mice after oral chloroform doses of 1.5 mmol/kg body weight (about 180 mg/kg body weight). The body and kidney weights of the treated animals were unchanged relative to those of the controls (Potter *et al.* 1996).

After oral administration of the substance, tests for covalent binding to DNA yielded weak positive results in the kidneys and liver of the rat, and negative results in the mouse

(Pereira *et al.* 1982). The doses given were 48 mg/kg body weight in the rat and 119 mg/kg body weight in the mouse. In a second investigation, no covalent binding was detected in the liver of the mouse after intraperitoneal doses of 15 mg/kg body weight (Diaz-Gomez and Castro 1980). In addition, it was reported in a third publication that no DNA binding in the liver and kidneys of mice was detected (Reitz *et al.* 1982). The radioactivity was determined in all investigations of covalent binding. DNA adducts were not identified.

Cytogenetic tests

In a chromosomal aberration test in bone marrow cells of the Chinese hamster, the effects of chloroform doses of 0, 40, 120 or 400 mg/kg body weight were investigated. The substance was administered by gavage and preparation was carried out 6, 24 and 48 hours later. In the highest dose group, in which the animals exhibited reduced spontaneous activity and eyelid closure and did not consume any food, a slight but statistically significant increase in the incidence of aberrations was determined (higher than in the controls by a factor of 1.9 and 4.5) 6 and 24 hours after administration; according to the authors, this was, however, within the range of the historical controls of the laboratory. The authors reported the occasional occurrence of cells with severe chromosomal damage: one cell with an unspecified exchange aberration in three of the four treated groups of animals and one cell with an undefined multiple aberration in two of the four treated groups (1000 cells evaluated in each case). The authors did not regard these results as positive, but do not exclude the possibility that chloroform has clastogenic potential (Hoechst AG 1988). In bone marrow cells of the rat, chromosomal aberrations were induced even after single, intraperitoneal doses of 1.2 mg/kg body weight. Also five oral doses of 119.4 mg/kg body weight led to the induction of chromosomal aberrations (Fujie *et al.* 1990). There are no data for toxicity available. Chromosomal aberration tests in bone marrow cells of the mouse produced inconsistent results. Positive results were obtained after subcutaneous doses of 100 or 200 mg/kg body weight (Sharma and Anand 1984), negative results after intraperitoneal administration of up to 1000 mg/kg body weight (Shelby and Witt 1995). Seen in this light, the positive results for chromosomal aberrations obtained in the rat after intraperitoneal injection of the low dose are not explicable and are therefore not included in the present evaluation. There are five publications available about micronucleus tests in the bone marrow of mice. In most cases high doses were used. Data is lacking, however, for the occurrence of systemic-toxic effects in the treated animals. Also the ratio of normochromatic to polychromatic erythrocytes, which would provide information about cytotoxic effects and entry of the substance into the bone marrow, is not given in any of the publications available. Intraperitoneal injection of chloroform doses up to 800 mg/kg body weight carried out on three consecutive days did not cause the induction of micronuclei. In a second experiment with the same dose pattern, in which doses of 400 and 600 mg/kg body weight were tested, but not 800 mg/kg body weight, weak positive effects were obtained (Shelby and Witt 1995). In the micronucleus tests that yielded negative results, two intraperitoneal injections of chloroform in doses of up to 90 mg/kg body weight (Tsuchimoto and Matter 1981), of up to 952 mg/kg body weight (Gocke *et al.* 1981) and of around 80 % of the LD_{50} (Salamone *et al.* 1981) were tested. In a publication, which cannot be included in the evaluation as a result of inadequate description of the method, a positive result was obtained with 700 mg/kg body weight. Doses of 100 to 600 mg/kg body weight and 800 and 900 mg/kg body weight yielded negative results (San Agustin and Lim-Sylianco 1978). Data are lacking for the route of administration, and the sex and number of animals, and there was no positive control.

Male rats, from which a kidney had been removed and which were given folic acid to stimulate cell proliferation after nephrectomy, were treated with oral chloroform doses of 4 mmol/kg body weight (about 480 mg/kg body weight). According to the authors, this dose is about 50 % of the LD_{50}. In the kidney cells, the frequency of cells containing micronuclei increased from a control value of 0.13 % to 0.44 % after administration of chloroform (Robbiano *et al.* 1998). The method used in this investigation is very unusual, and in addition only one concentration was tested. This test can therefore only be included in the evaluation with reservations.

Germ cells

In two studies, the test for sex-linked recessive lethal mutations with *Drosophila melanogaster* yielded negative results after administration of chloroform with the diet (Gocke *et al.* 1981; Vogel *et al.* 1981).

A test for morphologically abnormal sperms in mice treated via inhalation with chloroform revealed an increase in the number of morphologically abnormal sperms; a test after intraperitoneal treatment yielded negative results (Land *et al.* 1981; Topham 1980). Changes in sperm morphology are not, however, a reliable indicator of mutations; the relevance of the effects for germ cell mutagenicity is questionable (ICPEMC 1983; Salamone 1988; Wild 1984).

The available *in vivo* genotoxicity tests with mammals indicate a weak clastogenic potential, but only at high and usually toxic doses.

Manifesto (MAK value/classification)

Chloroform is not mutagenic in bacteria. In eukaryotes, the substance was found to have a weak mutagenic and clastogenic potential *in vitro* only in cytotoxic concentrations. There was no evidence of mutagenic effects in somatic and germ cells of *Drosophila melanogaster*. A gene mutation test with transgenic LacI mice yielded negative results. There are no investigations available of the mutagenic effects in the germ cells of mammals. The available *in vivo* genotoxicity tests with mammals indicate a weak clastogenic potential, but only at high and usually toxic doses. On the basis of this data, chloroform has therefore not been classified in a category for germ cell mutagens.

References

Ammann P, Kedderis GL (1997) Chloroform-induced DNA double-strand breaks in freshly isolated male B6C3F$_1$ mouse and F-344 rat hepatocytes. *Toxicologist 36*: 223

Butterworth BE, Kedderis GL, Conolly RB (1998) The chloroform cancer risk assessment: a mirror of scientific understanding. *CIIT Act 18*: 1–12

Callen DF, Wolf CR, Philpot RM (1980) Cytochrome P-450 mediated genetic activity and cytotoxicity of seven halogenated aliphatic hydrocarbons in *Saccharomyces cerevisiae. Mutat Res 77*: 55–63

Caspary WJ, Daston DS, Myhr BC, Mitchell AD, Rudd CJ, Lee PS (1988) Evaluation of the L5178Y mouse lymphoma cell mutagenesis assay: interlaboratory reproducibility and assessment. *Environ Mol Mutagen 12, Suppl 13*: 195–229

Diaz-Gomez MI, Castro JA (1980) Covalent binding of chloroform metabolites to nuclear proteins – no evidence for binding to nucleic acids. *Cancer Lett 9*: 213–218

DiRenzo AB, Gandolfi AJ, Sipes IG (1982) Microsomal bioactivation and covalent binding of aliphatic halides to DNA. *Toxicol Lett 11*: 243–252

Fujie K, Aoki T, Wada M (1990) Acute and subacute cytogenetic effects of the trihalomethanes on rat bone marrow cells *in vivo. Mutat Res 242*: 111–119

Fujie K, Aoki T, Ito Y, Maeda S (1993) Sister-chromatid exchanges induced by trihalomethanes in rat erythroblastic cells and their suppression by crude catechin extracted from green tea. *Mutat Res 300*: 241–246

Gocke E, King MT, Eckhardt K, Wild D (1981) Mutagenicity of cosmetic ingredients licensed by the European communities. *Mutat Res 90*: 91–109

Hoechst AG (1987) *Chloroform: Detection of gene mutations in somatic cells in culture. HGPRT-test with V79 cells.* Report No. 870692, Frankfurt, unpublished report

Hoechst AG (1988) *Chromosome aberrations in Chinese hamster bone marrow cells.* Report No. 880445, Frankfurt, unpublished report

ICPEMC (1983) Committee 1 final report: screening strategy for chemicals that are potential germ cell mutagens in mammals. *Mutat Res 114*: 117–177

Iijima S, Morimoto K, Koizumi A (1982) Induction of sister chromatid exchanges in mouse bone marrow cells by inhaled chloroform (Japanese). *Igaku No Ayumi 122*: 978–980

Kirkland DJ, Smith KC, Van Abbe NJ (1981) Failure of chloroform to induce chromosome damage or sister chromatid exchanges in cultured human lymphocytes and failure to induce reversion in *E coli. Food Cosmet Toxicol 19*: 651–656

Land PC, Owen EL, Linde HW (1981) Morphologic changes in mouse spermatozoa after exposure to inhalational anesthetics during early spermatogenesis. *Anesthesiology 54*: 53–56

Larson JL, Sprankle CS, Butterworth BE (1994) Lack of chloroform-induced DNA repair *in vitro* and *in vivo* in hepatocytes of female B6C3F$_1$ mice. *Environ Mol Mutagen 23*: 132–136

Mirsalis JC, Tyson CK, Butterworth BE (1982) Detection of genotoxic carcinogens in the *in vivo-in vitro* hepatocyte DNA repair assay. *Environ Mutagen 4*: 553–562

Mirsalis JC, Tyson CK, Steinmetz KL, Loh EK, Hamilton CM, Bakke JP, Spalding JW (1989) Measurement of unscheduled DNA synthesis and S-phase synthesis in rodent hepatocytes following *in vivo* treatment: testing of 24 compounds. *Environ Mol Mutagen 14*: 155–164

Mitchell AD, Myhr BC, Rudd CJ, Caspary WJ, Dunkel VC (1988) Evaluation of the L5178Y mouse lymphoma cell mutagenesis assay: methods used and chemicals evaluated. *Environ Mol Mutagen 12, Suppl 13*: 37–101

Morimoto K, Koizumi A (1983) Trihalomethanes induce SCE in human lymphocytes *in vitro* and mouse bone marrow cells *in vivo. Environ Res 32*: 72–79

Pereira MA, Lin LHC, Lippitt JM, Herren SL (1982) Trihalomethanes as initiators and promoters of carcinogenesis. *Environ Health Perspect 46*: 151–156

Potter CL, Chang LW, DeAngelo AB, Daniel FB (1996) Effects of four trihalomethanes on DNA strand breaks, renal hyaline droplet formation and serum testosterone in male F-344 rats. *Cancer Lett 106*: 235–242

Reitz RH, Fox TR, Quast JF (1982) Mechanistic considerations for carcinogenic risk estimation: chloroform. *Environ Health Perspect 46*: 163–168

Robbiano L, Mereto E, Morando AM, Pastore P, Brambilla G (1998) Increased frequency of micronucleated kidney cells in rats exposed to halogenated anaesthetics. *Mutat Res 413*: 1–6

Salamone MF (1988) Summary report on the performance of the sperm assays. in: Ashby J, de Serres FJ, Shelby MD, Margolin BH, Ishidate Jr M, Becking GC (Eds) *Evaluation of short-term tests for carcinogens, report of the International Programme on Chemical Safety's collaborative study on in vivo assays*, Vol. 2, Cambridge University Press, 2.229–2.234

Salamone MF, Heddle JA, Katz M (1981) Mutagenic activity of 41 compounds in the *in vivo* micronucleus assay. in: Ashby J, de Serres FJ (Eds) (1981) *Progress in mutation research*, Vol. 1, Evaluation of short-term tests for carcinogens, Elsevier, New York, 686–697

San Agustin J, Lim-Sylianco CY (1978) Mutagenic and clastogenic effects of chloroform. *Bull Philipp Biochem Soc 1*: 17–23

Sharma GP, Anand RK (1984) Two halogenated hydrocarbons as inducers of chromosome aberrations in rodents. *Proc Natl Acad Sci India 54*: 61–67

Shelby MD, Witt KL (1995) Comparison of results from mouse bone marrow chromosome aberration and micronucleus tests. *Environ Mol Mutagen 25*: 302–313

Sina JF, Bean CL, Dysart GR, Taylor VI, Bradley MO (1983) Evaluation of the alkaline elution/rat hepatocyte assay as a predictor of carcinogenic and mutagenic potential. *Mutat Res 113*: 357–391

Sturrock JE (1977) Lack of mutagenic effect of halothane or chloroform on cultured cells using the azaguanine test system. *Br J Anaesth 49*: 207–210

Suzuki H (1987) Assessment of the carcinogenic hazard of 6 substances used in dental practices. *Shigaku 74*: 1385–1403

Topham JC (1980) Do induced sperm-head abnormalities in mice specifically identify mammalian mutagens rather than carcinogens? *Mutat Res 74*: 379–387

Tsuchimoto T, Matter BE (1981) Activity of coded compounds in the micronucleus test. in: Ashby J, de Serres FJ (Eds) *Progress in mutation research*, Vol. 1, Evaluation of short-term tests for carcinogens, Elsevier, New York, 705–711

Vogel E, Blijleven WG, Kortselius MJH, Zijlstra JA (1981) Mutagenic activity of 17 coded compounds in the sex-linked recessive lethal test in *Drosophila melanogaster*. in: Ashby J, de Serres (Eds) *Progress in mutation research*, Vol. 1, Evaluation of short-term tests for carcinogens, Elsevier, New York, 660–65

Vogel EW, Nivard MJM (1993) Performance of 181 chemicals in a *Drosophila* assay predominantly monitoring interchromosomal mitotic recombination. *Mutagenesis 8*: 57–81

White AE, Takehisa S, Eger EI, Wolff S, Stevens WC (1979) Sister chromatid exchanges induced by inhaled anesthetics. *Anesthesiology 50*: 426–430

Wild D (1984) The sperm morphology test, a rapid *in vivo* test for germinal mutations? in: Baß R, Glocklin V, Grosdanoff P, Henschler D, Kilbey B, Müller D, Neubert D (Eds) *Critical evaluation of mutagenicity tests*, bga-Schriften 3/84, MMV Medizin Verlag München, 299–306

completed 28.11.2002

Dimethylformamide[1]

Supplement 2001

MAK value (1992)	**10 ml/m^3 (ppm) ≙ 30 mg/m^3**
Peak limitation (2001)	**Category II, excursion factor 2**
Absorption through the skin (1969)	**H**
Sensitization	**–**
Carcinogenicity	**–**
Prenatal toxicity (1989)	**Pregnancy Risk Group B**
Germ cell mutagenicity	**–**
BAT value (1993)	**35 mg *N*-methylformamide/l urine**
Synonyms	*N,N*-dimethylmethanamide DMF DMFA *N*-formyldimethylamine formic acid dimethylamide
Chemical name (CAS)	*N,N*-dimethylformamide
CAS number	68-12-2

Peak Limitation Category

Hepatotoxicity is the critical effect for which the metabolites of dimethylformamide (methyl isocyanate, *S*-(*N*-methylcarbamoyl)cysteine, *S*-(*N*-methylcarbamoyl)glutathione) are considered to be responsible (Mráz *et al.* 1989; Pearson *et al.* 1991). The half-time for renal elimination of dimethylformamide is about 5 hours within the first 15 hours after end of exposure (Walter *et al.* 1998). The half-time for the formation and elimination of the acetylated cysteine adducts is 23 hours (Mráz and Nohová 1992). Slight changes in liver parameters can occur even at a concentration of 20 ml/m^3 at the workplace. Carcinogenic effects in humans cannot be substantiated at this time, although there is evidence from epidemiological studies.

---

[1] A more recent supplement follows.

The MAK-Collection Part I: MAK Value Documentations, Vol. 26. DFG, Deutsche Forschungsgemeinschaft

ISBN: 978-3-527-32306-7

Marked irritation of the mucous membranes in the respiratory tract and the eyes was reported to have occurred at workplace concentrations of 8 to 58 mg/m^3 (average 22 mg/m^3) (Cirla *et al.* 1984). However, the authors themselves do not rule out a synergistic effect with other irritating substances acting at the same time. Possibly the irritation was caused by higher exposure for a brief period.

An excursion factor of 2 has been provisionally established on the basis of dimethylformamide's long half-time, its critical metabolites and the uncertainty about classifying the irritant effects; it will be reviewed after a new evaluation of the carcinogenicity of dimethylformamide.

References

Cirla AM, Pisati G, Invernizzi E, Torricelli P (1984) Epidemiological study on workers exposed to low dimethylformamide concentrations. *G Ital Med Lav 6*: 149–156

Mráz J, Nohová H (1992) Absorption, metabolism and elimination of N,N-dimethylformamide in humans. *Int Arch Occup Environ Health 64*: 85–92

Mráz J, Cross H, Gescher A, Threadgill MD, Flek J (1989) Differences between rodents and humans in the metabolic toxification of N,N-dimethylformamide. *Toxicol Appl Pharmacol 98*: 507–516

Pearson PG, Slatter JG, Rashed MS, Han DH, Baillie TA (1991) Carbamoylation of peptides and proteins *in vitro* by S-(N-methylcarbamoyl)glutathione and S-(N-methylcarbamoyl)cysteine, two electrophilic S-linked conjugates of methyl isocyanate. *Chem Res Toxicol 4*: 436–444

Walter D, Jochim C, Knecht U (1998) Toxikokinetische Untersuchungen zur Reevaluierung des BAT-Wertes von Dimethylformamid (Toxicological studies for the reevaluation of the BAT value for dimethylformamide) (German). In: Hallier E, Bünger J (Eds) Dokumentationsband 38. Jahrestagung der Deutschen Gesellschaft für Arbeitsmedizin und Umweltmedizin e.V., Rindt Verlag, Fulda, 619–620

completed 01.03.2001

Dimethylformamide

Supplement 2006

MAK value (2005)	**5 ml/m^3 (ppm) ≙ 15 mg/m^3**
Peak limitation (2005)	**Category II, excursion factor 4**
Absorption through the skin (1969)	**H**
Sensitization	**–**
Carcinogenicity	**–**
Prenatal toxicity (1989)	**Pregnancy Risk Group B**
Germ cell mutagenicity	**–**
BAT value (2001)	**35 mg *N*-methylformamide/l urine**
Synonyms	*N,N*-dimethylmethanamide DMF DMFA *N*-formyldimethylamine formic acid dimethylamide
Chemical name (CAS)	*N,N*-dimethylformamide
CAS number	68-12-2
1 ml/m^3 (ppm) ≙ 3.03 mg/m^3	**1 mg/m^3 ≙ 0.329 ml/m^3 (ppm)**

Since the documentation from 1992 ("Dimethylformamide", Volume 8, present series), further studies have been published that make a reassessment of the toxicity of *N,N*-dimethylformamide (dimethylformamide, DMF) necessary.

1 Toxic Effects and Mode of Action

Dimethylformamide causes irritation of the mucous membranes. Oedema and scaling of the skin are observed after long-term exposure. However, local irritation is not the main effect at dimethylformamide concentrations in the range of the previous MAK value of 10 ml/m^3.

The MAK-Collection Part I: MAK Value Documentations, Vol. 26. DFG, Deutsche Forschungsgemeinschaft

ISBN: 978-3-527-32306-7

Dimethylformamide is a solvent with a high vapour pressure, which is readily absorbed and distributed in the organism after inhalation or dermal exposure. The ability of dimethylformamide to penetrate the skin very readily contributes significantly to endogenous exposure at the workplace. Monitoring the BAT value is therefore of great importance.

Numerous studies of exposure in humans and animals have shown the liver to be a specific target organ, with increased activities of liver-specific enzymes and hepatocellular hyperplasia. Eosinophilic foci and necrosis were also found in animal experiments. Exposure to dimethylformamide in combination with alcohol can lead to intolerance reactions with flushing, similar to those observed with disulfiram in combination with alcohol. Such alcohol intolerance reactions after exposure to dimethylformamide may occur even some hours after the end of the work shift.

In animal experiments, dimethylformamide caused impairment of fertility in mice after doses of 800 mg/kg body weight and day and above. Embryotoxic effects were observed in rats after doses as low as 100 mg/kg body weight and day, and skeletal variations were found after 200 mg/kg body weight and day. In inhalation studies, the thresholds for embryotoxic effects were between 50 ml/m^3 (NOAEL: no observed adverse effect level) and 150 ml/m^3 (LOAEL: lowest observed adverse effect level) in rabbits, and between 18 and 30 ml/m^3 (NOAEL) and 172 and 300 ml/m^3 (LOAEL) in rats.

There is no clear evidence that dimethylformamide has genotoxic, germ cell mutagenic or carcinogenic effects.

2 Mechanism of Action

The liver is the target organ of dimethylformamide in humans and animals. Increased serum concentrations of liver-specific enzymes, such as alkaline phosphatase (AP), the transaminases aspartate aminotransferase (AST) and alanine aminotransferase (ALT), or γ-glutamyl transpeptidase (γ-GT), are detectable in humans (see Tables 1 and 2). In animal experiments, increased total cholesterol and phospholipid concentrations were found, and fatty degeneration of the liver was observed. The hepatotoxic effects are attributed to dimethylformamide itself, but in particular to the reactive metabolite *N*-acetyl-*S*-(*N*-methylcarbamoyl)cysteine (AMCC), which is formed by conjugation with glutathione (“Dimethylformamide”, Volume 8, present series; Wrbitzky 1999).

In rats, dimethylformamide is transformed by cytochrome P450-dependent enzymes, in particular CYP2E1, into free radicals, which react with the haem, thereby inactivating the enzyme (Tolando *et al.* 2001).

Exposure to dimethylformamide in combination with alcohol can produce an alcohol intolerance reaction of the disulfiram type with flushing of the facial skin and other objective and subjective symptoms. The reason for such intolerance reactions is an inhibition of alcohol dehydrogenase and aldehyde dehydrogenase. This leads to the accumulation of acetaldehyde, a metabolite of ethanol, which causes the symptoms described and results in liver damage (see the documentation for ethanol from 1990, 1998 ("Ethanol", Volume 12, present series), 2001 and 2002 ("Ethanol", this volume). Differences in tolerance to alcohol have been known for a long time (Chan 1986) and are based on a genetic polymorphism of the enzymes alcohol dehydrogenase and aldehyde dehydrogenase. In the European population, the atypical ADH1B*2 allele is found in only approximately 5 % of persons, whereas in Asiatic populations it is found in up to 90 %. Bearers of the atypical ADH1B*2 allele are more sensitive to alcohol (Quertemont 2004).

3 Toxicokinetics and Metabolism

3.1 Inhalation

3.1.1 Animal studies

After whole-body inhalation exposure of rats and mice to dimethylformamide concentrations of 0, 10, 250 or 500 ml/m^3 up to 10 times for 6 hours a day, there was evidence that dimethylformamide metabolism had become saturated after single exposures to the two high concentrations. After repeated exposure, induction of dimethylformamide metabolism occurred at the highest concentration (Hundley *et al.* 1993a).

After whole-body exposure of cynomolgus monkeys to dimethylformamide concentrations of 0, 30, 100 or 500 ml/m^3 over 13 weeks for 6 hours a day on 5 days a week, saturation of the metabolism could be seen at the two highest concentrations. At 30 ml/m^3, dimethylformamide was transformed very quickly into *N*-hydroxymethyl-*N*-methylformamide (HMMF) (56 %–95 %), which was detected in urine as *N*-methyl-formamide (NMF) after hydrolysis. The half-times of dimethylformamide in plasma are reported to be 1 to 2 hours, and that of NMF to be 4 to 15 hours (Hundley *et al.* 1993b).

3.1.2 Volunteers

After exposure of groups of up to 10 volunteers to dimethylformamide concentrations of 0, 10, 30 or 60 mg/m^3 (0, 3.3, 10, 20 ml/m^3) for 8 hours in an inhalation chamber, retention of the substance in the respiratory tract was 90 % at a pulmonary respiration rate of 10 l/min. Elimination of dimethylformamide and of the metabolites HMMF (detected as NMF), *N*-hydroxymethyl-formamide (detected as formamide, FA) and AMCC was determined in the urine. After exposure to dimethylformamide concentrations of 10 ml/m^3, the proportion of the total dose in urine was 0.3 % for dimethylformamide, 22.3 % for NMF, 13.2 % for FA and 13.4 % for AMCC. The elimination half-times were 3.8 hours for NMF, 6.9 hours for FA, and 23 hours for AMCC. The long half-time for AMCC is attributed to a reversible protein binding. Elimination was faster, however, after oral intake of 20 mg AMCC. After exposure to dimethylformamide concentrations of 10 ml/m^3 for 8 hours a day for 5 days, there was no accumulation of NMF, slight accumulation of FA, and marked accumulation of AMCC (Mráz and Nohová 1992b).

Thirteen volunteers were exposed once via the skin in an inhalation chamber to a dimethylformamide concentration of 6.2 ± 1.0 ml/m^3 for 4 hours at a temperature of 27°C and a relative humidity of 44 %. The volunteers breathed fresh air via a mask. In another experiment, the volunteers were exposed for 4 hours to a dimethylformamide concentration of 7.1 ± 1.0 ml/m^3 exclusively via a breathing mask. The amount of NMF eliminated with the urine within 24 hours was 3.25 mg after dermal exposure and 3.93 mg after inhalation exposure. From these data, absorption rates of 40.4 % after dermal exposure and of 59.6 % after inhalation exposure were calculated. The half-time for NMF in urine after dermal exposure (4.75 ± 1.63 hours) was longer than that after respiratory exposure (2.42 ± 0.63 hours) (Nomiyama *et al.* 2001a).

3.1.3 Workers

Recent studies have confirmed there is relevant dermal absorption of the substance in addition to the amount taken in by inhalation, as has been known for a long time from workplaces with exposure to dimethylformamide.

The amount of dimethylformamide absorbed by 3 workers was calculated by determining the levels of the metabolite NMF in urine both after an 8-hour working day without protective equipment and after an 8-hour working day with a respiratory protective mask. The concentrations of dimethylformamide in the air were between 12 and 40 mg/m^3 (4 and 13 ml/m^3). The workers did not perspire and their skin was dry. The NMF eliminated with the urine over 15 hours was approximately three times higher in the work shift without protective equipment (inhalation and dermal absorption) than in the shift during which a protective mask had been worn (dermal absorption only) (3.73, 8.36 and 11.13 mg, compared with 1.69, 1.92 and 3.44 mg NMF) The authors point out that dermal absorption contributes significantly to the total amount of dimethylformamide absorbed (Miyauchi *et al.* 2001).

A study was carried out with 178 workers exposed via inhalation and dermally to dimethylformamide (6.27 ml/m^3; NMF in urine: 24.39 mg/l) and with 37 workers exposed to dimethylformamide mainly via inhalation (2.07 ml/m^3; NMF in urine: 8.23 mg/l). Correlation analyses revealed that, based on a dimethylformamide concentration of 10 ml/m^3, the amount of NMF eliminated in the urine (45.3 mg/l) was higher in workers with inhalation and dermal exposure to dimethylformamide than in workers with mainly inhalation exposure (NMF in urine, 37.7 mg/l) (Yang *et al.* 2000).

3.2 Dermal absorption

After four hours dermal exposure of volunteers to a dimethylformamide concentration of 51 mg/m^3 in an exposure chamber while they inhaled fresh air via a respiration mask, absorption increased with increasing temperature and relative humidity. The amount of HMMF (detected as NMF) eliminated with the urine over 24 hours was 27 µmol at 21°C and 50 % relative humidity, 44 µmol at 28°C and 70 % relative humidity and 95 µmol at 30°C and 100 % relative humidity. In contrast, the amount of NMF eliminated with the urine during 24 hours after inhalation and dermal exposure to 52 mg/m^3 was 219 µmol (Mráz and Nohová 1992a).

In 4 volunteers, the absorption rate of dimethylformamide after dipping one hand up to the wrist in pure dimethylformamide for 2 to 20 minutes was determined as 9.4 ± 4.0 mg/cm^2 × hour. After dipping one hand in the substance for 15 minutes, 930 µmol HMMF (detected as NMF), 606 µmol *N*-hydroxymethylformamide (detected as FA) and 597 µmol AMCC were eliminated with the urine over five days. The elimination half-times were given as 7.8 hours for NMF, 9.9 hours for FA and 23.9 hours for AMCC. The metabolite levels corresponded approximately to those found after exposure for 8 hours to a dimethylformamide concentration of 60 mg/m^3 (20 ml/m^3). After inhalation of dimethylformamide, however, the half-times for NMF and FA were shorter (4 and 6.9 hours, respectively); the half-time for AMCC of 25.1 hours was somewhat longer (Mráz and Nohová 1992a).

After occlusive application of 2 mmol dimethylformamide (146 mg) to the forearm (100 cm^2) for 8 hours in a patch test, the amount of HMMF (detected as NMF) eliminated with the urine within 24 hours was only 7.6 % in the first 4 volunteers, and 8.7 % in the other 4 volunteers. In contrast, 16 % to 18 % was eliminated as NMF after inhalation exposure (Mráz and Nohová 1992a).

3.3 Metabolism

The metabolism of dimethylformamide (see Figure 1) has already been described in detail in the 1992 MAK documentation. The metabolites of dimethylformamide in urine are *N*-hydroxymethyl-*N*-methylformamide (HMMF), *N*-methylformamide (NMF), *N*-hydroxymethylformamide and formamide (FA). Cytochrome P450-dependent transformation of HMMF and NMF is thought to form the reactive and toxic metabolite methyl isocyanate, which together with glutathione forms *S*-(*N*-methylcarbamoyl)-glutathione (NMG). Further transformation leads to the cysteine adduct AMCC (*N*-acetyl-*S*-(*N*-methylcarbamoyl)cysteine). More AMCC is formed in humans than in rodents (Mráz *et al.* 1989).

In vitro it was shown that dimethylformamide can inhibit the formyl oxidation of HMMF and NMF to NMG (Mráz *et al.* 1993).

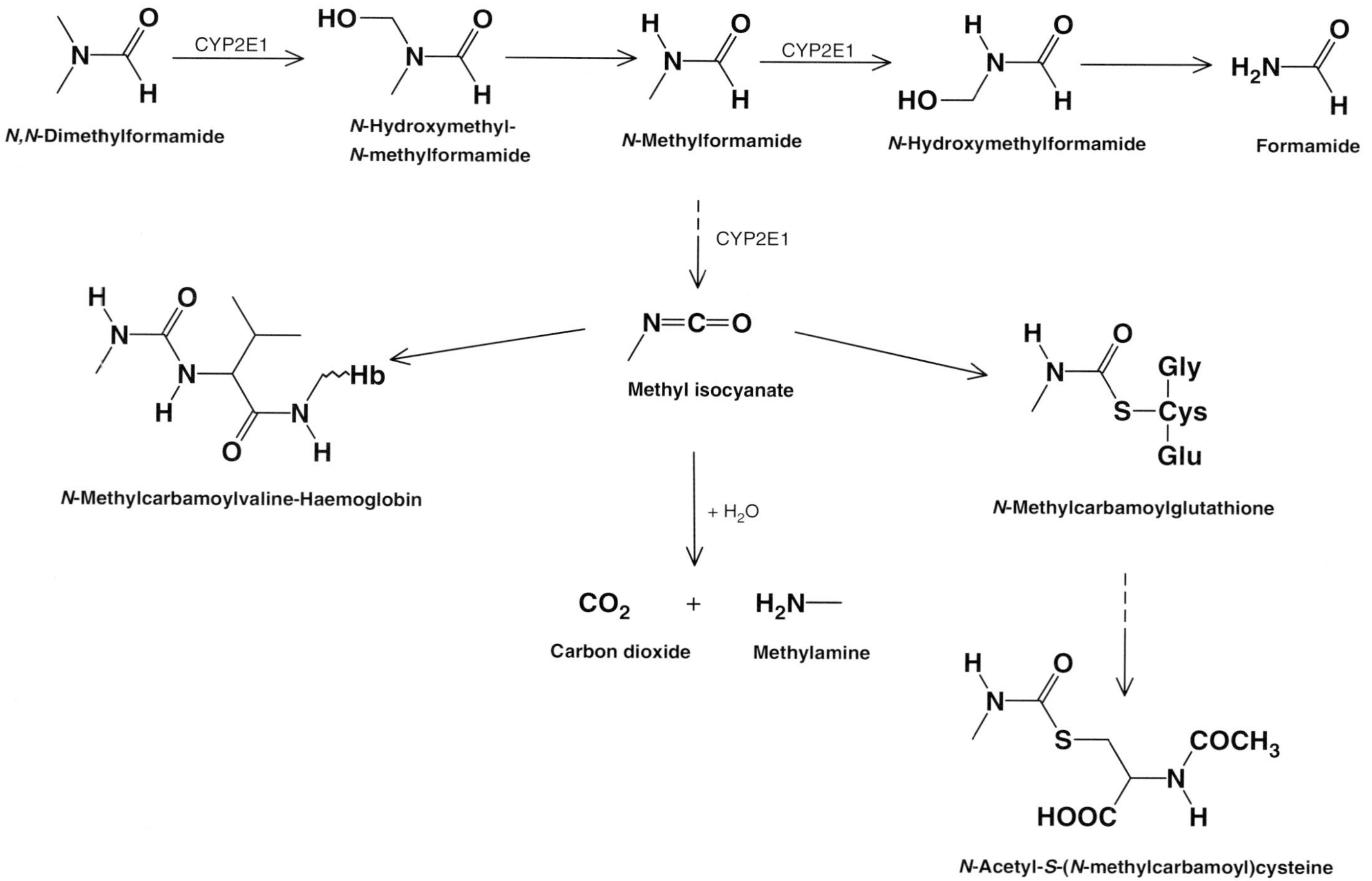

Figure. 1. Known metabolism of *N,N*-dimethylformamide in humans (see Drexler and Greim 2006)

In vitro studies of human liver microsomes revealed that CYP2E1 is a key enzyme in dimethylformamide metabolism (Amato *et al.* 2001; Mráz *et al.* 1993). The CYP isoforms 1A1, 1A2, 2B6, 2C10 and 3A4 proved to be inactive *in vitro* (Amato *et al.* 2001).

Investigation of CYP2E1 PstI/RsaI polymorphism in 123 volunteers, however, did not indicate any significant differences in the elimination half-time of NMF in urine (Nomiyama *et al.* 2001b).

4 Effects in Humans

4.1 Single exposures

In a case report, increased concentrations of dimethylformamide were determined in the serum of a suicide patient following the intake of a veterinary euthanasia drug (T-61®) with dimethylformamide as solvent. These concentrations decreased in a linear fashion and were no longer elevated after 30 hours. On admission of the patient to hospital, serum ALT and AST values were normal. AST increased over the subsequent period, reaching a maximum on day four after intoxication. ALT was increased from day two onwards, reaching a maximum on the sixth day after intoxication. A transient increase in serum bilirubin was found, and the prothrombin time was decreased. The alkaline phosphatase value was within the normal range. γ-GT was not determined. The authors attribute the hepatotoxic effects to dimethylformamide (Buylaert *et al.* 1996).

4.2 Repeated exposures

Where humans are concerned, differences are made between investigations conducted in Asians and in Europeans because of genetic polymorphism (see above).

4.2.1 Asian area

Case reports

In a biomonitoring study of 9 workers exposed to dimethylformamide concentrations of up to 5 ml/m^3 and who eliminated 19 mg NMF per 24 hours in the urine (corresponding to about 19 mg/l urine), the values for AST, ALT, alkaline phosphatase and γ-GT were within the normal range. Six of the workers reported that they were not able to tolerate as much alcohol as before starting work with dimethylformamide (Yonemoto and Suzuki 1980). In another biomonitoring study of 10 workers exposed to dimethylformamide concentrations of 2.5 to 10.4 ml/m^3 and who had 24.7 mg NMF per gram creatinine in their urine (corresponding to about 30 mg/l urine), the values for ALT, AST and alkaline phosphatase were also within the normal range (Sakai *et al.* 1995).

One abstract reports of 13 workers exposed to peak dimethylformamide concentrations of 10 to 42 ml/m^3 for 15 minutes. Seven of them complained of abdominal colic, liver function (no other details) was abnormal in 3 and 3 suffered facial flushing (Yang *et al.* 1994).

After exposure to dimethylformamide at the workplace for 5 months, one man suffered from hepatic dysfunction with increased AST, ALT and γ-GT values. On the basis of a urinary NMF concentration of 42.8 mg/l, a dimethylformamide concentration in the air of 10 to 30 ml/m^3 was calculated. The man returned to his workplace after two months and, eighteen days later, the liver values were once again increased (Nomiyama *et al.* 2001c).

A 42-year-old male Taiwanese exposed for 3 years to dimethylformamide at his workplace suffered from vomiting and abdominal pain over six months. The values for AST, ALT, γ-GT and alkaline phosphatase in previous routine investigations were within the normal range. The man drank no alcohol. On admission to hospital, the values for bilirubin, ALT, alkaline phosphatase and γ-GT were above the normal values, some of them markedly so. Slight fatty degeneration of the liver was diagnosed. Liver biopsies revealed oedematous portal tracts with proliferation of the bile ducts, minimal proliferation of the ductus, and slight acute and chronic inflammation-related cell infiltration. Fatty vacuoles, increased liver cell regeneration, focal hypertrophy of Kupffer's cells and small neutrophil aggregates were observed in the acinus. Ten days after admission, the patient was discharged free of symptoms, although the ALT value was still increased (Huang *et al.* 1998). No data are available for the levels of exposure to dimethylformamide or metabolite analyses in urine.

Cohort studies in workers

Findings in collectives of Asian workers are shown in Table 1.

In Chinese workers exposed to mean dimethylformamide concentrations of more than 3.9 ml/m^3 (range 2.0–4.9 ml/m^3), a high proportion of alcohol intolerance reactions (71 %) was reported. In workers with exposures of up to 1.9 ml/m^3, the incidence of alcohol intolerance was 14 % compared with 25 % in the control group. Only in 2 workers were the values for AST and ALT above the normal range. γ-GT was not increased above the normal range in any of the workers (Cai *et al.* 1992). This study indicates that alcohol intolerance reactions in Chinese workers may occur more frequently at dimethylformamide concentrations of more than 2 ml/m^3, on average at 3.9 ml/m^3. Biomonitoring was not performed, however, so that the contribution of dermally absorbed dimethylformamide cannot be estimated.

Table 1. Effects in workers exposed to dimethylformamide in Asian countries (NMF in urine was not determined in any of the studies)

Collective studied		DMF concentration in air (ml/m^3)[1]	Result	References
143	controls	–	alcohol intolerance (25 %)	Cai *et al.* 1992 (China)
207	workers	**4.5** 8-hour TWA	Total cohort: symptoms (e.g. nasal irritation, unusual taste, dizziness, nausea, dry mouth); biochemical parameters above the normal values only in individual cases	
59	laboratory B	0.2	no increased alcohol intolerance	
23	laboratory A	0.4		
17	production of shoe soles	0.7		
65	polyurethane production	**3.9** (2–4.9)	increased alcohol intolerance	
43	leather production	**9.1**	symptoms (nausea, vomiting, abdominal pain); increased alcohol intolerance	
176	workers in a plant for synthetic leather and synthetic resins	**11.6 ± 13.8 (0.1–86.6)** TWA		Luo *et al.* 2001 (Taiwan)
74		2.9 ± 1.1 (0.5–5)	abnormal liver function (AST, ALT and γ-GT) (18 %)	
37		6.4 ± 0.7 (5.9–8.2)	abnormal liver function (27 %); adjusted OR 1.62 (0.61–4.28) in comparison with the low-exposure group	
65		24.6 ± 15.6 (11.2–86.6)	abnormal liver function (37 %); adjusted OR 2.93 (1.27–6.8) in comparison with the low-exposure group; effects on individual liver enzymes (AST, γ-GT)	

Table 1. continued

Collective studied		DMF concentration in air (ml/m³)[1]	Result	References
183	workers in a plant for synthetic leather[2]	–	Total cohort: symptoms (dizziness, anorexia, nausea, epigastric pains) increasing with increasing exposure; AST and γ-GT not significantly changed	Wang *et al.* 1991 (China)
76	administration and maintenance	< 10	ALT (14 ± 6 IU/l); 6.4 % of the workers with alcohol consumption > 24 g/day	
83	dry process material mixing and separation	**10–40** (besides DMF, methylethylketone, toluene)	ALT increased (18 ± 7 IU/l); 3.7 % of the workers with alcohol consumption > 24 g/day	
24	wet processing and mixing	**25–60**	ALT increased (25 ± 6 IU/l); significantly increased proportion of workers with high ALT (> 35 IU/l); no workers with alcohol consumption > 24 g/day	

AST: aspartate aminotransferase; ALT: alanine aminotransferase; DMF: *N,N*-dimethylformamide; γ-GT: γ-glutamyltranspeptidase; OR: odds ratio; TWA: time-weighted average
[1] stationary sampling [2] workers with increased prevalence of liver damage

In a Taiwanese plant producing synthetic leather, an increase in the percentage of workers with abnormal liver function values (AST, ALT and γ-GT) was found in the middle (6.4 ± 0.7 ml/m³) and high (24.6 ± 15.6 ml/m³) dimethylformamide exposure groups in comparison with the group exposed to 2.9 ± 1.1 ml/m³. The values for individual liver enzymes (AST, γ-GT) were also significantly increased in the high exposure group (Luo *et al.* 2001). Since there was no control group without exposure to dimethylformamide, no statement can be made as to whether an increase in abnormal liver values could also occur at dimethylformamide concentrations below 3 ml/m³. Biomonitoring was not performed.

Another study was carried out with Chinese workers at a plant producing synthetic leather, who were found to have an increased prevalence of liver damage. In workers exposed to dimethylformamide concentrations above 10 ml/m³, the ALT values were increased, and the proportion of workers with a daily alcohol consumption of more than 24 g was decreased compared with workers exposed to dimethylformamide concentrations below 10 ml/m³ (Wang *et al.* 1991). Since there was also no control group without exposure to dimethylformamide, a statement whether the ALT values were also increased at dimethylformamide concentrations below 10 ml/m³ is not possible. Biomonitoring was also not performed.

4.2.2 European area

Many publications are available in which increased liver transaminase values, alcohol intolerance reactions and histological liver changes have been reported in workers after exposure to dimethylformamide ("Dimethylformamide", Volume 8, present series; BUA 1992; Fleming *et al.* 1990; Redlich *et al.* 1988, 1990).

Case reports

Flushing was described in 19 of 102 workers exposed to dimethylformamide who drank alcoholic beverages after work. It was reported that even one glass of beer was sufficient to induce flushing. Episodes of intolerance reactions were reported particularly during the months in which exposure concentrations of dimethylformamide were above 20 ml/m^3 (Lyle *et al.* 1979).

In 13 workers at an Italian factory manufacturing synthetic leather, in which dimethylformamide concentrations between 14 and 60 mg/m^3 (5–20 ml/m^3) were determined in air, 11 workers complained of irritation of the eyes and upper respiratory tract, 8 of nausea, a further 8 of alcohol intolerance reactions, and 4 of liver problems (Tomasini *et al.* 1983).

In one man with episodic reddening of the face, upper thorax and forearms after alcohol consumption or stress, these symptoms also occurred after the introduction of dimethylformamide at his workplace. It is reported that about 10 % of the other workers had similar symptoms (Cox and Mustchin 1991). No data were given for the concentrations of dimethylformamide in the air or of NMF in the urine.

In one publication, 30 cases of poisoning were listed after exposure to dimethylformamide at the workplace. Eye irritation was reported and, after dermal contact usually lasting for several hours, skin irritation. After repeated or long-term skin contact, alcohol intolerance and abdominal pains were reported as symptoms, and increased γ-GT values were determined. According to the authors, a number of cases were reported even after observance of the threshold limit value in air of 30 mg/m^3 (10 ml/m^3) and avoidance of skin contact; no details of the analysis were given and biomonitoring was not carried out (Garnier *et al.* 1992).

Cohort studies in workers

Table 2 shows the reports of collectives exposed to dimethylformamide in Europe.

The most valuable study was performed by the Wrbitzky research group (Wrbitzky 1999; Wrbitzky and Angerer 1998). In this German study, 125 workers were exposed on average to dimethylformamide concentrations of 4.1 ml/m^3 (median 1.2 ml/m^3). The exposure was determined by personal sampling. The average dimethylformamide concentrations in the separate work areas were 1.4, 2.5, 6.4 and 7.3 ml/m^3. The workers were divided into groups without exposure to dimethylformamide (n = 54 or 53), with low exposure (n = 55 or 53) and with high exposure (n = 71 or 70). According to the details given by Wrbitzky and Angerer (1998; see Table 2), the low exposure group comprised the workers in the finishing area (n = 55; DMF: 1.4 ml/m^3), the high exposure group (n = 70) those in the dyeing (DMF: 2.5 ml/m^3), dry spinning (DMF: 6.4 ml/m^3) and wet spinning areas (DMF: 7.3 ml/m^3).

In the total collective of workers exposed to dimethylformamide, significantly more complaints of flushing, increased γ-GT and AST values, and reduced alcohol consumption since starting employment at the factory were reported.

Table 2. Effects in workers exposed to dimethylformamide in Europe

Collective		DMF concentration in air (ml/m^3)	NMF concentration in urine	Result	References
126	workers, acrylic fibre production	**4.1** ± 7.4 (< 0.1–37.9)[1] 8-hour TWA	**14.9 ± 18.7** (0.9–100) **mg/l**; 9.1 ± 11.4 (0.4–62.3) mg/g creatinine	Total cohort: effects on liver enzymes (AST, γ-GT); reports of alcohol intolerance	Wrbitzky 1999, Wrbitzky and Angerer 1998 (Germany)
55	finishing	1.4 ± 2.2 (< 0.1–13.7)	4.5 ± 4.3 (0.6–19.9) mg/g creatinine	low exposure group workers **with** alcohol consumption: effects on liver index (ALT, AST and γ-GT)	
12	dyeing	2.5 ± 3.1 (0.1–9.8)	6.7 ± 5.4 (0.8–17.2) mg/g creatinine	high exposure group workers **with** alcohol consumption: effects on liver index (ALT, AST and γ-GT); reduced alcohol consumption; workers **without** alcohol consumption: no effects on liver index (ALT, AST and γ-GT)	
28	dry spinning	6.4 ± 9.6 (0.8–36.9)	11.6 ± 13.1 (0.9–62.3) mg/g creatinine		
30	wet spinning	7.3 ± 10.2 (0.3–37.9)	16.0 ± 15.9 (0.4–54.0) mg/g creatinine		
54	controls	–	–		
22	workers, acrylic fibre factory	**4.5** (0.4–15.3)[2]	20–63 mg/g creatinine	no effects on liver enzymes (ALT, AST, AP, γ-GT); alcohol intolerance in some workers (especially after peak exposures)	Lauwerys *et al.* 1980 (Belgium)
28	controls	–	–		

Table 2. continued

Collective	DMF concentration in air (ml/m³)	NMF concentration in urine	Result	References
28 workers, acrylic fibre plant, spinning department	**6** (4–8); 8-hour TWA	**22.3 mg/l**	no effects on liver enzymes (ALT, AST, γ-GT, AP)	Catenacci *et al.* 1984 (Italy)
26 polymer department	1 (0.6–1.6) ml/m³	7 mg/l		
54 controls (matched)	–	–		
75 workers, factory for synthetic leather	**7** ± 0.7 (1.6–13)[2] **6** ± 0.6 (0.7–12)[2], 8-hour TWA	[13.6 ± 3.3 mg/l; 13.4 ± 3.2 mg/g creatinine][3]	effects on liver enzymes (ALT, AST, γ-GT, AP); abnormal liver function (23 % of exposed workers, 4 % of controls); alcohol intolerance (40 % of exposed workers); high alcohol consumption (20–40 g/d) reduced	Fiorito *et al.* 1997 (Italy)
75 controls (matched)	–	–		
100 workers, factory for synthetic leather	**7** (3–19)[1]	–	symptoms (headache, nausea, reduced power of concentration, dizziness, eye irritation, gastritis); increased abnormal liver enzyme values (γ-GT significant, AST and ALT not significant); alcohol intolerance (even after the intake of small quantities of wine several hours after work); alcohol consumption reduced	Cirla *et al.* 1984 (Italy)
100 controls (matched)	–	–		

AP: alkaline phosphatase; AST: aspartate aminotransferase; ALT: alanine aminotransferase; BUN: blood urea nitrogen; DMF: *N,N*-dimethylformamide; γ-GT: γ-glutamyltranspeptidase; NMF: *N*-methylformamide

[1] stationary sampling

[2] personal sampling

[3] analytic method produces lower values than found e.g. in Wrbitzky and Angerer (1998)

To investigate the influence of alcohol on the liver, a liver index was determined (ALT, AST and γ-GT combined), and the workers and control persons were each divided into three groups: no alcohol consumption; < 50 g alcohol per day; and > 50 g alcohol per day. 61.8 % of the persons exposed and 62.3 % of the control persons drank alcohol at work. Of the 31 workers exposed to dimethylformamide who drank no alcohol, 23 from the high exposure group were found to have liver indices of up to 7.3 ± 10.2 ml/m³ on average, which did not constitute a significant increase. Workers who consumed alcohol already had a higher liver index even at dimethylformamide concentrations as low as 1.4 ml/m^3, which continued to increase with increasing alcohol consumption. Additional dimethylformamide exposure with alcohol consumption produced an additional slight increase in the liver index. As there was no significant difference in alcohol consumption between the control group and exposed persons, the authors attribute the increase in γ-GT and AST in the exposed group to dimethylformamide. A variance analysis confirmed that alcohol consumption has a marked influence on the liver index, whereas dimethylformamide only has a slight influence at the investigated exposure concentrations. According to the authors, the apparently contradictory finding that the liver index was increased only in the low exposure group but not in the high exposure group is explained by the fact that alcohol consumption was reduced in the low exposure group, probably because of intolerance reactions. The authors also suspect a healthy-worker effect, so that workers with alcohol intolerance and functional impairments changed workplace (Wrbitzky 1999; Wrbitzky and Angerer 1998). **The study concluded that increased liver enzyme values do not occur in workers exposed to dimethylformamide without alcohol consumption at mean dimethylformamide concentrations in the air of up to 7.3 ± 10.2 ml/m^3 in the wet spinning area (corresponding to NMF concentrations of up to 16 ± 16 mg/g creatinine). The liver index determined in workers with alcohol consumption was increased and increased further to a slight extent as a result of dimethylformamide exposure, which the authors attributed to a synergistic effect of alcohol.** It is not possible to make a quantitative statement from this study about the exposure conditions under which alcohol intolerance reactions occur, as the data for alcohol intolerance were obtained anamnestically and cannot be correlated with the determined exposures. According to the authors, no alcohol intolerance reactions were reported in the week during which the study was performed. No intolerance reactions were observed in an additional investigation at the end of the shift in 17 workers with average NMF concentrations in the urine of 19 ± 24.9 mg/l (range 1.07–99.96 mg/l urine) after the intake of 0.66 litres beer (about 33 g alcohol). The workers cited daily beer consumption of 0.0 litres (n = 3); 0.5 litres (n = 4); 1.0 litre (n = 1); 1.5 litres (n = 2); 2.0 litres (n = 2); 2.5 litres (n = 3); 3.0 litres (n = 2) (Angerer and Drexler 2005, personal communication).

The following studies are of limited usefulness on account of methodological shortcomings.

A Belgian study was carried out with 22 workers exposed to dimethylformamide in the spinning department of an acrylic fibre factory and in 28 controls who, because of confounders, were not comparable. The mean dimethylformamide concentration in the air, obtained by stationary sampling, was given as 13 mg/m^3 (4.3 ml/m^3); the mean exposure concentrations in the different work areas were between 1.3 and 46.6 mg/m^3 (0.4–15.3 ml/m^3). The integrated daily exposure on the 5 days investigated was given as 70.6, 92.6, 75.4, 50.2 and 51.7 mg × h/m^3. This corresponds to dimethylformamide concentrations of 11.8, 15.4, 12.6, 8.4 and 8.6 mg/m^3 or of 3.9, 5.1, 4.1, 2.8 and 2.8 ml/m^3 in a daily working period of 6 hours. As the dimethylformamide concentration was only established by stationary sampling, it must be assumed that personal exposures were higher. The NMF values in urine at the end of the shift during the 5 days of the investigation were on average 40 ± 14, 63 ± 16, 43 ± 17, 21 ± 5 and 23 ± 8 mg/g creatinine. The workers wore gloves with long sleeves, but no protective mask. According to the authors, only one worker drank alcohol (more than 3 glasses of beer per day). No information was provided on the alcohol consumption in the control group. Liver function tests were performed in the workers exposed to dimethylformamide and control persons on Monday and Friday mornings. No significant differences in AST, ALT and γ-GT were found between Monday and Friday, or between exposed persons and controls. The authors concluded that "a concentration of NMF in urine samples collected at the end of the workshift not exceeding 40–50 mg/g creatinine indicates an exposure which is probably safe with regard to the acute and long term (5 years) action of dimethylformamide on liver function". In spite of this, the authors did not exclude the possible occurrence of alcohol intolerance reactions. They mention that some workers reported alcohol intolerance reactions at the end of the day when, for example, they were exposed to high dimethylformamide peak concentrations during cleaning work (no other details). Assuming the avoidance of dermal absorption, the authors correlated NMF concentrations in urine of between 40 and 50 mg/g creatinine with dimethylformamide concentrations in air of 13 mg/m^3, corresponding to 4.3 ml/m^3 (not 45 ml/m^3 as cited in the publication) (Lauwerys *et al.* 1980).

In a short publication by an Italian research group, it is shown in a table that there was no statistical difference in the group comparison as regards the values for ALT, AST, γ-GT or alkaline phosphatase between two groups of a total of 54 workers in the spinning and polymer departments of an acrylic fibre factory with mean dimethylformamide concentrations in air of 6 and 1 ml/m^3 and mean NMF concentrations in urine of 22.3 and 7 mg/l, compared with the values in 54 matched controls. The workers had been exposed for at least 5 years (Catenacci *et al.* 1984). It should be noted that no statement was given about the alcohol consumption of the workers, although the controls were presumably matched as regards their alcohol consumption. Nothing is mentioned about alcohol intolerance reactions and no details are given about the method of dimethylformamide analysis.

The prevalence of liver changes in 75 workers in a synthetic leather factory was investigated in an Italian cross-sectional study. Seventy-five controls were included for comparison, who had been matched with regard to age, sex, social status and place of living. There was no significant difference in the daily quantity of alcohol consumed between exposed persons and controls, although more of those exposed (n = 42) abstained from alcohol than the controls (n = 29) (Larese 2004, personal communication). Workers with alcohol consumption of more than 50 g/day were excluded. The geometric mean values for the dimethylformamide concentrations in air obtained by stationary sampling in two work areas were 21.5 mg/m^3 (7 ml/m^3) and 18.7 mg/m^3 (6 ml/m^3), with upper ranges of up to 40 and 35 mg/m^3, respectively (13 or 12 ml/m^3). The NMF concentrations in urine were determined (at the end of the shift) in 22 of the 75 workers, yielding a geometric mean of 13.6 ± 3.3 mg/l or 13.4 ± 3.2 mg/g creatinine. The maximum value was 126 mg/g creatinine. The NMF values were obtained from the workers of one shift, and were regarded by the authors as being representative of all workers. On account of the great variability of the NMF concentrations in urine, the authors suspected that "overexposure" occurred occasionally (no further details). In addition to inhalation exposure, they assumed there was dermal absorption of dimethylformamide both through the gloves and through unprotected skin. Questioning the workers revealed that 50 % had gastrointestinal problems and 40 % alcohol intolerance reactions. There was a high rate of worker turnover, which the authors attributed to the frequency of effects. The average length of employment in this plant was 3.8 years. Serum analyses showed that ALT, AST, γ-GT and alkaline phosphatase values were significantly higher in those exposed than in the controls. Values were above the normal range in 17 of the 75 exposed workers. Multivariate analyses revealed that the transaminase values (AST, ALT) did not correlate with the daily alcohol consumption, but primarily with the duration of employment in the plant, and also with the body mass index and the cholesterol level. According to the authors, the resultant AST/ALT ratio of < 1 indicates alcohol-independent liver changes. The γ-GT values were increased in subjects with alcohol consumption (Fiorito *et al.* 1997). Problematical in this study is that the dimethylformamide concentrations were obtained by stationary sampling only and not by personal sampling; stationary analyses usually yield lower concentrations than personal sampling. Also, the method used for determining the NMF values in urine results in lower values than the validated method used by Wrbitzky and Angerer (1998). Thus, markedly higher exposure can be assumed for these workers, probably as a result of excessive dermal exposure. The high maximum value given suggests this was the case.

The MAK documentation from 1992 cites a study with 100 male workers (mean age 36 years) in another Italian synthetic polyurethane leather factory and 100 control persons (Cirla *et al.* 1984). The control persons were carefully selected as regards sex, age, alcohol consumption, cigarette and coffee consumption, socio-economic status, place of living and eating habits. Only workers continuously exposed to dimethylformamide were included; workers who were possibly exposed earlier to higher levels were excluded. The mean calculated from the 8-hour values determined by personal sampling was 7 ml/m^3, with a range of 3 to 19 ml/m^3. No details were given about the analytical method or the number of determinations; attention was merely drawn to a symposium presentation. As frequent throat irritations were reported (see below), it cannot be excluded that the method used resulted in lower values than the validated method of Wrbitzky and Angerer (1998). According to the authors, the production technology had not changed in the 15 years before the study and did not change during it. The authors assume that dermal absorption of dimethylformamide contributed significantly to the exposure of the individual; biological monitoring was, however, not performed. The mean exposure duration was 5 years with a range of 1 to 15 years. Symptoms that occurred frequently during the years prior to performing the study were throat irritation, coughing, headaches, dyspepsia, nausea, hepatic impairment and increased γ-GT values; frequent alcohol intolerance reactions with flushing were recorded in 39 of the 100 exposed persons. The alcohol intolerance reactions occurred usually after the intake of "small amounts of wine" (no other details) some hours after work. Of the 100 workers, 23 drank no alcohol, 51 workers had moderate consumption (< 1 litre wine per day), and 26 workers heavy alcohol consumption (> 1 litre wine per day). Twenty-two workers with wine consumption of originally more than one litre per day changed their drinking habits to lower wine consumption or ceased drinking altogether. Twenty-five of those exposed were found to have abnormal γ-GT values, compared with 10 of 100 control persons. The fact that the proportion of persons with abnormal γ-GT values was higher in exposed persons without any or with moderate alcohol consumption than in controls with comparable drinking behaviour was attributed by the authors to an effect of dimethylformamide. The proportion of persons with abnormal AST and ALT values was slightly increased in the exposed persons, but the increase was not statistically significant; no average values for the groups are presented for the respective enzyme values, however, and it is thus not clear whether the increases in transaminases in the exposed persons compared with those in the control persons were within the normal range (Cirla *et al.* 1984). As no biomonitoring was performed in this study and the contribution of dermal absorption to the total exposure cannot be estimated, the results of this study are practically not usable.

Summary

Although the preceding investigations are not always consistent, the following facts can be concluded:

1) In addition to inhalation exposure, the dermal absorption of dimethylformamide contributes considerably to the overall exposure to dimethylformamide. Thus, with an inhaled air volume of 10 m^3 over an 8-hour work shift, a dimethylformamide concentration of 30 mg/m^3 is equivalent to a daily intake of 300 mg or 0.3 g dimethylformamide. At a density of 0.94 g/ml, this corresponds to 0.3 ml or 300 μl dimethylformamide. This means that dermal exposure to a few drops of dimethylformamide, which is absorbed at a level of 100 %, already results in an intake exceeding the maximum permissible quantity of dimethylformamide absorbed per day. Dermal absorption of dimethylformamide through the gloves is possible if the material of the gloves is unsuitable. The dermal absorption of dimethylformamide in vapour form through unprotected areas of skin is also relevant. Inevitably, an air threshold value therefore always includes a share of dermally absorbed dimethylformamide which, depending on the work conditions, can vary greatly. Surveillance of the threshold value in biological material is thus of great importance.

2) As has long been known, the liver is the target organ for the toxic effects of dimethylformamide. Nausea, gastrointestinal complaints, increases in the serum transaminases ALT and AST and in γ-GT are observed. Alcohol intolerance of the disulfiram type is a frequent finding (see below). Depending on their severity, dimethylformamide-induced effects on the liver seem to be reversible relatively rapidly, as shown by empirical findings in workers (Lauwerys *et al.* 1980) and case reports.

3) An increase in the transaminases AST and ALT is a suitable parameter for detecting dimethylformamide-induced effects on the liver. Transaminase values above the normal range were reported in some publications (Fiorito *et al.* 1997; Huang *et al.* 1998; Nomiyama *et al.* 2001c). Whether increased transaminase values in a group exposed to dimethylformamide are within the normal range can be determined only through statistical comparison with a matched control group, such as was performed, for example, by Wrbitzky (1999). Thus, the transaminases in the serum of exposed workers were not changed compared with the levels in controls at average dimethylformamide concentrations in air of 7.3 ± 10.2 ml/m^3 and excluding alcohol consumption. The dimethylformamide concentrations in air cited correlate with NMF concentrations in urine of 16 ± 16 mg/g creatinine (Wrbitzky and Angerer 1998). Also in Asian workers, biochemical parameters were above the normal range only in individual cases at dimethylformamide concentrations of up to 9.1 ml/m^3 (Cai *et al.* 1992), or no effects on liver enzymes were found up to 5 ml/m^3 (Yonemoto and Suzuki 1980) or 10 ml/m^3 (Sakai *et al.* 1995).

However, there are also studies available in which effects are evident in this concentration range. For example, Wrbitzky (1999) found that the AST was also increased in addition to γ-GT, an indicator in particular of alcohol consumption, in the total cohort of exposed workers (mean DMF concentration 4.1 ± 7.4 ml/m^3). Effects were reported also in workers with presumably high dermal absorption of dimethylformamide (Fiorito *et al.* 1997).

Although some of the studies did not reveal significant impairment of liver function at dimethylformamide concentrations in air of 7 to 10 ml/m^3, this cannot be excluded in other studies that report such impairment even below 10 ml/m^3, particularly with excessive dermal absorption of dimethylformamide or in combination with alcohol consumption.

4) Alcohol intolerance reactions of the disulfiram type have been reported after repeated or long-term skin contact with dimethylformamide (Garnier *et al.* 1992) and after cleaning work with high peak concentrations of dimethylformamide in air (Lauwerys *et al.* 1980). Alcohol intolerance reactions are mentioned in other publications as well ("Dimethylformamide", Volume 8, present series; Cirla *et al.* 1984; Fiorito *et al.* 1997; Wrbitzky 1999).

It is not possible at present to give exact quantitative data for the conditions under which intolerance reactions occur. The occurrence of such reactions appears to be dependent on the level of exposure to dimethylformamide resulting from inhalation and dermal absorption, individual sensitivity and the quantity of alcohol consumed.

In an earlier study with high dimethylformamide concentrations (maximum 200 ml/m^3), it was suspected that flushing could occur even without alcohol consumption (Lyle *et al.* 1979). Generally, however, flushing has been reported after the intake of alcoholic beverages after work. Two publications mention that "small quantities of alcohol" were already sufficient to induce flushing (Cirla *et al.* 1984), for example a glass of beer or wine (Lyle *et al.* 1979). More accurate data on the quantity of alcohol that induces flushing are not available.

The reported flushing occurred, however, only in some of those exposed to dimethylformamide. This can probably be attributed to genetic polymorphism (see Section 2).

4.3 Local effects on skin and mucous membranes

Liquid dimethylformamide causes degreasing of the skin with oedema and subsequent desquamation after long-term exposure (BUA 1992).

4.4 Allergenic effects

Despite the widespread use of dimethylformamide in industry, there are no reports of sensitization reactions.

4.5 Reproductive toxicity

Fertility

Both microscopic investigation and computer-assisted analysis revealed reduced sperm motility in 12 exposed workers (DMF concentrations of 11.4 ml/m^3; NMF concentrations of 17.9 mg/l urine) compared with in 8 control persons. The motility parameters correlated in a dose-dependent manner with the elimination of NMF in urine, but not with the dimethylformamide concentration in air. The authors concluded that NMF could be responsible for the changes in sperm function (Chang *et al.* 2004). Taking into account the small group size of only 12 workers, and the fact that the number of exposed persons with alcohol consumption (8/12; 66.7 %) was significantly above that of the control persons (3/8; 37.5 %), this study is practically not usable. The findings should be clarified in further investigations, as multigeneration studies (see below) yielded evidence of reduced sperm concentrations in young male F_1 animals (but not reduced sperm motility), albeit after the administration of high doses of 400 mg/kg body weight and day and above.

4.6 Genotoxicity

In the MAK documentation from 1992, a study was described in which the incidence of sister chromatid exchange (SCE) increased slightly with the dimethylformamide concentration (0.3, 0.7, 5.8 ml/m^3) in 22 non-smoking women employed in manufacturing synthetic leather, compared with in 22 control persons. The SCE values of the exposed women were 5.67 ± 1.35, 7.24 ± 1.53 and 8.26 ± 1.76, and were significantly higher than in the corresponding controls (6.57 ± 1.12, 4.66 ± 0.56 and 5.63 ± 1.56). The medium-level exposure group was also exposed to toluene concentrations of 0.9 ml/m^3. The exposure duration lasted between 1.1 and 9.9 years (Seiji *et al.* 1992). As the increase in SCEs was only slight, its biological importance is questionable despite the statistical significance.

In contrast, exposure of 85 male workers to median dimethylformamide concentrations of 10 to 24.8 ml/m^3 with an exposure duration of 9.2 ± 5.7 years was not associated with an increase in the frequency of SCE (Cheng *et al.* 1999).

The incidences of chromosomal aberrations, SCE and DNA repair synthesis were increased in 26 workers exposed to maximum dimethylformamide concentrations of 8 ml/m^3 and maximum acrylonitrile concentrations of 17.6 mg/m^3, compared with in 26 matched control persons (Major *et al.* 1998). This investigation cannot be used in the assessment of dimethylformamide because of the co-exposure to acrylonitrile.

4.7 Carcinogenicity

Case reports and epidemiological studies yielded evidence of possible carcinogenic effects of dimethylformamide as regards testicular tumours. The data were presented comprehensively in the MAK documentation from 1992 ("Dimethylformamide", Volume 8, present series). Further studies have not been published since. The available data were interpreted by the International Agency for Research on Cancer (IARC, 1999) to the effect that three cases of testicular tumours in an aircraft industry repair unit (Ducatman *et al.* 1986) and three cases in a leather tannery (Levin *et al.* 1987) have been reported which indicate a possible association with dimethylformamide. Possible confounders were not taken into account in these studies, nor were further investigations able to confirm this relationship. No other cases could be identified when screening was carried out in the leather tannery where the three persons with tumours were employed (Calvert *et al.* 1990). In the factories already investigated (Chen *et al.* 1988a, 1988b) and in two other factories, studies of cancer incidence and mortality (Chen *et al.* 1988a, 1988b) and one case–control study (Walrath *et al.* 1989) of testicular and other tumours did not reveal a convincing relationship between dimethylformamide exposure and testicular tumours. The IARC therefore judged the data in humans to be "inadequate evidence" for carcinogenic effects of dimethylformamide.

5 Animal Experiments and *in vitro* Studies

5.1 Subacute, subchronic and chronic toxicity

5.1.1 Inhalation

Groups of 10 male and 10 female F344 rats and BDF_1 mice were exposed in whole-body chambers for 6 hours a day on 5 days a week for 2 weeks to dimethylformamide concentrations of 0, 100, 200, 400, 800 or 1600 ml/m^3 and for 13 weeks to 0, 50, 100, 200, 400 or 800 ml/m^3. Three male and seven female rats died during the two-week exposure to 1600 ml/m^3. Even in the lowest exposure group, reduced body weight gains, increased relative liver weights, centrilobular hepatocellular hypertrophy, increased total cholesterol and increased phospholipid concentrations were observed in male mice after 13 weeks exposure. Focal necrosis was seen in the liver of both sexes after concentrations of 100 ml/m^3 and above. The absolute liver weights were significantly increased in rats after concentrations of 50 ml/m^3 and above, the relative liver weights after

100 ml/m^3 (males) and 200 ml/m^3 (females) and above. Centrilobular hypertrophy and necrosis were observed after concentrations of 200 ml/m^3 and above. **In this study, 50 ml/m^3 was found to be the NOAEL for rats and the LOAEL for mice** (Senoh *et al.* 2003).

Groups of 10 male and 10 female F344 rats and B6C3F$_1$ mice were exposed in whole-body chambers for 6 hours a day on 5 days a week for 13 weeks to dimethylformamide concentrations of 0, 50, 100, 200, 400 or 800 ml/m^3. In both species, the relative liver weights were significantly increased even at the lowest exposure concentration of 50 ml/m^3. In addition, male mice were found to have centrilobular hepatocellular hypertrophy, whereas this effect was not observed in female animals until the next highest exposure concentration of 100 ml/m^3. In rats, the cholesterol values were also increased in all exposure groups (this parameter was not investigated in mice). Centrilobular hepatocellular necrosis was diagnosed in rats after concentrations of 400 ml/m^3 and above (Lynch *et al.* 2003; NTP 1992). **The LOAEL in this study was 50 ml/m^3 for rats and mice.**

Groups of 60 male and 60 female CD rats were exposed in whole-body chambers for 2 years, and groups of 60 male and 60 female CD-1 mice for 18 months, to dimethylformamide concentrations of 0, 25, 100, or 400 ml/m^3 for 6 hours a day on 5 days a week. Liver changes, such as centrilobular hepatocellular hypertrophy, accumulation of lipofuscin and haemosiderin in Kupffer's cells, centrilobular necrosis and eosinophilic liver foci, were observed in the mice, especially in male animals, of all exposure groups and increased in a concentration-dependent manner. In rats, the corresponding liver changes were observed after concentrations of 100 ml/m^3 and above (Malley *et al.* 1994). **In this study, 25 ml/m^3 was found to be the NOAEL for rats and the LOAEL for mice.** On the basis of histopathologic findings in mice (Table 3), a benchmark calculation was performed by the Commission (Figure 2). Centrilobular hepatocellular hypertrophy was used as the most sensitive relevant end point in male and female mice. The calculation was performed with the benchmark dose software from EPA (version 1.3.2) using a log-probit model with a benchmark response of 5 % extra risk. Since there is only little difference between the incidences for male and female animals, they were calculated together. **A value of 7.8 ml/m^3 was obtained for the benchmark dose lower confidence limit (BMDL) and of 14.7 ml/m^3 for the benchmark dose (BMD)** (Figure 2). With an AIC (Akaike Information Criterion) of 388 and a chi square of 1.35, the adjustment of the model was not optimal. However, benchmark calculations for the end points necrosis and Kupffer's cell hypertrophy were even more problematical because of the high control values, while the end point foci is not sensitive enough.

Table 3. Incidence of substance-related morphological observations in the liver of mice after exposure to dimethylformamide for 18 months (Malley *et al.* 1994)

		DMF (ml/m³)			
		0	25	100	400
Number of animals whose liver was investigated	♂	60	62	60	59
	♀	61	63	61	63
	♂+♀	121	125	121	122
Centrilobular hepatocellular hypertrophy	♂	0 (0 %)	5[1] (8 %)*	25 (41 %)*	31 (52 %)*
	♀	0 (0 %)	4 (6 %)	12 (19 %)*	34 (54 %)*
	♂+♀	0 (0 %)	9 (7 %)*	37 (31 %)*	65 (53 %)*
Single cell necrosis	♂	14 (24 %)	37 (59 %)*	41 (68 %)*	51 (87 %)*
	♀	18 (29 %)	28 (44 %)*	43 (70 %)*	48 (76 %)*
Kupffer's cell hyperplasia/ pigment accumulation	♂	13 (22 %)	32 (52 %)*	36 (60 %)*	51 (86 %)*
	♀	31 (51 %)	36 (57 %)	43 (71 %)*	61 (96 %)*
Foci of alterations	♂	0 (0 %)	2 (3 %)	8 (13 %)*	11 (19 %)*
	♀	0 (0 %)	0 (0 %)	2 (3 %)	2 (3 %)

* statistically significant $p < 0.05$
[1] The incidences of the number of animals with changes were calculated or estimated from the percentages given in the publication.

Exposure of cynomolgus monkeys to a dimethylformamide concentration of 500 ml/m³ for 6 hours a day on 5 days a week for 2 weeks did not produce any adverse effects (Hurtt *et al.* 1991). Nor were changes observed after exposure of groups of 3 cynomolgus monkeys in whole-body chambers to dimethylformamide concentrations of 0, 30, 100 or 500 ml/m³ for 6 hours a day on 5 days a week for 13 weeks. The investigations included clinical observations, analyses of the sperm quality and quantity, histopathological investigations, haematology and the determination of serum parameters including transaminases (Hurtt *et al.* 1992). **The NOAEL for monkeys was 500 ml/m³ in this study.** Since the studies with humans revealed dimethylformamide-related effects after concentrations as low as < 10 to 20 ml/m³, these investigations with monkeys are not relevant for assessing the risk in humans and are not used for the derivation of a MAK value.

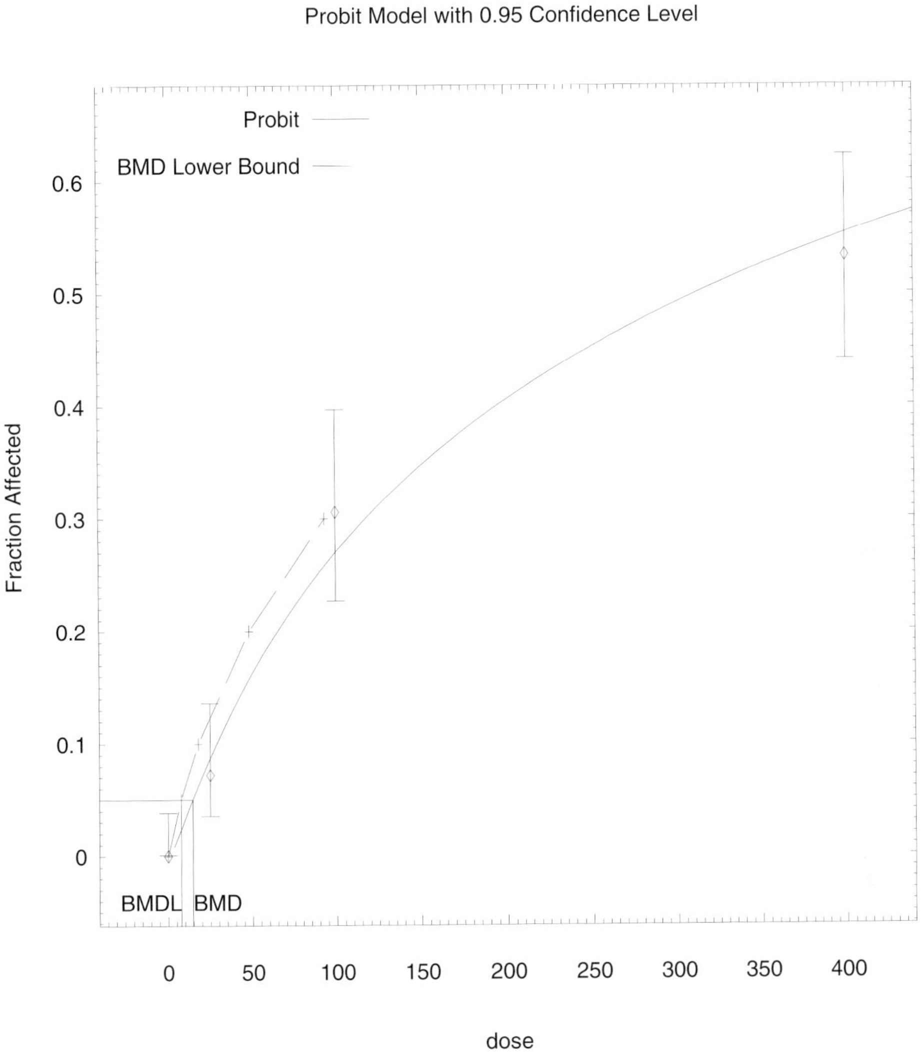

Figure 2. Benchmark dose calculation according to the log-probit model with a benchmark response of 5 % extra risk using the data for centrilobilar hepatocellular hypertrophy in mice from the 18-month inhalation study from Table 3 (Malley *et al.* 1994)

5.1.2 Ingestion

A **NOAEL** of 200 ppm dimethylformamide in the diet was obtained (about **12 mg/kg body weight and day**) in a 13-week feeding study with groups of 40 male and 40 female Sprague-Dawley rats. The relative liver weights were slightly increased after doses of 1000 ppm in the diet (about 60 mg/kg body weight and day). There was no histological correlate, but cholesteraemia and increased phospholipid values were found in the female animals. In addition, leukocytosis and a reduced erythrocyte count were seen. After doses of 5000 ppm in the diet (about 300 mg/kg body weight and day), food intake and body weight gains were reduced in both sexes. The relative liver weights were increased and histological investigations revealed slight liver damage (OECD 2003).

5.2 Local effects on skin and mucous membranes

5.2.1 Skin

After semi-occlusive exposure of 4 rabbits to 0.5 ml dimethylformamide for 20 hours, slight erythema was observed in 1 of the 4 rabbits after removal of the occlusive patch, but no longer on day two (OECD 2003).

No irritation was apparent after occlusive application of the undiluted substance to the abraded skin of rats for 24 hours (OECD 2003).

There was no indication of irritation in rabbits whose skin was treated with a dimethylformamide dose of 2 g/kg body weight for 6 hours, 15 times within 4 weeks. No skin irritation was found in rats and guinea pigs; skin irritation in mice was slight (BUA 1992).

5.2.2 Eyes

Marked irritation was observed in the rabbit eye after instillation of 0.1 ml undiluted dimethylformamide and of a 75 % solution of the product. Aqueous dilutions of 50 % to 25 % induced slight and no irritation, respectively. Irritation of the rabbit eye has also been reported by other authors (BUA 1992).

5.3 Allergenic effects

A maximization test in guinea pigs revealed no evidence of sensitizing effects (WHO 1991, 2001).

Several authors judged dimethylformamide to be a suitable vehicle for the local lymph node assay in the mouse (Hilton *et al.* 1998; Robinson and Cruze 1996; Ryan *et al.* 2002; Warbrick *et al.* 1999) and the epicutaneous test in the guinea pig (van't Erve *et al.* 1998), since it is not sensitizing and readily penetrates the skin.

5.4 Reproductive toxicity

5.4.1 Fertility

In a generation study with continuous mating, dimethylformamide was administered to Swiss mice in the drinking water in concentrations of 0, 1000, 4000 or 7000 ppm. At the lowest concentration of 1000 ppm and above (about 200 mg/kg body weight and day), liver weights were increased in mice of both sexes (F_0, F_1) and prostate weights reduced in male F_1 animals. The birth weights of the F_2 generation were reduced. After 4000 ppm (about 800 mg/kg body weight and day) and above, fertility (F_0, F_1), body weights (F_1) and postnatal survival (F_1) were reduced and the ovarian weights increased (F_1). Pups exposed to dimethylformamide (F_1 and F_2) and adult F_1 animals were found to have craniofacial malformations. At 7000 ppm (about 1300 mg/kg body weight and day), body weight gains were reduced in female F_0 animals. Crossover matings revealed that the reduced fertility was caused by the dams. However, sperm concentration was reduced in the male F_1 animals; sperm motility was not affected. A NOAEL could not be determined (Fail *et al.* 1998).

5.4.2 Developmental toxicity

One developmental toxicity study with oral administration has been published since the MAK documentation from 1992. In this study Sprague-Dawley rats were given gavage doses of dimethylformamide of 0, 50, 100, 200 or 300 mg/kg body weight and day on days 6 to 20 of gestation. Maternal toxicity was evident in the reduced food intake and reduced body weight gains after doses of 100 mg/kg body weight and above. At this dose and above, foetal body weight gains were reduced as well. After doses of 200 and 300 mg/kg body weight and day, skeletal variations were increased. The NOAEL for maternal toxicity and developmental toxicity was 50 mg/kg body weight and day in this study (Saillenfait *et al.* 1997).

5.5 Genotoxicity

The genotoxicity of dimethylformamide *in vitro* and *in vivo* has been thoroughly investigated ("Dimethylformamide", Volume 8, present series; OECD 2003; IARC 1999). There is no evidence of genotoxic effects.

5.6 Carcinogenicity

Groups of 60 male and 60 female CD rats were exposed for 2 years, and groups of 60 male and 60 female CD-1 mice for 18 months, to dimethylformamide concentrations of 0, 25, 100, or 400 ml/m^3 for 6 hours a day on 5 days a week. Effects on the liver were observed in the mice even at the lowest concentration of 25 ml/m^3, while in the rats not until 100 ml/m^3 and above (see Section 5.1, "Subacute, subchronic and chronic toxicity"). Tumour incidences were not increased (Malley *et al.* 1994).

6. Manifesto (MAK value, classification)

A large number of studies *in vitro* and *in vivo* indicate that dimethylformamide has no genotoxic or germ cell mutagenic effects. Nor were carcinogenic effects observed in animal experiments. Evidence of testicular tumours in workers could not be confirmed in subsequent studies. The classification of dimethylformamide in a category for carcinogenic substances or germ cell mutagenicity is therefore not necessary.

In 1992, the MAK value was established at 10 ml/m^3 to protect exposed workers against changes in liver parameters; this took into consideration the fact that the substance penetrates the skin readily. In the meantime, one relevant study has been published which did not reveal any significantly increased liver index values (ALT, AST and γ-GT combined) at dimethylformamide concentrations of up to 7.3 ± 10.2 ml/m^3 (NMF concentrations of 16 ± 16 mg/g creatinine) in workers without alcohol consumption. In workers with alcohol consumption, the drinking of alcohol led to an increase in the liver index, which was already slightly increased after exposure to dimethylformamide concentrations as low as 1.4 ml/m^3. Therefore, the authors do not exclude a synergistic effect of alcohol (Wrbitzky 1999; Wrbitzky and Angerer 1998). It is difficult to derive a threshold concentration in the air from the available studies carried out at the workplace since the contribution of the dermal uptake of dimethylformamide cannot be assessed. Therefore, animal experiments were used for the derivation of a MAK value as exposure conditions are better to control. Thus, the NOAEL for rats cited in the 2-year study was 25 ml/m^3, a concentration at which effects on the liver of mice were still seen (Malley *et al.* 1994). A benchmark dose calculation by the Commission yielded a BMDL of 7.8 ml/m^3 and a BMD of 14.7 ml/m^3. The MAK value has therefore been reduced to 5 ml/m^3. Even when this reduced MAK value is observed, alcohol intolerance reactions in persons exposed to dimethylformamide cannot be excluded.

The BAT value (Drexler and Greim 2006) has to be observed to take into account the amount of dimethylformamide taken up by inhalation and absorption through the skin.

The systemic effects of dimethylformamide are decisive for establishing the peak limitation category, not so much the local irritant effects. Therefore, Peak Limitation Category II has been retained. As dimethylformamide has a half-time of up to two hours, an excursion factor of 4 can be set.

The ability of dimethylformamide to penetrate the skin very readily makes it necessary to retain the designation with an "H".

Since there is no evidence of a sensitizing effect of dimethylformamide, the substance is not designated with an "Sa" or "Sh".

As a result of the small margin between prenatal toxicity and the MAK value, dimethylformamide has been classified in Pregnancy Risk Group B. Whether reclassification in Pregnancy Risk Group C is possible as a result of the reduction of the MAK value from 10 to 5 ml/m^3 is being investigated at present. Classification in Pregnancy Risk Group B has been retained provisionally.

7. References

Amato G, Grasso E, Longo V, Gervasi PG (2001) Oxidation of N,N-dimethylformamide and N,N-diethylformamide by human liver microsomes and human recombinant P450s. *Toxicol Lett 124*: 11–19

Angerer J, Drexler H (2005) Faserwerk Kehlheim – Untersuchung am 21.07.1995 (Kehlheim factory investigation on July 21, 1995) (German). Report to the Commission dated January 26, 2005

BUA (Beratergremium für umweltrelevante Altstoffe der Gesellschaft Deutscher Chemiker) (1992) *N,N-Dimethylformamid*, BUA-Stoffbericht 84 (Dezember 1991), also available in English. VCH, Weinheim

Buylaert W, Calle P, De Paepe P, Verstraete A, Samyn N, Vogelaers D, Vandenbulcke M, Belpaire F (1996) Hepatotoxicity of N,N-dimethylformamide (DMF) in acute poisoning with the veterinary euthanasia drug T-61. *Hum Exp Toxicol 15*: 607–611

Cai SX, Huang MY, Xi LQ, Li YL, Qu JB, Kawai T, Yasugi T, Mizunuma K, Watanabe T, Ikeda M (1992) Occupational dimethylformamide exposure. 3. Health effects of dimethyl formamide after occupational exposure at low concentrations. *Int Arch Occup Environ Health 63*: 461–468

Calvert GM, Fajen JM, Hills BW, Halperin WE (1990) Testicular cancer, dimethylformamide, and leather tanneries. *Lancet 336*: 1253–1254

Catenacci G, Grampella D, Terzi R, Sala A, Pollini G (1984) Hepatic function in subjects exposed to environmental concentrations of DMF lower than the actually proposed TLV. *Med Lav 6*: 157–158

Chan AW (1986) Racial differences in alcohol sensitivity. *Alcohol Alcoholism 21*: 93–104

Chang HY, Shih TS, Guo YL, Tsai CY, Hsu PC (2004) Sperm function in workers exposed to N,N-dimethylformamide in the synthetic leather industry. *Fertil Steril 81*: 1589–1594

Chen JL, Fayerweather WE, Pell S (1988a) Cancer incidence of workers exposed to dimethylformamide and/or acrylonitrile. *J Occup Med 30*: 813–818

Chen JL, Fayerweather WE, Pell S (1988b) Mortality study of workers exposed to dimethylformamide and/or acrylonitrile. *J Occup Med 30*: 819–821

Cheng TJ, Hwang SJ, Kuo HW, Luo JC, Chang MJ (1999) Exposure to epichlorohydrin and dimethylformamide, glutathione S-transferases and sister chromatid exchange frequencies in peripheral lymphocytes. *Arch Toxicol 73*: 282–287

Cirla AM, Pisati G, Invernizzi E, Torricelli P (1984) Epidemiological study on workers exposed to low dimethylformamide concentrations. *Med Lav 6*: 149–156

Cox NH, Mustchin CP (1991) Prolonged spontaneous and alcohol-induced flushing due to the solvent dimethylformamide. *Contact Dermatitis 24*: 69–70

Drexler H, Greim H (2006) Biologische Arbeitsstoff-Toleranz-Werte (BAT-Werte) und Expositionsäquivalente für krebserzeugende Arbeitsstoffe (EKA) (Biological Tolerance Values (BAT values) and exposure equivalents for carcinogenic materials (EKAs)) (German), 14. Lieferung, *N,N-Dimethylformamid*, Wiley-VCH, Weinheim

Ducatman AM, Conwill DE, Crawl (1986) Germ cell tumors of the testicle among aircraft repairmen. *J Urol 136*: 834–836

van't Erve EH, Wijnand E, Bol M, Seinen W, Pieters RH (1998) The vehicle modulates cellular and humoral responses in contact hypersensitivity to oxazolone. *Toxicol Sci 44*: 39–45

Fail PA, George JD, Grizzle TB, Heindel JJ (1998) Formamide and dimethylformamide: reproductive assessment by continuous breeding in mice. *Reprod Toxicol 12*: 317–332

Fiorito A, Larese F, Molinari S, Zanin T (1997) Liver function alterations in synthetic leather workers exposed to dimethylformamide. *Am J Ind Med 32*: 255–260

Fleming LE, Shalat SL, Redlich CA (1990) Liver injury in workers exposed to dimethylformamide. *Scand J Work Environ Health 16*: 289–292

Garnier R, Chataigner D, Perez-Trigalou B, Efthymiou ML (1992) Dimethylformamide poisoning after occupational exposure – report of 30 cases (French). *Arch Mal Prof 53*: 111–120

Hilton J, Dearman RJ, Harvey P, Evans P, Basketter DA, Kimber I (1998) Estimation of relative skin sensitizing potency using the local lymph node assay: a comparison of formaldehyde with glutaraldehyde. *Am J Contact Dermatitis 9*: 29–33

Huang JF, Kuo HT, Ho CK, Chen TJ, Chang WY (1998) Dimethylformamide-induced occupational liver injury – a case report. *Kaohsiung J Med Sci 14*: 655–658

Hundley SG, Lieder PH, Valentine R, Malley LA, Kennedy Jr GL (1993a) Dimethylformamide pharmacokinetics following inhalation exposures to rats and mice. *Drug Chem Toxicol 16*: 21–52

Hundley SG, McCooey KT, Lieder PH, Hurtt ME, Kennedy Jr GL (1993b) Dimethylformamide pharmacokinetics following inhalation exposure in monkeys. *Drug Chem Toxicol 16*: 53–79

Hurtt ME, McCooey KT, Placke ME, Kennedy GL (1991) Ten-day repeated-exposure inhalation study of dimethylformamide (DMF) in cynomolgus monkeys. *Toxicol Lett 59*: 229–237

Hurtt ME, Placke ME, Killinger JM, Singer AW, Kennedy GL Jr (1992) 13-Week inhalation toxicity study of dimethylformamide (DMF) in cynomolgus monkeys. *Fundam Appl Toxicol 18*: 596–601

IARC (International Agency for Research on Cancer) (1999) *Dimethylformamide*. IARC Monographs on the Evaluation of Carcinogenic Risks to Humans, Volume 71, Part 2, IARC, Lyon, France, 545–574

Larese F (2004) Tabellen mit Daten zum Alkoholkonsum und zu Leberenzymwerten der Studie Fiorito *et al.* (1997) (Tables with data on alcohol consumption and liver enzyme values in the study by Fiorito *et al.* (1997) (German). Report to the Commission dated December 20, 2004

Lauwerys RR, Kivits A, Lhoir M, Rigolet P, Houbeau D, Buchet JP, Roels HA (1980) Biological surveillance of workers exposed to dimethylformamide and the influence of skin protection on its percutaneous absorption. *Int Arch Occup Environ Health 45*: 189–203

Levin SM, Baker DB, Landrigan PJ, Monoghan SV, Frumin E, Braithwaite M, Towne W (1987) Testicular cancer in leather tanners exposed to dimethylformamide. *Lancet 330*: 1153

Luo JC, Kuo HW, Cheng TJ, Chang MJ (2001) Abnormal liver function associated with occupational exposure to dimethylformamide and hepatitis B virus. *J Occup Environ Med 43*: 474–482

Lyle WH, Spence TW, McKinneley WM, Duckers K (1979) Dimethylformamide and alcohol intolerance. *Br J Ind Med 36*: 63–66

Lynch DW, Placke ME, Persing RL, Ryan MJ (2003) Thirteen-week inhalation toxicity of N,N-dimethylformamide in F344/N rats and $B6C3F_1$ mice. *Toxicol Sci 72*: 347–358

Malley LA, Slone TW Jr, Van Pelt C, Elliott GS, Ross PE, Stadler JC, Kennedy GL Jr (1994) Chronic toxicity/oncogenicity of dimethylformamide in rats and mice following inhalation exposure. *Fundam Appl Toxicol 23*: 268–279

Major J, Hudak A, Kiss G, Jakab MG, Szaniszlo J, Naray M, Nagy I, Tompa A (1998) Follow-up biological and genotoxicological monitoring of acrylonitrile- and dimethylformamide-exposed viscose rayon plant workers. *Environ Mol Mutagen 31*: 301–310

Miyauchi H, Tanaka S, Nomiyama T, Seki Y, Imamiya S, Omae K (2001) N,N-Dimethylformamide (DMF) vapor absorption through the skin in workers. *J Occup Health 43*: 92–94

Mráz J, Cross H, Gescher A, Threadgill MD, Flek J (1989) Differences between rodents and humans in the metabolic toxification of N,N-dimethylformamide. *Toxicol Appl Pharmacol 98*: 507-516

Mráz J, Nohová H (1992a) Percutaneous absorption of N,N-dimethylformamide in humans. *Int Arch Occup Environ Health 64*: 79–83

Mráz J, Nohová H (1992b) Absorption, metabolism and elimination of N,N-dimethylformamide in humans. *Int Arch Occup Environ Health 64*: 85–9

Mráz J, Jheeta P, Gescher A, Hyland R, Thummel K, Threadgill MD (1993) Investigation of the mechanistic basis of N,N-dimethylformamide toxicity. Metabolism of N,N-dimethylformamide and its deuterated isotopomers by cytochrome P450 2E1. *Chem Res Toxicol 6*: 197–207

Nomiyama T, Nakashima H, Chen LL, Tanaka S, Miyauchi H, Yamauchi T, Sakurai H, Omae K (2001a) N,N-Dimethylformamide: significance of dermal absorption and adjustment method for urinary N-methylformamide concentration as a biological exposure item. *Int Arch Occup Environ Health 74*: 224–228

Nomiyama T, Nakashima H, Sano Y, Chen LL, Tanaka S, Miyauchi H, Yamauchi T, Sakurai H, Omae K (2001b) Does the polymorphism of cytochrome P-450 2E1 affect the metabolism of N,N-dimethylformamide? Comparison of the half-lives of urinary N-methylformamide. *Arch Toxicol 74*: 755–759

Nomiyama T, Uehara M, Miyauchi H, Imamiya S, Tanaka S, Seki Y (2001c) Causal relationship between a case of severe hepatic dysfunction and low exposure concentrations of N,N-dimethylformamide in the synthetics industry. *Ind Health 39*: 33–36

NTP (National Toxicology Program) (1992) NTP technical report on the toxicity studies of N,N-dimethylformamide (CAS No. 68-12-2) administered by inhalation to F344/N rats and B6C3F1 mice. *NTP Toxic Rep Ser 22*

OECD (Organisation for Economic Co-operation and Development) (2003) SIDS Initial Assessment Report (SIAR) Dimethylformamide, Update 2003, July 2003, www.inchem.org/documents/sids/sids/DIMETHYLFORM.pdf

Quertemont E (2004) Genetic polymorphism in ethanol metabolism: acetaldehyde contribution to alcohol abuse and alcoholism. *Mol Psychiatry 9*: 570–581

Redlich CA, Beckett WS, Sparer J, Barwick KW, Riely CA, Miller H, Sigal SL, Shalat SL, Cullen MR (1988) Liver disease associated with occupational exposure to the solvent dimethylformamide. *Ann Intern Med 108*: 680–686

Redlich CA, West AB, Fleming L, True LD, Cullen MR, Riely CA (1990) Clinical and pathological characteristics of hepatotoxicity associated with occupational exposure to dimethylformamide. *Gastroenterology 99*: 748–757

Robinson MK, Cruze CA (1996) Preclinical skin sensitization testing of antihistamines: guinea pig and local lymph node assay responses. *Food Chem Toxicol 34*: 495–506

Ryan CA, Cruse LW, Skinner RA, Dearman RJ, Kimber I, Gerberick GF (2002) Examination of a vehicle for use with water soluble materials in the murine local lymph node assay. *Food Chem Toxicol 40*: 1719–1725

Saillenfait AM, Payan JP, Beydon D, Fabry JP, Langonne I, Sabate JP, Gallissot F (1997) Assessment of the developmental toxicity, metabolism, and placental transfer of N,N-dimethylformamide administered to pregnant rats. *Fundam Appl Toxicol 39*: 33–43

Sakai T, Kageyama H, Araki T, Yosida T, Kuribayashi T, Masuyama Y (1995) Biological monitoring of workers exposed to N,N-dimethylformamide by determination of the urinary metabolites, N-methylformamide and N-acetyl-S-(N-methylcarbamoyl) cysteine. *Int Arch Occup Environ Health 67*: 125–129

Seiji K, Inoue O, Cai SX, Kawai T, Watanabe T, Ikeda M (1992) Increase in sister chromatid exchange rates in association with occupational exposure to N,N-dimethylformamide. *Int Arch Occup Environ Health 64*: 65–67

Senoh H, Katagiri T, Arito H, Nishizawa T, Nagano K, Yamamoto S, Matsushima T (2003) Toxicity due to 2- and 13-wk inhalation exposures of rats and mice to N,N-dimethylformamide. *J Occup Health 45*: 365–375

Tolando R, Zanovello A, Ferrara R, Iley JN, mano M (2001) Inactivation of rat liver cytochrome P450 (P450) by N,N-dimethylformamide and N,N-dimethylacetamide. *Toxicol Lett 124*: 101–111

Tomasini M, Todaro A, Piazzoni M, Peruzzo GF (1983) Patologia da dimethylformamide: osservazioni su 14 casi (Pathology of dimethylformamide: observations in 14 cases (Italian). *Med Lav 4*: 217–220

Walrath J, Fayerweather WE, Gilby PG, Pell S (1989) A case-control study of cancer among Du Pont employees with potential for exposure to dimethylformamide. *J Occup Med 31*: 432–438

Wang JD, Lai MY, Chen JS, Lin JM, Chiang JR, Shiau SJ, Chang WS (1991) Dimethylformamide-induced liver damage among synthetic leather workers. *Arch Environ Health 46*: 161–166

Warbrick EV, Dearman RJ, Basketter DA, Kimber I (1999) Influence of application vehicle on skin sensitization to methylchloroisothiazolinone/methylisothiazolinone: an analysis using the local lymph node assay. *Contact Dermatitis 41*: 325–329

WHO (World Health Organization) (1991) *Dimethylformamide*. International Programme on Chemical Safety, IPCS, No. 114, WHO, Geneva

WHO (World Health Organization) (2001) *N,N-Dimethylformamide*. International Programme on Chemical Safety, IPCS, Concise International Chemical Assessment (CICAD) No. 31, WHO, Geneva

Wrbitzky R (1999) Liver function in workers exposed to N,N-dimethylformamide during the production of synthetic textiles. *Int Arch Occup Environ Health 72*: 19–25

Wrbitzky R, Angerer J (1998) N,N-Dimethylformamide – influence of working conditions and skin penetration on the internal exposure of workers in synthetic textile production. *Int Arch Occup Environ Health 71*: 309–316

Yang C, Ger J, Lin S, Yang G, Deng J (1994) Abdominal colic occurred in workers in a dye manufacturing plant. *Vet Hum Toxicol 36*: 345 (Abstract 28)

Yang JS, Kim EA, Lee MY, Park IJ, Kang SK (2000) Biological monitoring of occupational exposure to N,N-dimethylformamide – the effects of co-exposure to toluene or dermal exposure. *Int Arch Occup Environ Health 73*: 463–470

Yonemoto J, Suzuki S (1980) Relation of exposure to dimethyl formamide vapor and the metabolite, methylformamide, in urine of workers. *Int Arch Occup Environ Health 46*: 159–165

completed 09.03.2005

Ethanol[1]

Supplement 2001

MAK value (1998)	**500 ml/m^3 (ppm) ≙ 950 mg/m^3**
Peak limitation (2001)	**Category II, excursion factor 2**
Absorption through the skin	**–**
Sensitization	**–**
Carcinogenicity (1998)	**Category 5**
Prenatal toxicity (1994)	**Pregnancy Risk Group C**
Germ cell mutagenicity (1998)	**Category 2**
BAT value	**–**
Chemical name (CAS)	ethanol
CAS number	64-17-5

Peak Limitation Category

Carcinogenicity induced by acetaldehyde is the critical effect at high concentrations. Volunteers developed irritation after concentrations of 1900 ml/m^3 and above, but not at a concentration of 1000 ml/m^3 (see "Ethanol", Volume 12, present series). Therefore, an excursion factor of 2 is established for ethanol.

completed 06.12.2000

[1] A more recent supplement follows.

The MAK-Collection Part I: MAK Value Documentations, Vol. 26. DFG, Deutsche Forschungsgemeinschaft

ISBN: 978-3-527-32306-7

Ethanol

Supplement 2002

MAK value (1998)	**500 ml/m^3 (ppm) ≙ 960 mg/m^3**
Peak limitation (2001)	**Category II, excursion factor 2**
Absorption through the skin	**–**
Sensitization	**–**
Carcinogenicity (1998)	**Category 5**
Prenatal toxicity (1994)	**Pregnancy Risk Group C**
Germ cell mutagenicity (2002)	**Category 5**
BAT value	**–**
Chemical name (CAS)	ethanol
CAS number	64-17-5

In 1998, ethanol was classified in Category 5 for carcinogenic substances and the MAK value was lowered to 500 ml/m^3. At the same time, ethanol was classified in Category 2 for germ cell mutagens, since ethanol leads to aneuploidy in the germ cells of the mouse and induces dominant lethal mutations in the rat and the mouse ("Ethanol", Volume 12, present series). It was necessary to re-evaluate ethanol when the new categories for germ cell mutagens were introduced ("Germ cell mutagens", Volume 17, present series).

Genotoxicity

The *in-vitro* and *in-vivo* data for the genotoxic effects of ethanol were described in detail in the MAK documentation from 1998 ("Ethanol", Volume 12, present series). No new data about the effects of ethanol on germ cells have been published since.

From the wealth of information available it is apparent that ethanol has a weak genotoxic potential. Clastogenic effects are observed *in vitro* only after metabolic activation. Acetaldehyde and radicals with genotoxic properties are formed during biotransformation.

The MAK-Collection Part I: MAK Value Documentations, Vol. 26. DFG, Deutsche Forschungsgemeinschaft

ISBN: 978-3-527-32306-7

All positive results *in vivo* were obtained only with extremely high, systemically toxic doses. The results reveal that the mutagenic effects *in vivo* are specific for species and strain ("Ethanol", Volume 12, present series).

Manifesto

The positive results in germ cell tests with mice and rats led to the classification of ethanol in Category 2 for germ cell mutagenicity in 1998. As stated in the documentation, the effects on germ cells in animal studies (aneuploidy in the germ cells of the mouse and dominant lethal mutations in the mouse and rat) occurred only at very high, already clearly toxic doses of ethanol. Since there is no significant increase in the lifetime body burden from ethanol provided the MAK value is observed, also the potential for germ cell mutagenicity should be negligible under these conditions.

The introduction of Category 5 for germ cell mutagens provided a new aspect for the classification of ethanol. Category 5 is reserved for germ cell mutagens for which the potency is considered to be so low that their contribution to genetic risk for humans is expected not to be significant.

Inhalation exposure to ethanol at the workplace should be kept at such a low level that the lifetime body burden of the genotoxic metabolites does not significantly increase ("Ethanol", Volume 12, present series). In analogy to the assessment of cancer risk, ethanol is now classified in Category 5 for germ cell mutagens.

completed 29.11.2001

Hydrogen bromide[1]

Supplement 2000

MAK value (1996)	**2 ml/m³ (ppm) ≙ 6.7 mg/m³**
Peak limitation (2000)	**Category I, excursion factor 1**
Absorption through the skin	–
Sensitization	–
Carcinogenicity	–
Prenatal toxicity (1989)	**Pregnancy Risk Group D[2]**
Germ cell mutagenicity	–
BAT value	–
Synonyms	hydrobromic acid
Chemical name (CAS)	anhydrous hydrobromic acid
CAS number	10035-10-6

Hydrogen bromide is one of the strongest acids and is corrosive to the skin and mucous membranes of the eyes, nose and respiratory tract. To establish a MAK value (see "Hydrogen bromide", Volume 13, present series), 6 volunteers were exposed to the substance in a chamber; irritation of the nose and throat was observed after concentrations of 3 ml/m^3 and above (Stokinger 1981). In another study, exposure to 5 ml/m^3 led after a short time (no other details) to irritation of the respiratory passages (Kühn and Birett 1977).

Because of the strong local irritation, an excursion factor of 1 has been set for Peak Limitation Category I.

[1] A more recent supplement follows.

[2] The definitions of the pregnancy risk groups were revised with the *List of MAK and BAT Values 2006.*

The MAK-Collection Part I: MAK Value Documentations, Vol. 26. DFG, Deutsche Forschungsgemeinschaft

ISBN: 978-3-527-32306-7

References

Kühn R, Birett K (1977) Bromwasserstoff (Hydrogen bromide) (German). In: Ecomed Verlagsgesellschaft (Ed.) *Merkblätter Gefährliche Arbeitsstoffe*, Verlag Moderne Industrie, München, Sheet No. B 35

Stokinger HE (1981) Bromine. In: Clayton GD, Clayton FE (Eds) *Patty's Industrial Hygiene and Toxicology*, Vol IIB, Wiley-Interscience, New York, 2970

completed 25.04.2000

Hydrogen bromide

Supplement 2004

MAK value (1996)	**2 ml/m^3 (ppm) ≙ 6.7 mg/m^3**
Peak limitation (2000)	**Category I, excursion factor 1**
Absorption through the skin	**–**
Sensitization	**–**
Carcinogenicity	**–**
Prenatal toxicity (1989)	**Pregnancy Risk Group D[1]**
Germ cell mutagenicity	**–**
BAT value	**–**
Synonyms	hydrobromic acid
Chemical name (CAS)	anhydrous hydrobromic acid
CAS number	10035-10-6

In 1996 the MAK value for hydrogen bromide was set at 2 ml/m^3; in 2000, exposure to peak concentrations was limited according to Category I with an excursion factor of 1. In 1989 the substance was classified for prenatal toxicity in Section IIc of the *List of MAK and BAT Values*, which was revised to Pregnancy Risk Group D in 2006.

When the hydrogen halide compounds were reviewed recently, it was checked whether any new data for hydrogen bromide have become available. There were no studies in the IUCLID database after 1996 (ECB 2000). A literature search for publications since 1996 yielded two new case reports of short-term exposure to hydrogen bromide in humans, but they did not contain any exposure data (Burns and Linden 1997; Orlando *et al.* 1997). Therefore, no new data for hydrogen bromide relevant to the evaluation has been published since 1996.

The MAK value, the Peak Limitation Category and the classification for prenatal toxicity in Pregnancy Risk Group D have been retained.

[1] The definitions of the pregnancy risk groups were revised with the *List of MAK and BAT Values 2006.*

The MAK-Collection Part I: MAK Value Documentations, Vol. 26. DFG, Deutsche Forschungsgemeinschaft

ISBN: 978-3-527-32306-7

References

Burns MJ, Linden CH (1997) Another hot tub hazard. Toxicity secondary to bromine and hydrobromic acid exposure. *Chest 111*: 816–819

ECB (European Chemicals Bureau) (2000) *Hydrogen bromide*. IUCLID Dataset, 18.02.2000, ECB, Ispra, Italy

Orlando JP, de Haro L, Leroyer S (1997) Reactive airways dysfunction syndrome and bronchiolitis obliterans after exposure to acid vapors. *Rev Pneumol Clin 53*: 339–342

completed 03.02.2004

Hydrogen peroxide

MAK value (2005)	**0.5 ml/m³ (ppm) ≙ 0.7 mg/m³**
Peak limitation (2000)	**Category I, excursion factor 1**
Absorption through the skin	–
Sensitization	–
Carcinogenicity (2005)	**Category 4**
Prenatal toxicity (2005)	**Pregnancy Risk Group C**
Germ cell mutagenicity	–
BAT value	–
Synonyms	hydrogen dioxide hydroperoxide
Chemical name	dihydrogen dioxide
CAS number	7722-84-1
Molecular formula	H_2O_2
Molecular weight	34.02
Melting point	–0.04 to –0.43°C (EU 2003)
Boiling point at 1013 hPa	150–152°C (EU 2003)
Vapour pressure at 25°C	3 hPa (EU 2003)
log P_{ow}*	–1.57 (calculated) (SRC 2004)
1 ml/m³ (ppm) ≙ 1.4 mg/m³	**1 mg/m³ ≙ 0.71 ml/m³ (ppm)**

Hydrogen peroxide has a wide range of uses, for example as a bleaching agent in the textile, paper and cellulose industries, in foods and cosmetics, for synthesis and as a disinfectant (EU 2003). The present documentation is partly based on summaries of toxicological data by other organizations (ECETOC 1996; EU 2003; FIN Competent Authorities 1998).

* *n*-octanol/water partition coefficient

The MAK-Collection Part I: MAK Value Documentations, Vol. 26. DFG, Deutsche Forschungsgemeinschaft

ISBN: 978-3-527-32306-7

1 Toxic Effects and Mode of Action

The main effect of hydrogen peroxide is local irritation, which occurred in workers after daily exposure to concentrations of 1.7 to 3.4 mg/m^3 (1.2–2.4 ml/m^3) with peak values of 11.3 mg/m^3 (8 ml/m^3). Irritation of the eyes was reported in volunteers exposed to aqueous hydrogen peroxide solutions at concentrations of 0.04 % and above. In humans, the ingestion of hydrogen peroxide doses of 150 mg/kg body weight resulted in massive oxygen embolism, and about 600 mg/kg body weight was lethal.

The target organs in animal studies with repeated administration were local tissues, mainly the airways and the skin, and, after oral administration, the gastrointestinal tract. Exposure to hydrogen peroxide concentrations of 10 ml/m^3 for 28 days caused necrosis in the nose of rats. No effects occurred at 2 ml/m^3. Hydrogen peroxide was found to be genotoxic in a large number of *in vitro* test systems. *In vivo*, chromosomal aberrations were not induced, neither were micronuclei in polychromatic erythrocytes or DNA repair synthesis in liver cells. Two host-mediated assays yielded positive results. The available studies do not indicate a potential for adverse effects on fertility. No valid studies of developmental toxicity are available, so that no statement can be made about this end point. After oral administration, hydrogen peroxide was carcinogenic in the duodenum of mice with low catalase activity and caused papillomas in the forestomach of rats. Hydrogen peroxide has a tumour-promoting effect on the mucosa of hamsters at concentrations of 30 % and above, and a weak tumour-promoting effect on the skin of mice at 6 % and above. The local carcinogenic effects are dependent on the hydrogen peroxide concentration and the catalase activity of the animal strain used.

2 Mechanism of Action

Hydrogen peroxide is produced endogenously in aerobic cells and is converted in the organism to oxygen and water. The oxygen bubbles formed at higher concentrations after different routes of administration produce local mechanical damage. Thus, 1 ml of a 30 % hydrogen peroxide solution causes the release of approximately 100 ml oxygen. Reduced blood circulation in tissue caused by oxygen embolism in the capillaries results in a characteristic bleaching of the skin or other tissues and can lead to disturbances in organ functions (ECETOC 1996; EU 2003).

Endogenously active forms of oxygen which are cytotoxic can be produced from hydrogen peroxide (see Section 3.2). These are capable, for example, of causing lipid peroxidation and subsequently induced haemolysis. In humans, a genetically-determined deficiency in the activities of catalase (acatalasaemia) or glucose-6-phosphate dehydrogenase leads to reduced detoxification, which results in increased sensitivity to hydrogen peroxide (ECETOC 1996; EU 2003). The reactive oxygen species released are responsible for the cytotoxic and genotoxic effects.

A cytotoxic or genotoxic effect of hydrogen peroxide does not occur until the detoxification capacity of the organism is overloaded. One study showed that the primary event, at least in the *in vitro* cell culture system, is the necrotic effect, not the genotoxic effect (Fenech *et al.* 1999). This would mean that a cytotoxic effect, which is an irritant effect, occurs prior to the genotoxic effect.

3 Toxicokinetics and Metabolism

3.1 Absorption, distribution, elimination

Hydrogen peroxide is rapidly absorbed after inhalation and oral and dermal administration. Presumably it is completely absorbed, as the permeability coefficient for biological membranes is comparable with that of water. Due to its immediate and efficient metabolism, no absorption rates have been determined up to now. As the degradation product oxygen is transferred to the oxygen pool of the organism, elimination via the lungs is possible (ECETOC 1996; EU 2003).

In a valid study with 16 volunteers, the concentration of endogenous hydrogen peroxide in the expired air was ≤ 0.2 mg/m^3 (0.14 ml/m^3) (EU 2003). In an earlier study, which cannot be used for evaluation as the methods used are unclear, a level of $1–3 \times 10^{-8}$ M (about 0.2–0.7 ml/m^3) hydrogen peroxide was reported in the expired air of volunteers (Williams *et al.* 1982).

3.2 Metabolism

Hydrogen peroxide is an endogenous metabolite and its concentration is normally kept in equilibrium. Its degradation takes place via two enzymes: catalase, which dominates at high substrate concentrations, and glutathione peroxidase, which dominates at low substrate concentrations. The enzyme activity of catalase is high in the intestine, liver, kidneys, mucosae and other extensively vascularized tissues, and low in the brain, lungs and heart. The highest level of activity of glutathione peroxidase was found in the liver and the erythrocytes (ECETOC 1996; EU 2003).

The levels of catalase activity in the lungs and intestine of humans, mice and rats are shown in Table 1.

In the erythrocytes, glucose-6-phosphate dehydrogenase provides the NADPH pool for the regeneration of oxidized glutathione produced by the degradation of hydrogen peroxide (ECETOC 1996; EU 2003).

Table 1. Catalase activity in tissue samples of lung and intestine

Species, strain	Catalase activity [U/mg protein] (n)[c] Intestine	Lungs	References
humans	0.065 ± 0.0047 (3) jejunum	about 70 ± 20[b] (5)	Bryan and Jenkinson 1987; ECETOC 1996
rats	2.42 ± 0.6 (no other details) 1.60 ± 0.1 (no other details) jejunum	about 20 ± 10[b] (6)	
mouse, C3H/HeN[a]	0.0317 ± 0.0084 (11) duodenum	not determined	ECETOC 1996; Ito *et al.* 1984
mouse, B6C3F$_1$[a]	0.0102 ± 0.0012 (12) duodenum	not determined	
mouse, C57BL/6N[a]	0.0042 ± 0.0018 (8) duodenum	not determined	
mouse, C3H/Csb[a]	0.0024 ± 0.0006 (7) duodenum	not determined	

[a] see also Section 5.7.2
[b] values from graphic presentation
[c] number of tissue samples

In the presence of oxidizable metal ions such as Fe^{2+} or Cu^{+}, hydrogen peroxide can be converted to cytotoxic hydroxyl radicals via the Fenton or Haber-Weiss reaction (ECETOC 1996; EU 2003; IUPAC 2004).

Catalase: $2\ H_2O_2 \rightarrow 2\ H_2O + O_2$

Glutathione peroxidase: $H_2O_2 + 2\ GSH \rightarrow 2\ H_2O + GSSG$

Glutathione reductase, regeneration of the oxidized glutathione: $GSSG + NADPH + H^+ \rightarrow 2\ GSH + NADP^+$

Glucose-6-phosphate (G6P) dehydrogenase, regeneration of the NADPH pool: $G6P + NADP^+ \rightarrow$ 6P-glucono-δ-lactone + NADPH

Fenton reaction: $H_2O_2 + Cu^+/Fe^{2+} \rightarrow OH\bullet + OH^- + Cu^{2+}/Fe^{3+}$

Haber-Weiss reaction: $H_2O_2 + OH\bullet + \text{metal ions} \rightarrow H_2O + O_2^-\bullet + H^+$

$H_2O_2 + O_2^-\bullet + \text{metal ions} \rightarrow OH\bullet + OH^- + O_2$

4 Effects in Humans

4.1 Single exposures

4.1.1 Inhalation

The lowest observed adverse effect level (LOAEL) for airway irritation caused by hydrogen peroxide in 32 volunteers was 10 mg/m^3 (7.1 ml/m^3), independent of the duration of exposure of five minutes to four hours. The no observed adverse effect level (NOAEL) was 5 mg/m^3 (3.6 ml/m^3; no other details; Kondrashov 1977). As a result of the unclear methods, these data cannot be evaluated.

4.1.2 Ingestion

A number of lethal intoxications have been reported after ingestion of hydrogen peroxide. The lowest known lethal dose was about 600 mg/kg body weight in a child. Cases of severe but non-fatal poisonings have been reported after the intake of 25 to 500 ml hydrogen peroxide or 150 to 350 mg/kg body weight at different concentrations. The most frequent symptoms of intoxication were gas emboli in the heart, the lungs, the walls of the gastrointestinal tract, the brain and portal venous system; corrosion in oesophagus and stomach; and neurological deficits (ECETOC 1996; EU 2003).

4.1.3 Dermal absorption

Characteristic bleaching of the skin occurred after dermal application of a 20 % to 30 % hydrogen peroxide solution (ECETOC 1996).

4.2 Repeated exposures

In 1998 and 1999, six workers were exposed to hydrogen peroxide concentrations of 1.7 to 3.4 mg/m^3 (1.2–2.4 ml/m^3) with peak values of 8.5 mg/m^3 (6 ml/m^3) several times per hour during the working period (no other details) and to 11.3 mg/m^3 (8 ml/m^3) at the start of the work shift for about 1.5 hours daily. Three workers reported redness and burning in the eyes, a blocked nose, itching and dryness in the throat, coughing and asthma symptoms. Most of the symptoms increased towards the end of the working week. It is not evident from the studies to what extent these effects can also be attributed to the peak exposure. After two years, the three workers developed bronchitis and sinusitis. Treatment with antibiotics produced an improvement. The coughing and headaches disappeared only after treatment with corticosteroids and reduction of the exposure concentrations in the year 2000 to 0.5 to 0.7 mg/m^3 (0.36–0.5 ml/m^3; 8-hour mean values, no higher peak exposures). No symptoms were described in four workers at two other machines in the same working area, who had been exposed since 1998 to hydrogen peroxide concentrations of 0.2 to 0.6 mg/m^3 (0.14–0.43 ml/m^3; 8-hour mean values) without higher peak exposures (Table 2; Riihimäki *et al.* 2002; Riihimäki 2004).

In another study, workers reported airway irritation after exposure to hydrogen peroxide at a concentration of 10 mg/m^3 (7.1 ml/m^3; no other details; Kondrashov 1977). A communication from the FMC Corp (1990) revealed that hydrogen peroxide caused bleaching of the hair in workers at concentrations of 0.5 to 1 ml/m^3. These concentrations were apparently not irritating to the eyes, nose or throat. However, this information is not confirmed and cannot be used for evaluation.

One worker, who had been exposed to hydrogen peroxide for two hours per week for three years and subsequently every day for six months, developed interstitial lung disease with severely impaired gas exchange. A concentration of 41 mg/m^3 (29.1 ml/m^3) was determined at the machine, which had been in operation without a ventilation system for one year, and 12 mg/m^3 (8.5 ml/m^3) at floor level. Subsequent analyses after installation of a ventilation system revealed concentrations of 4.5 mg/m^3 (3.2 ml/m^3) and 1.5 mg/m^3 (1.06 ml/m^3). The study provides no information about when the exposure levels were determined in relation to the contraction of the disease by the worker, who had been smoking two packs of cigarettes daily for 25 years. Examination of the lungs revealed collapsed alveoli, thickened alveolar walls, infiltration of mononuclear cells and haemosiderin-loaded macrophages in the alveoli. The catalase activity of the erythrocytes was within the normal range. The authors ascribe the clinical findings to hydrogen peroxide exposure, particularly as the dyspnoea was reversible after 1.5 exposure-free months. The smoking habit of the worker was considered to be a contributing factor. The radiographic appearance of the lungs and the results of lung function tests returned to normal after oral administration of corticosteroids. The worker in question and six other colleagues reported irritation of the eyes and throat and bleached hair (Kaelin *et al.* 1988).

Table 2. Effects of hydrogen peroxide in humans after repeated inhalation

Number of persons	Duration of exposure; concentration	Peak values	Findings	References
workers n = not given	**no other details**; 0.5–1.0 ml/m^3		**0.5–1.0 ml/m^3 and above:** bleaching of hair, no irritation	FMC Corp 1990
workers n = not given	**no other details**; max. 1-hour value: ≤ 3.5 mg/m^3 (2.4 ml/m^3)	no other details	**≤ 2.4 mg/m^3:** slight irritation in the nose, white skin after contact	EU 2003
workers n = not given	**no other details**; 10 mg/m^3 (7.1 ml/m^3)		**7.1 ml/m^3 and above:** respiratory irritation	Kondrashov 1977
workers 6	**about 1 year**; 1) 0.5–0.7 mg/m^3 (0.36–0.5 ml/m^3) 2) 1.7–3.4 mg/m^3 (1.2–2.4 ml/m^3) per 8-hour time-weighted mean value	1) no peak exposure 2) at start-up (1.5 hours): 11.3 mg/m^3 (8 ml/m^3), during operation (6 hours) a number of times per hour: 8.5 mg/m^3 (6 ml/m^3); at 2nd machine: 3 ml/m^3	**1) 0.36–0.5 ml/m^3: NOAEL** **2) 1.2–2.4 ml/m^3**: reddened, burning eyes; blocked nose; itching, dry throat; coughing; asthma symptoms; symptoms increased during the work period. Bronchitis and sinusitis in 3/6. The other three workers reported only slight or no symptoms (no other details).	Riihimäki *et al.* 2002

Table 2. continued

Number of persons	Duration of exposure; concentration	Peak values	Findings	References
workers 7	**about 3 years, 2 hours/week, then daily for 6 months;** 1st year: 41 mg/m³ (29.1 ml/m³), then: 4.5 mg/m³ (3.2 ml/m³)	no other details	**3.2 ml/m³ and above:** irritation of eyes and throat; bleaching of hair; dyspnoea, collapsed alveoli; thickened alveolar walls; infiltration of mononuclear cells and haemosiderin-loaded macrophages in 1/7. Reversible after 1.5 months	Kaelin *et al.* 1988
workers n = not given	**3–5 years** ≤ 0.79 mg/m³ (0.56 ml/m³)	no other details	no findings in lung function tests	EU 2003
workers 43, control group: 31 workers	**2–11 years;** **1997:** 0.48 mg/m³ (0.30–0.94 mg/m³; 0.21–0.67 ml/m³) **1998:** 0.43 mg/m³ (0.27–0.72 mg/m³; 0.19–0.51 ml/m³) **1999:** 0.30 mg/m³ (0.10–0.95 mg/m³; 0.07–0.67 ml/m³) **2000:** 0.15 mg/m³ (0.11–0.30 mg/m³; 0.08–0.21 ml/m³) **2001:** 0.32 mg/m³ (0.23–0.53 mg/m³; 0.16–0.34 ml/m³) 8-hour mean values from single annual determinations at ≥ 5 workplaces	no other details	no findings in lung function tests for 1993–2002; no information about how many years the workers were exposed	Mastrangelo *et al.* 2005
workers 110	n = 80, >10 years n = 18, 5–10 years n = 12, 1–5 years 1991–1999: ≤ 1.4 mg/m³ (1 ml/m³); before 1991 presumably > 1 ml/m³	5 mg/m³ (3.6 ml/m³); in accident situations > 10 mg/m³ (7.1 ml/m³)	after accidental contact: skin irritation and bleaching; bleaching of hair (presumably higher concentrations in the 1970s); no findings in lung function tests, one complaint of throat irritation	Degussa-Hüls 1999

A study in an Italian plant did not reveal any adverse findings in lung function tests carried out annually between 1993 and 2002 in a total of 43 workers exposed to hydrogen peroxide for two to eleven years. From the annual analyses of the levels of exposure at at least five different workplaces, 8-hour mean values of 0.10 to 0.95 mg/m^3 (0.07–0.67 ml/m^3) were determined for the years 1997 to 2001 (Mastrangelo *et al.* 2005).

Lung function tests carried out in workers (no other details) who had been employed in hydrogen peroxide production for three to five years yielded no adverse findings. The "typical" (no other details) hydrogen peroxide concentrations were in the range between "not detectable" and 0.79 mg/m^3 (0.56 ml/m^3). Cases of hair bleaching, nose bleeds and irritation of the eyes and respiratory tract had been reported in the past by workers at the same production plant. Nasal irritation (no other details) occurred in workers exposed to 1-hour mean values of up to 3.5 mg/m^3 (2.4 ml/m^3) (EU 2003).

Lung function tests also yielded no adverse findings in 110 workers, of whom 80 had been employed in the production of hydrogen peroxide for 10 to 40 years, 18 for 5 to 10 years and 12 between 1 and 5 years. Health monitoring was carried out annually in 95 of the workers cited, and every three years in the remainder. Between 1991 and 1999, the mean concentration during a work shift in the production area was below 1.4 mg/m^3 (1 ml/m^3; no other details); short-term concentrations in separate areas, such as the pumping station, were up to about 5 mg/m^3 (3.6 ml/m^3) and above 10 mg/m^3 (7.1 ml/m^3) in the case of one accident. Skin irritation and skin bleaching were reported after accidental contact. Cases of hair bleaching were reported during the 1970s at one of the four production sites. Attention is drawn to the fact that the exposure level of the workers was presumably higher in earlier years (Degussa-Hüls 1999). As more detailed information on the duration of exposure is not available, these data cannot be used for evaluation.

4.3 Local effects on skin and mucous membranes

4.3.1 Skin

The hands of 32 volunteers were exposed to hydrogen peroxide vapour. The LOAEC (lowest observed adverse effect concentration) for skin irritation was 20 mg/m^3 (14.2 ml/m^3) after four hours and 180 mg/m^3 (128 ml/m^3) after five minutes. The amount of hydrogen peroxide absorbed through the skin was determined immediately after the end of exposure and was 1.1 to 1.7 mg/dm^2 at the LOAEC (Kondrashov 1977). The study cannot be used for evaluation because of the inadequate documentation.

4.3.2 Mucous membranes

Several studies showed that hydrogen peroxide causes irritation of the oral mucosa. Rinsing of the mouth several times a day with 3 % to 12.5 % hydrogen peroxide solutions produced lesions in the gingiva and tongue. Cases of inflammatory intestinal disease were described following endoscopic examinations with instruments cleaned with a 3 % hydrogen peroxide solution (ECETOC 1996).

4.3.3 Eyes

The mean LOAEC for eye irritation after direct instillation of a hydrogen peroxide solution into the eyes of 10 volunteers was 0.08 % (0.04–0.15 %; McNally 1990). Reports of hydrogen peroxide solutions used as antibacterial agent in the eyes describe reversible inflammatory reactions at and above 0.5 % and corneal changes at and above 3 %; the latter were described as being reversible in some cases (ECETOC 1996; EU 2003). In studies of doubtful validity dating from 1913 and 1965, the instillation of hydrogen peroxide solutions of up to 20 % into the eye for the treatment of ulcers was described as having no further effects (ECETOC 1996).

4.4 Allergenic effects

The allergic reaction to hydrogen peroxide was investigated in three studies involving persons who had been exposed to hydrogen peroxide hair dyes. In one of the studies, 2 of the 158 persons tested produced positive reactions. They were also sensitized to other additives in the hair dyes (Aguirre *et al.* 1994).

All patch tests with hydrogen peroxide conducted between 1974 and 1993 in 35 female hairdressers with allergic dermatitis (Leino *et al.* 1998a) and in 54 female hairdressers from a cohort totalling 355 patients (Leino *et al.* 1998b) yielded negative results.

In two other skin sensitization studies, carried out between 1991 and 1997, none of 130 and 59 dermatitis patients produced a positive result. Of a total of 29800 recorded cases of allergic dermatitis reported in the Finnish Register for Occupational Diseases between 1975 and 1997, 4 were caused by hydrogen peroxide. The authors concluded from this that no relevant sensitizing potential exists for hydrogen peroxide (Kanerva *et al.* 1998).

4.5 Reproductive toxicity

There are no data available for the reproductive toxicity of hydrogen peroxide.

4.6 Genotoxicity

There are no data available for the genotoxicity of hydrogen peroxide.

4.7 Carcinogenicity

A case-control study investigated the relationship between exposure to 293 workplace substances and the development of carcinomas in oesophagus, stomach, intestine, rectum, pancreas, lungs, prostate, bladder, kidneys, and lymph nodes, and melanomas of the skin. Hydrogen peroxide produced no increase in tumour incidence. However, as regards interpretation of these negative results, attention was drawn to the low number of investigated persons exposed to hydrogen peroxide (hairdressers, textile bleachers and furriers) and the presumably low level of exposure (IARC 1999).

5 Animal Experiments and *in vitro* Studies

5.1 Acute toxicity

5.1.1 Inhalation

Studies of inhalation toxicity were carried out with hydrogen peroxide in aerosol or vapour form. With a 4-hour LC_{50} of 2000 mg/m^3, the acute inhalation toxicity of hydrogen peroxide vapour for rats was low (1420 ml/m^3; Kondrashov 1977). In other studies, exposure for 8 hours to a saturated vapour atmosphere (about 2880 ml/m^3) was not lethal for rats (ECETOC 1996). The LOEC (lowest observed effect concentration) in rats was 60 mg/m^3 (42.6 ml/m^3) after exposure to hydrogen peroxide vapour for 4 hours. At this concentration, an increase in NAD diaphorase was determined in the bronchial epithelium (Kondrashov 1977). Mice were more sensitive with a 4-hour LC_{50} between 159 and 274 mg/m^3 (113 and 194 ml/m^3, presumably vapour/aerosol mixture; Svirbely 1961). In the case of aerosols consisting of a 90 % hydrogen peroxide solution, the 2-hour LC_{50} for mice was between 920 and 2000 mg/m^3 (653 and 1420 ml/m^3). After exposure for 30 minutes to an aerosol consisting of a 70 % hydrogen peroxide solution, the RD_{50} for mice was 665 mg/m^3 (472 ml/m^3; EU 2003).

5.1.2 Ingestion

The oral LD_{50} in the rat was between 800 mg/kg body weight after administration of a 70 % hydrogen peroxide solution and over 1500 mg/kg body weight after administration of a 9.6 % hydrogen peroxide solution in studies conducted in accordance with present-day requirements (ECETOC 1996; EU 2003).

5.1.3 Dermal absorption

The studies of acute dermal toxicity are insufficiently described. After application of a 90 % hydrogen peroxide solution, the dermal LD_{50} was between 700 and 5000 mg/kg body weight in various species, of which the rat was the least sensitive (ECETOC 1996; EU 2003). The most sensitive species was the rabbit, in which hydrogen peroxide doses of 700 mg/kg body weight were lethal for 6 of 12 animals (no other details; Hrubetz *et al.* 1951). After application of a 70 % hydrogen peroxide solution in doses of 6500 or 13000 mg/kg body weight, the dermal LD_{50} was 9200 mg/kg body weight in four male rabbits. The lowest dose of 6500 mg/kg body weight was not lethal, but produced oedema and skin corrosion at the site of application (FMC Corp 1979).

5.2 Subacute, subchronic and chronic toxicity

5.2.1 Inhalation

In a 28-day study carried out according to OECD Test Guideline 412, 5 Wistar rats per sex and concentration group were exposed to hydrogen peroxide vapour in concentrations of 0, 2, 10 or 25 ml/m^3 for 6 hours per day and 5 days per week. The symptoms after concentrations of 10 ml/m^3 and above were reddened noses and stains around the nose, and additionally in the highest concentration group salivation, piloerection and rales. The incidence and the intensity of the symptoms increased with exposure concentration and duration. Body weight gains in the male animals of the highest concentration group were reduced by 8.2 %. A slight decrease in haemoglobin concentration, erythrocyte count and erythrocyte volume was considered to be not relevant taking historical controls into account. After concentrations of 10 ml/m^3 and above, minimal to slight necrosis and inflammation were observed in the anterior region of the nasal cavity in all animals and in the respiratory epithelium in one animal. Inflammation of the nasal mucosa, infiltration of mononuclear cells and epithelial erosion in the larynx occurred after 25 ml/m^3. In all concentration groups, but not in the controls, increased perivascular neutrophil infiltration in the lungs was observed, as well as haemorrhages in the two lower concentration groups. The latter effects were evaluated as being not substance-related, as they were not concentration-dependent. According to the authors, the NOAEC (no observed adverse effect concentration) was 2 ml/m^3 (CEFIC 2002).

Evidence of local effects on the skin, such as thickening, hair loss and bleaching, and of damage to the airways was obtained in earlier studies with rats and dogs after whole-body exposure to hydrogen peroxide concentrations of 10 mg/m^3 (7.1 ml/m^3) for 5 to 6 hours per day on 4 to 5 days per week for 4 to 6 months. In the dogs, hyperplastic muscular coats were found in the bronchioles. In addition, areas with atelectasis and emphysema were observed. Red-staining, circular areas, which consisted of collagen and contained occasional muscle cells and strands of elastic tissue, could be observed particularly in areas with fragmented alveolar walls (EU 2003).

In rats, a reduction in succinate dehydrogenase, monoamine oxidase and acid phosphatase and an increase in alkaline phosphatase were found in the lungs in the highest concentration group after 4 months of exposure to hydrogen peroxide concentrations of 0.1, 1 or 10 mg/m^3 (0.07–7.1 ml/m^3; 5 hours/day, 5 days/week). No effects were observed after 1 mg/m^3 (0.71 ml/m^3) (Kondrashov 1977). This study cannot be included in the evaluation as the methods and analyses used are unclear.

Irritation of the airways was reported after exposure of mice, rats and rabbits to hydrogen peroxide concentrations of 31, 93 or 107 mg/m^3 (22, 66 or 76 ml/m^3) in studies of limited usefulness. No significant histopathological changes in the trachea or the lungs were observed (EU 2003).

5.2.2 Ingestion

Groups of Wistar rats were given gavage doses of hydrogen peroxide of 0, 52.6, 168.7 or 506 mg/kg body weight and day in the form of a 0.5 % solution for 6 days a week for 12 weeks. After doses of 52.6 mg/kg body weight and day and above, there was a dose-dependent reduction in enzyme activities in the liver, and changes in haematological parameters occurred (EU 2003; FIN Competent Authorities 1998).

After oral administration of hydrogen peroxide doses of 6 to 60 mg/kg body weight and day (0.06–0.6 % solution) by gavage to 6 Wistar rats per dose group for 100 days, the catalase activity in the plasma was reduced at doses of 30 mg/kg body weight and day and above. Higher hydrogen peroxide doses produced reduced body weights and a decrease in the haematocrit and plasma protein concentration. The NOAEL was given as 20 mg/kg body weight and day (EU 2003; FIN Competent Authorities 1998).

No effects were observed in Wistar rats given hydrogen peroxide doses of 0.6 to 6 mg/day with the feed (about 3–30 mg/kg body weight and day) for 90 days. No statement was made about the stability of hydrogen peroxide in the feed (FIN Competent Authorities 1998).

The NOAEL after administration of hydrogen peroxide for six months was 0.005 mg/kg body weight and day in an insufficiently documented study with rats and rabbits. At higher dose levels (no other details), changes in haematological parameters and enzyme activities in the gastrointestinal tract were reported. The study could not be included for evaluation as a result of the insufficient documentation (FIN Competent Authorities 1998).

Body weights were reduced in rats and mice after the administration of hydrogen peroxide with the drinking water in concentrations of 0.15 % and 0.3 %, respectively. This corresponds to about 150 or 600 mg/kg body weight and day in the case of normal water uptake; reduced water uptake was described in most other studies, however, so that the actual dose might have been lower (EU 2003; FIN Competent Authorities 1998).

In a 90-day study with strain C57BL/6NCrlBr mice (with low catalase activity) carried out according to OECD Test Guideline 408, the NOAEL was 0.01 % hydrogen peroxide in the drinking water (26 mg/kg body weight for males and 37 mg/kg body weight for females). Effects at the LOAEL of 0.03 % (about 76 mg/kg body weight or 103 mg/kg body weight) were reduced feed and water intake in the females and mucosal hyperplasia of the duodenum in the males. In females, mucosal hyperplasia of the duodenum occurred only after 0.1 % (328 mg/kg body weight) and above, and was reversible in both sexes in the 6-week recovery period (EU 2003; FIN Competent Authorities 1998).

5.2.3 Dermal exposure

There are no data available for the dermal effects of hydrogen peroxide.

5.3 Local effects on skin and mucous membranes

5.3.1 Skin

In studies carried out according to OECD Test Guideline 404, a 10 % hydrogen peroxide solution had a slight irritant effect on the skin of rabbits, a 35 % solution was irritating, and 50 % and 70 % solutions were corrosive (ECETOC 1996; FIN Competent Authorities 1998).

In rats, the NOAEC for effects of hydrogen peroxide on the dorsal skin was 0.1 mg/m^3 (0.07 ml/m^3) after whole-body exposure for four months. Hydrogen peroxide concentrations of 1 mg/m^3 (0.71 ml/m^3) increased the activity of monoamine oxidase and NAD diaphorase in the epidermis after two months; after four months succinate and lactate dehydrogenase also increased and a functional disturbance of the stratum corneum was observed (Kondrashov 1977). The results of the study are not of use as the description is inadequate and the methods unconventional.

5.3.2 Eyes

In studies carried out according to OECD Test Guideline 405, a 5 % hydrogen peroxide solution was weakly irritating to the rabbit eye, a 10 % hydrogen peroxide solution was highly irritating and caused damage to the cornea. A 35 % hydrogen peroxide solution was corrosive (ECETOC 1996; FIN Competent Authorities 1998).

5.4 Allergenic effects

In a test for skin sensitization in the guinea pig, a 3 % hydrogen peroxide solution was slightly irritating in 3 of 10 animals one hour after topical induction, and not irritating after 24 and 48 hours. The subsequent application of a single drop of test solution to the scarified skin was slightly irritating in 2 of 5 animals after one hour and reversible in 1 animal after 24 hours. After the intradermal injection of 0.1 ml of a 0.1 % solution, slight irritation was observed in 3 of the other 5 animals after one hour and after 24 hours. For induction, the last 2 applications were repeated a total of 6 times at intervals of 2 to 4 days. After a 2-week rest period, the animals were challenged with the same concentration according to the same pattern. There was no reaction after the topical challenge. After challenge by intradermal injection, a slight reddening occurred after 1 hour in 3 of 5 animals, which was found in all animals after 24 hours and produced a haemorrhagic reaction in one animal. Challenge by application to the scarified skin produced no reaction after one hour, but a slight reddening in 2 of 5 animals after 24 hours, which was reversible after 48 hours. The authors interpreted the results as negative (DuPont 1953). The study does not meet present-day requirements and is not sufficiently documented.

5.5 Reproductive and developmental toxicity

5.5.1 Fertility

The only investigations of fertility are three early drinking water studies. Administration of a 1 % hydrogen peroxide solution to male mice for 3 weeks yielded no evidence of reduced fertility. After administration of a 3 % hydrogen peroxide solution for 6 weeks, no adverse effects were found on the sperm of rabbits. Administration of 0.45 % hydrogen peroxide with the drinking water for 5 months had no influence on reproduction in female rats (ECETOC 1996; EU 2003).

Oral administration of hydrogen peroxide doses of 0.005 to 50 mg/kg body weight and day to male and female rats for 6 months resulted in reduced sperm motility and a reduced number of litters in the animals of the highest dose group. The maternal NOAEL in the study was 0.005 mg/kg body weight and day (see Section 5.2.2). Attention is drawn to the insufficient documentation of the methods and results (ECETOC 1996; EU 2003).

No hydrogen peroxide-induced effects were observed in the reproductive organs of female and male animals in the 90-day study performed according to OECD Test Guideline 408 with C57BL/6N mice with low catalase activity (see Section 5.2.2) and the carcinogenicity studies with C57BL/6N mice and F344 rats (see Section 5.7.2) (EU 2003).

In vitro studies showed that the sperm of rabbits were less sensitive to hydrogen peroxide than those of mice, dogs or humans. The concentration which produced a 50 % reduction in motility in human sperm *in vitro* was between 30 and 300 mg/l. The sensitivity increased in the absence of endogenous catalase (ECETOC 1996).

5.5.2 Developmental toxicity

In an inadequate study with Wistar rats, in which hydrogen peroxide was administered for 1 week “during the critical period of pregnancy” (no other details), a concentration-dependent increase in foetotoxicity was found at concentrations of 0.2 to 10 mg/kg feed (0.02–1 mg/kg body weight and day), and delayed skeletal ossification, but no teratogenicity at 0.2 mg/kg body weight and day and above. Foetal resorption was increased and foetal body weights were decreased at the highest concentration; the body weights of the dams were only slightly decreased. As a result of methodological shortcomings as regards exposure, and suspected degradation of essential nutrients in the feed, the study was described as not relevant for inclusion in the evaluation (EU 2003).

5.6 Genotoxicity

5.6.1 *In vitro*

Hydrogen peroxide was clearly genotoxic in a series of *in vitro* test systems. The corresponding studies are therefore presented here only in summarized form, with reference to the secondary literature. Hydrogen peroxide induced DNA repair in bacteria, among others in the SOS chromotest with *Escherichia coli*. In the mutagenicity test with *Salmonella typhimurium*, hydrogen peroxide was particularly mutagenic in those strains that are sensitive to hydroxyl radicals. Hydrogen peroxide induced sister chromatid exchange and DNA repair in various mammalian cell lines. In the UDS (unscheduled DNA synthesis) test, hydrogen peroxide induced DNA repair synthesis in rat hepatocytes and human foetal lung cells. DNA single strand breaks were produced in various mammalian cell lines, DNA double strand breaks in human leukocytes and V79 cells, but not in rat hepatocytes. Hydrogen peroxide induced chromosomal aberrations in a large number of cells, including V79, CHL (Chinese hamster liver) and CHO (Chinese hamster ovary) cells, but not in human D98/AHZ cells. The negative results in a micronucleus test with spleen cells of mice (C57BL/6J mice) were ascribed to methodological shortcomings. Hydrogen peroxide induced gene mutations in mammalian cells in the HPRT and $TK^{+/-}$ test (ECETOC 1996; EU 2003).

In general, bacterial test systems without catalase were particularly sensitive, and the mammalian cell lines R-8 and R-10 of the Chinese hamster were highly resistant to hydrogen peroxide as a result of their high catalase activity. A number of tests with metabolic activation indicated that the enzymes contained in the S9 mix contribute to the detoxification of hydrogen peroxide in the genotoxicity tests (ECETOC 1996; EU 2003).

5.6.2 *In vivo*

A 3 % hydrogen peroxide solution was not mutagenic in the test for X-chromosomal recessive lethal mutations (SLRL) in *Drosophila melanogaster* (EU 2003). In a valid *in vivo/in vitro* UDS test with rat liver cells from five male Wistar rats, hydrogen peroxide was not genotoxic after intravenous injection of 0, 25 or 50 mg/kg body weight (infusion of a 0, 0.1 or 0.2 % solution, for 30 minutes at a rate of 0.2 ml/minute). DNA repair synthesis was increased in the orally administered positive controls 2-acetylamino-fluorene and dimethyl nitrosamine (EU 2003).

Hydrogen peroxide induced no micronuclei in polychromatic erythrocytes (PCE) of groups of 10 male and 10 female C57BL/6NCrlBr mice (low catalase activity) after 14 days administration of 0, 200, 1000, 3000 or 6000 mg/l in the drinking water (males: 0, 42.4, 164, 415 or 536 mg/kg body weight and day; females: 0, 48.5, 198, 485 or 774 mg/kg body weight and day). 2000 PCE from each animal of the negative controls, of the cyclophosphamide positive control and of the animals in the highest dose group were examined. The positive control substance induced the formation of micronuclei

(EU 2003). It is not certain whether hydrogen peroxide reached the target cells, as the ratio between PCE and normochromatic erythrocytes (NCE) remained unchanged.

Hydrogen peroxide did not increase the incidence of micronuclei in PCE in groups of 5 male and 5 female Swiss OF1 mice given single intraperitoneal doses of 250, 500 or 1000 mg/kg body weight (1 %, 2 % or 4 % solution, 25 ml/kg body weight). 2000 PCE from each animal were examined. The ratio of PCE to NCE was significantly reduced in all dose groups in the samples taken after 24 hours, but only after doses of 250 and 1000 mg/kg body weight in the samples taken after 48 hours. One male died after administration of 1000 mg/kg body weight; hypoactivity and piloerection were observed in the other males. The cyclophosphamide positive control induced the formation of micronuclei. From the reduced PCE to NCE ratios, the authors concluded that hydrogen peroxide induces effects in the bone marrow, though the mechanism is not clear (EU 2003).

Two other insufficiently documented micronucleus tests with mice are available. Oral administration of 0.003 % to 3.0 % hydrogen peroxide in milk to Swiss mice for 32 hours or a single intraperitoneal dose of 1/2, 1/5, 1/25 or 1/100 of the LD_{50} (no other details) of hydrogen peroxide in mice (strain not known) did not increase the formation of micronuclei (EU 2003).

No increase in chromosomal aberrations was observed in the bone marrow cells of Wistar rats after exposure to hydrogen peroxide (no other details) (ECETOC 1996).

After intraperitoneal injection of *Salmonella typhimurium* TA1530 or G46 as indicator organisms, 2 oral doses of 0.5 ml of a 0.3 % solution produced an increase in mutation frequency within two hours in a host-mediated assay in Swiss mice (EU 2003).

In another host-mediated assay, intraperitoneal injection of 1 ml of 0.01 to 0.5 M solutions of hydrogen peroxide into 2 to 4 mice (inbred strain AB Jena Gat.) induced concentration-dependent chromosomal aberrations in the ascites tumour cells injected intraperitoneally 48 hours previously. The tumour cells themselves had very low catalase activity (Schöneich 1967). The cause of the positive results obtained in this host-mediated assay is thought to be the direct contact of hydrogen peroxide with the target cells and their low catalase activity.

In a 4-week study with Sencar mice, the application of up to 200 μl of 70 % hydrogen peroxide to the skin two times a week did not cause increased formation of 8-OH-2′-deoxyguanine or mutations in the *c-Ha-ras* gene. Nor were dermal changes revealed upon examination of the sites of application for epidermal hyperplasia and cellular composition. No abnormal findings were obtained at necropsy of the animals. The positive and negative controls, dimethylbenzanthracene and ethanol, yielded the expected results (SPI 1997).

5.7 Carcinogenicity

5.7.1 Short-term studies

Initiation promotion studies

Sixteen F344 rats were given 1.5 % hydrogen peroxide in the drinking water for 4 weeks. Following this, methyl azoxymethanol acetate (MAM) doses of 25 mg/kg body weight were additionally injected intraperitoneally 3 times at 2-week intervals. Half of the animals continued to receive hydrogen peroxide in the drinking water, the other half received normal drinking water. The animals were examined 17 weeks after the beginning of the intraperitoneal injections. As controls, groups of 3 rats were given only hydrogen peroxide in the drinking water or remained untreated for 25 weeks. In the 8 F344 rats given normal drinking water after MAM treatment, 2 duodenal tumours, 2 jejunal tumours, and 1 colon tumour were observed. The subsequent administration of 1.5 % hydrogen peroxide in the drinking water induced 8 duodenal and 5 jejunal tumours. This increased tumour incidence was evaluated by the authors as evidence of a tumour-promoting effect of hydrogen peroxide. No tumours were produced after administration of 1.5 % hydrogen peroxide alone for 25 weeks. The administration of MAM alone was not investigated (Hirota and Yokoyama 1981). The lack of administration of MAM alone is not a deficit of the study: the tumours which developed without follow-up treatment with hydrogen peroxide can be attributed to MAM, as hydrogen peroxide alone does not produce any tumours. The initiation protocol is nevertheless unusual.

Another study with 20 Wistar rats did not reveal any tumour-promoting activity for intestinal tumours after oral intake of 1 % hydrogen peroxide with the drinking water for 32 weeks, but the substance did cause an increased incidence of forestomach papillomas (100 % after administration of *N*-methyl-*N'*-nitro-*N*-nitrosoguanidine (MNNG) and hydrogen peroxide, 50 % after administration of hydrogen peroxide alone; see also Table 3) and a significant increase in adenomatous hyperplasia (38 %) in the glandular stomach. The administration of MNNG alone produced neither forestomach papillomas nor hyperplasia in the glandular stomach (Takahashi *et al.* 1986).

After initiation with 9,10-dimethyl-1,2-benzanthracene (DMBA, 43 % incidence, 3/7 animals), tumour-promoting activity for buccal pouch carcinomas in the Syrian hamster was reported after twice weekly application of 30 % hydrogen peroxide for 22 weeks (100 % incidence, 5/5 animals), but not after 3 % hydrogen peroxide (55 % incidence, 6/11 animals). No carcinomas were produced by 30 % hydrogen peroxide alone (0/9 animals; Weitzman *et al.* 1986).

After initiation with DMBA, hydrogen peroxide had a weak tumour-promoting effect on the skin of Sencar mice after concentrations of 6 % and above. A 6 %, 10 %, 15 % or 30 % hydrogen peroxide solution (0.2 ml in acetone) induced papillomas after 25 weeks in 10, 10, 8 and 6 of 65 mice, respectively. After application of acetone alone, no papillomas were observed. The authors justified the absence of a dose-effect relationship with the necrotic effects of hydrogen peroxide concentrations of 15 % and above, which prevented survival of the initiated cells (Klein Szanto and Slaga 1982).

A 5 % hydrogen peroxide solution (0.2 ml in acetone) was not tumour-promoting on the skin of 20 Sencar mice after twice weekly application for 51 weeks after initiation with DMBA, and not carcinogenic when administered alone. In the promotion study, however, 45 % of the animals were found to have epidermal hyperplasia. After administration of hydrogen peroxide alone, epidermal hyperplasia was observed in only one animal (5 %). No information was given as to whether skin irritation or necrosis occurred (Kurokawa *et al.* 1984).

After initiation with DMBA, a 3 % hydrogen peroxide solution (0.2 ml in water) was not tumour-promoting after application to the skin of 30 female ICR Swiss mice 5 or 7 times per week for 56 or 40 weeks. No information was given as to whether skin irritation or necrosis occurred (Bock *et al.* 1975; Shamberger 1972).

5.7.2 Long-term studies

The results of the long-term studies of carcinogenicity in mice and rats are shown in Table 3.

Mouse

The administration of 0.1 % or 0.4 % hydrogen peroxide in the drinking water (about 300 or 1200 mg/kg body weight and day) for 100 weeks to groups of about 50 male and 50 female C57BL/6J mice (it is not clear whether the duodenal cells of this strain have low catalase activity like strain C57BL/6N; see Ito *et al.* (1982)) induced erosion and ulcers in the glandular stomach and hyperplasia, adenomas and carcinomas in the duodenum (see Table 3). Erosion in the stomach occurred after 40 weeks, duodenal hyperplasia after 55 weeks and duodenal carcinomas after 65 weeks. No metastases or other treatment-related tumours were observed. Body weights were significantly reduced in the high dose females after 15 months (Ito *et al.* 1981a, 1981b, 1982).

Further studies showed that tumours of the duodenum had the highest incidence in strains of mice with low catalase activity, and the lesions in the intestine caused by hydrogen peroxide (plaques and nodules) were reversible after the end of treatment. Hydrogen peroxide was administered in the drinking water (fresh preparations daily) in concentrations of 0 %, 0.1 % or 0.4 % to groups of 2 to 29 mice of the strains C57BL/6N, DBA/2N and BALB/cAnN (with low to high catalase activity) for up to 740 days. Examination of the C57BL/6N mice revealed damage to the forestomach (erosion and nodules) in 67 % of the animals after 120 days and damage to the duodenum (plaques and nodules) in 80 % of the animals after 60 days. When administration of hydrogen peroxide was discontinued for 10 to 30 days after exposure for 120 to 160 days, these findings were reversible. After 700 days, 34 % of the animals were found to have nodules in the forestomach and 100 % to have nodules in the duodenum. The nodules were not further differentiated. Comparison of the three mice strains (5–16 animals) after 210 days showed that the incidence of erosion and nodules in the forestomach or the plaques and nodules in the duodenum were not significantly different after administration of 0.4 % hydrogen peroxide, although the number of lesions per mouse were different (C57BL/6N > DBA/2N > BALB/cAnN), which therefore correlated inversely with the catalase activity. Data are lacking from control animals not exposed (Ito *et al.* 1982).

Table 3. Studies on the carcinogenicity of hydrogen peroxide in animal experiments

Author:	Ito *et al.* 1981a, 1981b, 1982
Species:	C57BL/6J mouse, about 50 ♂, 50 ♀ per group (no details of catalase activity)
Administration route:	drinking water
Concentration:	0, 0.1, 0.4 %; dose: about 0, 200, 800 mg/kg body weight and day [a]
Duration:	100 weeks
Toxicity:	at and above 0.1 %: body weights decreased (♀) at 0.4 %: body weights significantly decreased (–10 %, ♀)

		Exposure concentration (%)		
		0	0.1	0.4
Survivors:		54 %	61 %	63 %
Tumours:				
Glandular stomach:				
Erosion and ulcers	♂	2/48 (4.2 %)	13/51 (25 %)$^+$	19/50 (38 %)$^+$
	♀	2/50 (4.0 %)	7/50 (14 %)	29/49 (59 %)$^+$
	total	4/98 (4.1 %)	20/101 (20 %)**	42/99 (42 %)**
Hyperplasia	♂	2/48 (4.2 %)	6/51 (12 %)	3/50 (6.0 %)
	♀	5/50 (10 %)	7/50 (14 %)	7/49 (14 %)
	total	7/98 (7.1 %)	13/101 (13 %)	10/99 (10 %)
Duodenum:				
Hyperplasia	♂	2/48 (4.2 %)	16/51 (31 %)$^+$	30/50 (60 %)$^+$
	♀	7/50 (14 %)	24/50 (48 %)$^+$	31/49 (63 %)$^+$
	total	9/98 (9.2 %)	40/101 (40 %)**	61/99 (62 %)**
Adenomas	♂	0/48 (0 %)	2/51 (3.9 %)	2/50 (4.0 %)
	♀	1/50 (2.0 %)	4/50 (8.0 %)	0/49 (0 %)
	total	1/98 (1.0 %)	6/101 (5.9 %)	2/99 (2.0 %)
Carcinomas	♂	0/48 (0 %)	1/51 (1.9 %)	1/50 (2.0 %)
	♀	0/50 (0 %)	0/50 (0 %)	4/49 (8.2 %)
	total	0/98 (0 %)	1/101 (0.9 %)	5/99 (5.1 %)*

Table 3. continued

Author:	Ito *et al.* 1984				
Species:	Mouse, 9–12 ♂, 9–12 ♀ C3H/HeN (high catalase activity), $B6C3F_1$ (medium catalase activity), C57BL/6N (low catalase activity), C3H/Csb (hypocatalasaemia)				
Administration route:	drinking water				
Concentration:	0.4 %, no unexposed control group; dose: about 800 mg/kg body weight and day [a]				
Duration:	7 months in the C57BL/6N mice, 6 months in the other strains				
Toxicity:	not further specified				
Tumours:		Strain			
		C3H/HeN	$B6C3F_1$	C57BL/6N	C3H/Csb
Catalase activity in the duodenum (10^{-4} k/mg protein)[b]		5.3	1.7	0.7	0.4
Duodenum:					
Tumours:	♂, ♀	2/18 (11.1 %)	7/22 (31.8 %)	21/21 (100 %)	22/24 (91.7 %)
Tumours/mouse:	♂, ♀	0.11	0.36	3.91	2.63
Author:	Takayama 1980				
Species:	F344 rat, 50 ♂, 50 ♀ per group				
Administration route:	drinking water				
Concentration:	0.3 %, 0.6 %; dose: about 300, 600 mg/kg body weight and day [a]				
Duration:	78 weeks or for 2 years				
Toxicity:	body weight gains decreased				
Tumours:	no treatment-related tumours				
Author:	Takahashi *et al.* 1986				
Species:	Wistar rat, 10 ♂ per group				
Administration route:	drinking water				
Concentration:	1 %; dose: about 1000 mg/kg body weight and day [a]				
Duration:	32 weeks				
Toxicity:	not further specified				
Tumours:		Exposure concentration (%)			
		0	1		
Survivors:	♂	10/10 (100 %)	10/10 (100 %)		
Forestomach:					
Papillomas	♂	0/10 (0 %)	5/10 (50 %)*		

[+] $p < 0.01$ (subsequently calculated with Fisher's exact test, one-sided)
* $p < 0.05$, ** $p < 0.005$ (Ito *et al.* 1982)
[a] calculated for normal water intake (rat: 100 ml/kg body weight and day; mouse: 200 ml/kg body weight and day). Reduced water intake was described in most studies with repeated administration, so that the real doses were presumably lower.
[b] k = reaction constant

In another study with the four mice strains C3H/HeN (high catalase activity), $B6C3F_1$ (medium catalase activity), C57BL/6N (low catalase activity) and C3H/Csb (hypocatalasaemia), groups of 9 to 12 male and female mice were given 0.4 % hydrogen peroxide in the drinking water for up to 7 months. There was a negative correlation between the incidence of duodenal tumours and the number of tumours per mouse and the catalase activity of these strains (see Table 3). There were no data for controls (Ito *et al.* 1984).

Rat

Administration of 0.3 % or 0.6 % hydrogen peroxide in the drinking water to groups of 50 F344 rats per sex (about 217 or 433 mg/kg body weight and day for male animals and 339 or 677 mg/kg body weight for female animals) for 78 weeks or for 2 years induced no treatment-related tumours. The drinking water solutions were freshly prepared four times per week. Body weight gains were reduced in the treated animals. The tumour incidence in the treated animals was not increased compared with that in the controls (Takayama 1980). The study is valid and well performed, though not completely documented.

6 Manifesto (MAK value, classification)

Hydrogen peroxide was found to be genotoxic *in vitro* in a series of test systems; bacterial strains without catalase were particularly sensitive, compared with mammalian cells with high catalase activity, which were resistant. *In vivo*, however, DNA repair synthesis (UDS) was not observed in liver cells of rats, neither were chromosomal aberrations in bone marrow cells of rats. Nor were micronuclei observed in polychromatic erythrocytes of catalase-deficient mice or in the bone marrow cells of Swiss mice. *In vivo*, therefore, sufficient catalase or other detoxifying enzymes seem to be present for protection against the systemic genotoxic effects of hydrogen peroxide.

Local carcinogenic effects of hydrogen peroxide were also observed in animal experiments only when the capacities of detoxifying enzymes were overloaded. In mice, concentrations of 0.1 % and above in the drinking water produced a dose-dependent increase in duodenal carcinomas; the highest incidence occurred in the strains having the lowest catalase activity. In rats, which have considerably higher duodenal catalase activity than mice, no duodenal tumours occurred in a 2-year study with hydrogen peroxide concentrations of up to 0.6 %. In a short-term study with rats, papillomas of the forestomach did not occur before doses of 1 % hydrogen peroxide in the drinking water. Tumour-promoting effects were found in rats (papillomas of the forestomach after 1 % hydrogen peroxide, intestinal tumours after 1.5 % hydrogen peroxide), hamsters (buccal pouch carcinomas after 30 % hydrogen peroxide) and mice (skin papillomas after 6 %).

The mechanisms for the carcinogenic effects of hydrogen peroxide are seen as being, on the one hand, the local irritant effects causing tissue damage and, on the other hand, the genotoxic effects, both of which occur when the detoxifying enzymes, particularly catalase, are overloaded. If the endogenous detoxifying capacities are not overloaded, the genotoxic effects play no or only a minor role. Under these conditions, the cancer risk for humans is negligible. Hydrogen peroxide is consequently classified in Carcinogen Category 4. Since the tissue concentrations above which the detoxification of hydrogen peroxide in the respiratory tract becomes overloaded are not known, the sensory irritation of the eyes and nose is used to derive the MAK value. As this endpoint is based, like the carcinogenic effects, on the release of reactive oxygen species, the avoidance of sensory irritation also guarantees that the detoxifying enzymes are not overloaded and consequently that neither tissue-damaging nor genotoxic effects occur. The NOAEC from a 28-day inhalation study with rats was 2 ml/m^3. After concentrations of 10 ml/m^3 and above, histological changes in the anterior region of the nasal sinus and in the respiratory epithelium were observed (CEFIC 2002). Eye and throat irritation, a blocked nose, coughing and asthma symptoms are reported by workers who were exposed to hydrogen peroxide concentrations of 1.2 to 2.4 ml/m^3 with peak values of 8 ml/m^3. From these studies it is not evident whether these effects were produced by the peak exposure. No effects were described after exposure to concentrations of up to 0.5 ml/m^3 for 8 hours (Riihimäki *et al.* 2002). Lung function tests did not yield adverse findings after concentrations of up to 0.56 ml/m^3 and 0.67 ml/m^3 (EU 2003; Mastrangelo 2005). The MAK value has thus been established at 0.5 ml/m^3 on the basis of this NOAEC. Hydrogen peroxide is classified in Peak Limitation Category I on account of its local irritant effects. As the MAK value is based on the avoidance of both local irritation in the target tissues of the respiratory tract and carcinogenic effects, and the difference between the MAK value and the LOAEC is very small, an excursion factor of 1 has been set.

The permeability coefficient of hydrogen peroxide corresponds to that of water. There are no data available for its dermal penetration rate. It cannot be assumed from the available studies that hydrogen peroxide is absorbed in quantities that cause a systemic effect. Designation with an "H" is thus not necessary.

In view of the fact that there are only two published cases of sensitization in humans, compared with the widespread use of hydrogen peroxide, it can be concluded that there is no relevant sensitizing potential. In animal studies, the substance was not found to be sensitizing. Designation with "Sh" or "Sa" is thus not necessary.

Because of the tumour-promoting potential of hydrogen peroxide, the avoidance of long-term dermal exposure to hydrogen peroxide solutions at concentrations that cause irritant effects is recommended (classified by the EU as irritating at a concentration of 5 % and above).

As regards prenatal toxic effects, only one inadequate study in rats is available, which cannot be used for evaluation. As a result of its effective metabolism, hydrogen peroxide is rapidly degraded in the organism so that placental transfer with damage to the foetus in the uterus is unlikely if the MAK value is observed. For this reason, the substance is classified in Pregnancy Risk Group C.

As the genotoxic metabolites react very rapidly on site, exposure of germ cells is not expected, so that classification in a category for germ cell mutagens is not necessary.

7 References

Aguirre A, Zabala R, Sanz de Galdeano C, Landa N, Diaz-Pérez L (1994) Positive patch tests to hydrogen peroxide in 2 cases. *Contact Dermatitis 30*: 113

Bock FG, Myers HK, Fox HW (1975) Cocarcinogenic activity of peroxy compounds. *J Natl Cancer Inst 55*: 1359–1361

Bryan CL, Jenkinson SG (1987) Species variation in lung antioxidant enzyme activities. *J Appl Physiol 63*: 597–602

CEFIC (European Chemical Industry Council) (2002) *Hydrogen peroxide: 28-day inhalation study*. CTL study No. MR0211, Degussa Report No. 2002-0082-DKT, unpublished

Degussa-Hüls (1999) *Draft report of hydrogen peroxide production workers health survey*. Summary of workers health data of 4 production sites for hydrogen peroxide of Degussa-Hüls AG, Hanau, unpublished report

DuPont (1953) *Primary irritancy and skin sensitization tests, hydrogen peroxide 3 %*, Memphis medical research project, DuPont, Wilmington, NC, USA, unpublished report

ECETOC (European Centre for Ecotoxicology and Toxicology of Chemicals) (1996) *Hydrogen peroxide*, Special Report No. 10, Aug 1996, ECETOC, Brussels

EU (European Union) (2003) *Risk assessment report, hydrogen peroxide*, Vol 38, EU, Brussels

Fenech M, Crott J, Turner J, Brown S (1999) Necrosis, apoptosis, cytostasis and DNA damage in human lymphocytes measured simultaneously within the cytokinesis-block micronucleus assay: description of the method and results for hydrogen peroxide. *Mutagenesis 14*: 605–612

FIN Competent Authorities (1998) *HEDSET dataset, hydrogen peroxide*, 18th Sept. 1998

FMC Corp (1979) *Acute dermal toxicity of 70 % hydrogen peroxide in rabbits*. Study No. ICG/T79027-02, Princeton, NJ, USA, unpublished report

FMC Corp (1990) *Occupational exposure data hydrogen peroxide*. 1st June 1990, FMC, Princeton, NJ, USA, unpublished report

Hirota N, Yokoyama T (1981) Enhancing effect of hydrogen peroxide upon duodenal and upper jejunal carcinogenesis in rats. *Gann 72*: 811–812

Hrubetz MC, Conn LW, Gittes HR, MacNamee JK (1951) *The cause of increasing intravenous toxicity of 90 % hydrogen peroxide with progressive dilutions*. Chemical Corps Medical Laboratories, Army Chemical Center, Edgewood, MA, USA, Research Report No. 75, unpublished

IARC (International Agency for Research on Cancer) (1999) *Monographs on the evaluation of the carcinogenic risk of chemicals to humans*, Volume 71, IARC, Lyon, 671–689

Ito A, Naito M, Watanabe H (1981a) Implication of chemical carcinogenesis in the experimental animal. Tumorigenic effect of hydrogen peroxide in mice. *Annu Rep Hiroshima Univ Inst Nucl Med Biol 22*: 147–158

Ito A, Watanabe H, Naito M, Naito Y, Kawashima K (1981b) Induction of duodenal tumors in mice by oral administration of hydrogen peroxide. *Gann 73*: 174–175

Ito A, Naito M, Naito Y, Watanabe H (1982) Induction and characterization of gastro-duodenal lesions in mice given continuous oral administration of hydrogen peroxide. *Gann 73*: 315–322

Ito A, Watanabe H, Naito M, Naito Y, Kawashima K (1984) Correlation between induction of duodenal tumor by hydrogen peroxide and catalase activity in mice. *Gann 75*: 17–21

IUPAC (International Union of Pure and Applied Chemistry) (2004) http://www.iupac.org/goldbook/HT06787.pdf

Kaelin RM, Kapanci Y, Tschopp JM (1988) Diffuse interstitial lung disease associated with hydrogen peroxide inhalation in a dairy worker. *Am Rev Respir Dis 137*: 1233–1235

Kanerva L, Jolanki V, Riihimäki V, Kalimo K (1998) Patch test reactions and occupational dermatoses caused by hydrogen peroxide. *Contact Dermatitis 39*: 146

Klein Szanto AJP, Slaga TJ (1982) Effects of peroxides on rodent skin: epidermal hyperplasia and tumor promotion. *J Invest Dermatol 79*: 30–34

Kondrashov VA (1977) On comparative toxicity of hydrogen peroxide vapours with their inhalation and dermal modes of action. *Gig Tr Prof Zabol 10*: 22–25 (Russian)

Kurokawa Y, Takamura N, Matsushima Y, Imazawa T, Hayashi Y (1984) Studies on the promoting and complete carcinogenic activities of some oxidizing chemicals in skin carcinogenesis. *Cancer Lett 24*: 299–304

Leino T, Estlander T, Kanerva L (1998a) Occupational allergic dermatoses in hairdressers. *Contact Dermatitis 38*: 166–167

Leino T, Tammilehto L, Hytönen M, Sala E, Paakulainen H, Kanerva L (1998b) Occupational skin and respiratory diseases among hairdressers. *Scand J Work Environ Health 24*: 398–406

McNally OD (1990) Clinical aspects of topical application of dilute hydrogen peroxide solutions. *Contact Lens Assoc Ophthalmol J 16*: 46–51

Mastrangelo G, Zanibellato R, Fedeli U, Fadda E, Lange JH (2005) Exposure to hydrogen peroxide at TLV level does not induce lung function changes: a longitudinal study. *Int J Environ Health Res 15*: 313–317

Riihimäki V, Toppila A, Piirilä P, Kuosma E, Pfäffli P, Tuomela P (2002) Respiratory health in aseptic packaging with hydrogen peroxide: a report of two cases. *J Occup Health 44*: 433–438

Riihimäki V (2004) Mitteilung an das Kommissionssekretariat vom 5. November 2004 (Communication to the Secretariat of the Commission dated November 5, 2004) (German)

Schöneich J (1967) The induction of chromosomal aberrations by hydrogen peroxide in strains of ascites tumors in mice. *Mutat Res 4*: 384–388

Shamberger RJ (1972) Increase of peroxidation in carcinogenesis. *J Natl Cancer Inst 48*: 1491–1497

SPI (Society of the Plastics Industry) (1997) *Determination of the potential for organic peroxides to induce sustained skin hyperplasia and DNA damage*. Final report SPI/SPRD study #96-1. The University of Texas MD Anderson Cancer Center, Smithville, TX, USA, unpublished report

SRC (Syracuse Research Corporation) (2004) Physprop Databank, http:/www.syrres.com/phys/demo.htm

Svirbely JL, Dobrogorski OJ, Stokinger HM (1961) Enhanced toxicity of ozone–hydrogen peroxide mixtures. *Am Ind Hyg Assoc J 20*: 21–26

Takahashi M, Hasegawa R, Furukawa F, Toyoda K, Sato H, Hayashi Y (1986) Effects of ethanol, potassium, metabisulphite, formaldehyde and hydrogen peroxide on gastric carcinogenesis in rats after initiation with N-methyl-N′-nitro-N-nitrosoguanidine. *Jpn J Cancer Res 77*: 118–124

Takayama S (1980) *Report on a carcinogenicity study*. Research Group, Ministry of Health and Welfare, Japan. Cancer Institute of Japan, Foundation for Cancer Research, Tokyo

Weitzman SA, Weitberg AB, Stossel TP, Schwartz J, Shklar G (1986) Effects of hydrogen peroxide on oral carcinogenesis in hamsters. *J Periodontol 57*: 685–688

Williams MD, Leigh JS, Chance B (1982) Hydrogen peroxide in human breath and its probable role in spontaneous breath luminescence. *Ann N Y Acad Sci 386*: 478–483

completed 29.03.2006

Methacrylic acid

MAK value (2005)	**5 ml/m³ (ppm) ≙ 18 mg/m³**
Peak limitation (2005)	**Category I, excursion factor 2**
Absorption through the skin	–
Sensitization	–
Carcinogenicity	–
Prenatal toxicity (2005)	**Pregnancy Risk Group C**
Germ cell mutagenicity	–
BAT value	–
Synonyms	2-methylacrylic acid 2-methylpropenoic acid 2-methylene propionic acid
Chemical name	2-methyl-2-propenoic acid
CAS number	79-41-4
Structural formula	$H_2C{=}C(CH_3){-}COOH$
Molecular formula	$C_4H_6O_2$
Molecular weight	86.09
Melting point	14–16°C (EU 2002)
Boiling point at 1013 hPa	159–163°C (EU 2002)
Density at 20°C	1.015 g/cm³ (EU 2002)
Vapour pressure at 20°C	0.9 hPa (EU 2002)
$\log K_{ow}$*	0.93, 0.99 (EU 2002)
pK_a#	4.66 (EU 2002)
1 ml/m³ (ppm) ≙ 3.572 mg/m³	**1 mg/m³ ≙ 0.280 ml/m³ (ppm)**

---

* *n*-octanol/water partition coefficient
# acid dissociation constant

The MAK-Collection Part I: MAK Value Documentations, Vol. 26. DFG, Deutsche Forschungsgemeinschaft

ISBN: 978-3-527-32306-7

The documentation is based on the risk assessment for the EU's existing substances programme (EU 2002). Data for methyl methacrylate, which releases methacrylic acid during metabolism, have also been used for the assessment of some end points.

1 Toxic Effects and Mode of Action

Methacrylic acid has mainly local effects. It is corrosive to the skin and eyes. In a 90-day study with rats and mice, the nasal epithelial cells were the target tissues, with exudate and epithelial hyperplasia occurring in one of the two strains of rat even at the lowest concentration tested of 20 ml/m^3. Systemic effects, in the form of reduced relative and absolute liver weights, were observed only in male mice at concentrations of 300 ml/m^3, but there were no histopathological effects on this organ. There is no clear evidence of contact sensitization or any evidence of sensitization of the respiratory tract induced by methacrylic acid. A developmental toxicity study with rats provided no evidence of such an effect up to the highest concentration of 300 ml/m^3. No fertility studies are available. Methacrylic acid yielded negative results in a *Salmonella* mutagenicity test. Other studies of genotoxicity or carcinogenicity are not available. Comparison with the data for methyl methacrylate shows that in this respect methacrylic acid is not expected to present a hazard.

2 Mechanism of Action

The local effects can be explained by the acidic character of methacrylic acid. As shown for acrylic acid, the double bond can presumably also react with nucleophilic compounds (e.g. glutathione) (Esterbauer *et al.* 1975). This reaction is probably even slower than with acrylates, since methacrylates are more inert (Osman *et al.* 1988).

3 Toxicokinetics and Metabolism

3.1 Absorption, distribution, elimination

After methacrylic acid vapour was drawn through the isolated upper respiratory tracts of rats (unidirectional flow, i.e. only simulation of inhalation) at a concentration of 113 ml/m^3 for 30 to 60 minutes, 95 % of the substance was found to have been deposited

(Morris and Frederick 1995). It may be concluded from this result that methacrylic acid acts on the upper respiratory tract and hardly reaches the lungs. This assumption is supported by the results of the 90-day study (see Section 5.2.1).

By means of a hybrid model, which had initially been developed for acrylic acid (detailed discussion of the model in the 2006 MAK documentation "Acrylic acid", this volume), deposition of 78 % was predicted for the human nasal tract after exposure to methacrylic acid concentrations of 10 to 80 ml/m^3, and a two to three times less pronounced effect on the human olfactory epithelium compared with that on the rat epithelium. Unidirectional flow and a flow rate of 1 m^3/hour were assumed (Frederick *et al.* 1998). This model and the prediction were criticized because of several weak points: the parameters for rats were validated using only one concentration (about 130 ml/m^3), whereas the prediction was made for 0–75 ml/m^3. There is no model for mice (and only they developed lesions of the olfactory epithelium in the 90-day study). There are no experimental data for cyclic flow, nor is there a sensitivity analysis of the parameters. Clearance mechanisms and metabolism, which are slow compared with absorption and are thus only of minor relevance, were not taken into account. There are no human data which might substantiate the prediction. The prediction is a point estimate without confidence intervals. Toxicodynamic differences were not considered (EU 2002). Modelling is still too unreliable at present to substantiate that a lower dose of methacrylic acid deposits on the human olfactory epithelium. In analogy to acrylic acid, the effects on humans and rats are assumed to be of similar degree (see "Acrylic acid" in this volume).

3.2 Metabolism

There are no specific studies of the metabolism of methacrylic acid. It is assumed that methacrylic acid reacts with coenzyme A and is then converted into (*S*)-3-hydroxy-isobutyryl-CoA. It then enters the citric acid cycle (EU 2002).

Structure-effect comparisons with methyl methacrylate, which has been investigated in more detail, to predict local effects can be drawn only to a limited extent since the ester is more lipophilic and methacrylic acid is first released intracellularly by cleavage of the ester. Structural analogy with methyl methacrylate can, however, be used for predicting systemic effects, since the ester is completely cleaved and thus no major difference in the availability of the amount of methacrylic acid released from methyl methacrylate compared with the inhalation of methacrylic acid itself is to be expected.

4 Effects in Humans

4.1 Single exposures

There are no data available for effects in humans after single exposures.

4.2 Repeated exposures

No valid data are available for repeated exposure to methacrylic acid. A study of Russian workers published as an abstract (EU 2002) is not suitable for the assessment of methacrylic acid because the persons may have been exposed to a mixture of substances, the findings are not very plausible and the documentation is inadequate.

4.3 Local effects on skin and mucous membranes

There are no data available for methacrylic acid.

4.4 Allergenic effects

A patient with sensitization presumably caused by an implant made of a methyl methacrylate/methyl acrylate copolymer, who reacted to the polymer, some methacrylates, 0.1 % acrylonitrile and 0.1 % acrylic acid, also produced a 3+ reaction to 0.1 % methacrylic acid in petrolatum (Romaguera *et al.* 1985).

According to an inadequately documented study, 1 of 45 patients with proven sensitization to 1 of a total of 64 tested substances possibly occurring in shoe materials also reacted to methacrylic acid (no other details) (Grimalt and Romaguera 1975). Another inadequately documented study reports about a 46-year-old worker in a car factory with sensitization caused by an adhesive containing acrylate, who reacted to 1 % triethylene glycol dimethacrylate and 0.5 % methacrylic acid in petrolatum in a patch test (Daecke *et al.* 1994).

None of the six workers who had been sensitized by contact with anaerobic sealants and had, for example, reacted to 2 % hydroxyethyl methacrylate in a patch test reacted to 0.1 % methacrylic acid in petrolatum (Condé-Salazar *et al.* 1988). A 47-year-old cosmetician with sensitization caused by photobonding materials for artificial fingernails reacted to numerous methacrylates and some (di-)acrylates, but not to 0.1 % to 1 % methacrylic acid (Kanerva *et al.* 1996). Nor did a worker with sensitization to hydroxypropyl acrylate (Lovell *et al.* 1985) or three other workers with sensitization to sealants containing (meth)acrylate (Dempsey 1982) produce reactions to 0.1 % methacrylic acid in methyl ethyl ketone or 1 % methacrylic acid in petrolatum in a patch test.

4.5 Reproductive toxicity

There are no data available for the reproductive toxicity of methacrylic acid. Three studies reported about adverse effects on reproduction in male and female workers from Russia who were exposed to methyl methacrylate. These studies cannot be used for assessing methyl methacrylate or methacrylic acid because they were inadequately documented and exposure was insufficiently characterized (EU 2002).

4.6 Genotoxicity

There are no data available for the genotoxicity of methacrylic acid. A study of workers with high levels of exposure to methyl methacrylate revealed increased sister chromatid exchange (see "Methyl methacrylate" in this volume). In view of the negative results obtained in carcinogenicity studies in humans and animals, the relevance of this finding is unclear.

4.7 Carcinogenicity

There are no data available for the carcinogenicity of methacrylic acid. Epidemiological studies with methyl methacrylate do not provide evidence of carcinogenicity. Therefore no carcinogenic potential is expected for the acid, either (EU 2002).

5 Animal Experiments and *in vitro* Studies

5.1 Acute toxicity

5.1.1 Inhalation

A study carried out according to OECD Test Guideline 403 revealed a 4-hour LC_{50} of 7100 mg/m^3 (1988 ml/m^3). Loss of weight and gross-pathologically evident irritation of the respiratory tract were described (EU 2002).

The RD_{50} in mice (10-minute exposure) was 22000 ml/m^3 (EU 2002).

5.1.2 Ingestion

Most of the oral LD_{50} values for rats, mice and rabbits were below 2000 mg/kg body weight and slightly above 2000 mg/kg body weight after administration of diluted solutions (EU 2002).

5.1.3 Dermal absorption

A range-finding test was carried out with groups of 2 rabbits given methacrylic acid doses of 500, 1000 or 2000 mg/kg body weight as 50 % solutions applied to the skin (time not specified). The 500 mg/kg body weight doses were not lethal, whereas the other doses were lethal for all rabbits either overnight or within 2 hours. Severe skin irritation was specified as a clinical finding after 500 mg/kg body weight (EU 2002).

5.2 Subacute, subchronic and chronic toxicity

5.2.1 Inhalation

In a 90-day study, groups of 10 $B6C3F_1$ mice, SD rats and F344 rats per strain, sex and concentration were exposed to methacrylic acid concentrations of 0, 20, 100 or 300 ml/m^3 in whole body exposure chambers for 6 hours a day, on 5 days a week. Another 10 animals per strain, sex and concentration were exposed to the same concentration for 4 days and examined on day 5. The study complies with present-day requirements. Reduced body weight gains in male F344 rats and in both sexes in mice, and reduced feed consumption in male F344 rats (about 10 % in each case) were the findings obtained in the 300 ml/m^3 groups. The reduced body weight gains in rats are thus directly related to reduced feed consumption and cannot be regarded as a direct effect of systemic toxicity. Reduced leukocyte counts and increased alkaline phosphatase activity were determined in female mice. The blood urea nitrogen level was increased in male F344 rats. Since these effects were not consistent, a relationship with the exposure is not plausible. The absolute liver weights were reduced in the males of both strains of rat and in both sexes in mice. The relative liver weights were reduced in mice only, but significantly so only in males and without any histopathological changes. The reduction in absolute liver weights in rats therefore results from the loss of body weight and is not an effect in its own right. For systemic effects, the no observed adverse effect concentration (NOAEC) was thus 300 ml/m^3 in rats and 100 ml/m^3 in mice.

The histopathologically evident findings in the nose detected after 4 and 90 days of exposure are listed in the following tables (Tables 1 and 2). Histopathological signs of irritation were detected in the nasal region after only 4 days. In the 300 ml/m^3 groups, exudate, goblet cell hyperplasia, focal ulceration, hyperkeratosis and acute necrosis were found in the F344 rats after the 4-day exposure, while $B6C3F_1$ mice were found to have acute rhinitis, acute necrosis, exudate and ulceration. Concentrations of 20 ml/m^3 and above additionally caused rhinitis in F344 rats, while SD rats also had epithelial vesicles, acute rhinitis and hyperkeratosis.

Table 1. Findings in the nasal region of rats and mice in the 4-day study (CIIT 1984)

	Concentration [ml/m³]			
	0	20	100	300
Findings section level A:				
F344 rats				
acute rhinitis				
♂	0/10	4/10	2/10	9/10
♀	0/10	2/10	4/10	7/10
SD rats				
epithelial vesicles				
♂	1/9	2/10	4/10	1/10
♀	0/10	0/10	5/10	1/9
acute rhinitis				
♂	2/9	3/10	4/10	6/10
♀	0/10	2/10	4/10	6/9
exudate				
♂	0/9	1/10	0/10	3/10
♀	0/10	0/10	0/10	3/9
focal ulceration				
♂	0/9	0/10	0/10	1/10
♀	0/10	0/10	0/10	1/9
hyperkeratosis				
♂	0/9	1/10	2/10	2/10
♀	0/10	1/10	3/10	7/9

Table 2. Findings in the nasal region of rats and mice in the 90-day study (CIIT 1984)

	Concentration [ml/m³]			
	0	20	100	300
Findings section level A:				
F344 rats				
acute rhinitis				
♂	5/10 (1.2)*	6/9 (1.33)	4/10 (1.25)	9/9 (2.1)
♀	5/10 (1.2)	9/10 (1.44)	1/10 (2)	7/10 (1.5)
exudate				
♂	1/10	1/9	0/10	4/9
♀	0/10	0/10	0/10	4/10
SD rats				
acute rhinitis				
♂	5/10 (1)	6/10 (1.5)	10/10 (1.3)	8/10 (1)
♀	2/10 (1)	4/10 (2)	2/10 (1.5)	7/10 (1.14)
exudate				
♂	0/10	2/10	7/10	4/10
♀	0/10	2/10	2/10	4/10
epithelial hyperplasia				
♂	0/10	3/10	5/10	3/10
♀	0/10	1/10	1/10	3/10
$B6C3F_1$ mice				
acute rhinitis				
♂	0/10	0/10	0/10	4/9 (1)
♀	0/10	0/10	0/10	3/10 (1)
exudate				
♂	0/10	0/10	0/10	4/9
♀	0/10	0/10	0/10	2/10
ulceration				
♂	0/10	0/10	0/10	3/9
♀	0/10	0/10	0/10	2/10
B6C3F mice				
Findings section level B:				
degeneration of olfactory epithelium				
♂	0/10	0/10	1/10	1/10
♀	0/10	0/10	1/10	9/10
B6C3F mice				
Findings in section level C:				
degeneration of olfactory epithelium				
♂	0/10	0/9	1/10	8/10
♀	0/10	0/10	3/10	9/10

* severity (1–5; minimal to severe) in brackets

The findings were similar in the 90-day study: rhinitis, exudate and epithelial hyperplasia occurred in rats, but not in relation to the concentration. Rhinitis as an effect of methacrylic acid would have been plausible since it had already been observed after only 4 days exposure. However, since control animals were also affected in the 90-day study (and in the 4-day study with SD rats), it is difficult to interpret the data from the two rat strains. Whereas a NOAEC of 100 ml/m^3 was determined in F344 rats, exudate and epithelial hyperplasia were found in SD rats even at 20 ml/m^3, but not in relation to the concentration. It is unclear why rhinitis occurred in control SD rats even after 4-day exposure. These cases of rhinitis may be explained by a virus infection and the overall result may be due to the infection and the direct effects of the substance. In mice, rhinitis, exudate and ulceration were first observed at a concentration of 300 ml/m^3. After 90 days, degeneration of the mouse olfactory epithelium was found in section levels B and C in the middle and high concentration groups. Deposits of orange to pink-coloured material were seen in the cytoplasm of ciliated cells (sustentacular cells). These led to a loss of ciliated cells in severe cases. On the basis of the findings obtained in the olfactory epithelium, the NOAEC for local effects was 20 ml/m^3 in mice. In general, no reliable NOAEC can be derived from the study.

A number of other studies with inhalation exposure are available, but the documentation and the methods do not comply with present-day requirements (EU 2002). These studies are therefore not described here.

A long-term study with methyl methacrylate in rats and mice yielded a NOAEC of 25 ml/m^3 (EU 2002).

5.2.2 Ingestion

There are no data available for the effects after ingestion of methacrylic acid.

5.2.3 Dermal absorption

The application of 0.56 M methacrylic acid in water to 8 male rats 3 times a week did not result in any signs of skin irritation after 3 weeks. The same concentration in acetone was, however, slightly to moderately irritating. Concentrations of 1.12 M and 2.24 M in acetone induced more pronounced skin irritation. Systemic effects were not investigated (EU 2002).

5.3 Local effects on skin and mucous membranes

5.3.1 Skin

Undiluted methacrylic acid is corrosive to the rabbit skin after an exposure period of only 3 minutes (EU 2002).

5.3.2 Eyes

Undiluted methacrylic acid is corrosive to rabbit eyes (EU 2002).

5.4 Allergenic effects

Methacrylic acid was not found to produce contact sensitization in a modified Buehler test (induction 15–20 % and challenge 10 %). Since the concentration of 20 % applied in the first induction caused marked irritation, the second and third inductions were carried out with 15 %. This concentration was barely irritating (EU 2002).

In a maximization test, groups of five Dunkin-Hartley guinea pigs were treated with 0.001 %, 0.003 %, 0.01 %, 0.03 % or 0.1 % methacrylic acid in water for intradermal induction and with 0.3 % or 30 % methacrylic acid in water for epicutaneous induction. In the challenge treatment with 3 % methacrylic acid in water, only one animal from the groups which had been treated with 0.003 %/30 % and 0.1 %/0.3 % methacrylic acid reacted after 48 hours. One animal from the first group mentioned also produced a reaction after 72 hours. Two animals of the group pretreated with 0.001 %/0.3 % reacted only after 24 hours, as did two of the five control animals. Only one animal of the group pretreated with 0.03 %/30 % methacrylic acid reacted in the repeated challenge treatment (Boman *et al.* 1996).

A negative result was obtained also in a modified single injection adjuvant test according to Polak (footpad test). In this test, groups of male and female Hartley guinea pigs were given 0.1 ml of a 1:1 mixture of the substance (2 mg/ml in isotonic saline/ethanol, 4:1) and Freund's complete adjuvant by injection into the paws and 0.1 ml of this preparation by injection into the necks for induction. A total amount of about 11.5 µmol (about 1 mg) methacrylic acid was administered to the animals. On day 7, the open epicutaneous challenge treatment was carried out with 0.02 ml of a preparation of methacrylic acid in acetone/olive oil (4:1) (challenge concentration not specified; either 5 % or the maximum non-irritant concentration) (Parker and Turk 1983).

5.5 Reproductive toxicity

5.5.1 Fertility

No fertility studies with methacrylic acid are available. In the 90-day study, no evidence was found of impairment of the sex organs of the treated animals (EU 2002).

5.5.2 Developmental toxicity

A study with SD rats with methacrylic acid concentrations of 50, 100, 200 or 300 ml/m^3 (days 6–20, 6 hours a day, as in OECD Test Guideline 414) revealed reduced feed consumption and reduced body weight gains in the dams during exposure at the highest concentration. No clinical findings were reported. The examination of the foetuses for skeletal or visceral variations did not reveal any abnormalities (Saillenfait *et al.* 1999).

Numerous investigations with methyl methacrylate, including two carried out according to OECD Test Guideline 414 with concentrations up to 400 ml/m^3 in mice or 2000 ml/m^3 in rats, provided no evidence of developmental toxicity (see "Methyl methacrylate", this volume).

5.6 Genotoxicity

A *Salmonella* mutagenicity test with strains TA98, TA100, TA1535 and TA1537 with methacrylic acid concentrations up to 4 mg/plate yielded negative results. Higher concentrations were cytotoxic (EU 2002). Since there are no other studies with methacrylic acid, investigations with the structurally related ester methyl methacrylate were taken into consideration. These show that the ester has a clastogenic potential *in vitro* only at cytotoxic doses and that it is not genotoxic *in vivo* (dominant lethal test and micronucleus test) (see "Methyl methacrylate", this volume; EU 2002). Therefore, a clastogenic potential is expected for methacrylic acid *in vitro* only at cytotoxic doses. No genotoxic potential is assumed *in vivo* (EU 2002).

5.7 Carcinogenicity

There are no studies available of the carcinogenicity of methacrylic acid. Valid inhalation studies with methyl methacrylate in rats and hamsters provided evidence of local irritation, but not of a carcinogenic potential (see "Methyl methacrylate", this volume). Therefore carcinogenic potential is also unlikely for methacrylic acid (EU 2002).

6 Manifesto (MAK value, classification)

There are not sufficient data from humans for the establishment of a MAK value. Local toxicity is the main effect of methacrylic acid. In a 90-day study, histopathological evidence of irritation (exudate and hyperplasia) of the nasal epithelial cells was found in one of two rat strains even at the lowest concentration tested of 20 ml/m^3. For the other strain, 100 ml/m^3 was the NOAEC. It is not clear why rhinitis was also observed in the control animals of both strains of rat; the findings of exudate and hyperplasia are thus unreliable. Degeneration of the olfactory epithelium was found only in mice; here, the NOAEC was 20 ml/m^3.

A MAK value of 5 ml/m^3 has been established for methacrylic acid. This value reflects the fact that the lowest concentration obtained in one rat strain cannot be regarded as a definite NOAEC. Since the effects of the substance are probably mainly the direct result of its acidity and not a consequence of metabolism, specific differences between the species are not expected. However, a generally lower exposure of the olfactory epithelium of humans than of rodents, as in the case of esters, which release an acid after metabolism (e.g. methyl methacrylate), cannot be assumed at present. Comparison of the results obtained from the 4-day and 90-day studies is not expected to show a shift of the NOAEC to lower values after long-term exposure.

Methacrylic acid is classified in Peak Limitation Category I with an excursion factor of 2 since definite effects on the nasal epithelium were not observed in mice until concentrations of 100 ml/m^3. In analogy to methyl methacrylate, methacrylic acid is classified in Pregnancy Risk Group C. Methacrylic acid is corrosive to the skin.There are no studies which assess skin penetration. A study carried out to determine acute dermal toxicity reported mortality at 1000 mg/kg body weight, but absorption via injured skin cannot be ruled out and may have distorted the result. The systemic toxicity of methacrylic acid is relatively low, and the NOAEC obtained from a 13-week study was 100 ml/m^3. This would correspond to a dose of 3570 mg per day or 51 mg/kg body weight in humans. Assuming conditions similar to those for the structurally related acrylic acid (absorption 25 %, and a non-irritant concentration of 1 %), the no observed adverse effect level would correspond to dermal exposure to 1430 g of a 1 % solution for a person with a body weight of 70 kg. This level is so high that systemic toxicity resulting from dermal exposure to methacrylic acid is not expected and the substance is not designated with an "H".

There is no clear evidence of contact sensitization after exposure to methacrylic acid in humans. The results obtained in animal studies were negative. Methacrylic acid is therefore not designated with an "Sh". There are no studies available of sensitization to the respiratory tract caused by methacrylic acid. The substance is therefore not designated with an "Sa", either.

No data are available which would require classification of the substance in one of the categories for germ cell mutagens.

7 References

Boman A, Drying Jacobsen S, Klastrup S, Svendsen O, Wahlberg JE (1996) *Potency evaluation of contact allergens – II. Dose-response studies using the guinea pig maximization test*, Nordic Council of Ministers, TemaNord Kopenhagen, 1996: 570

CIIT (Chemical Industry Institute of Toxicology) (1984) *90-Day vapor inhalation toxicity study of methacrylic acid in B6C3F$_1$ mice, Sprague-Dawley rats and Fischer-344 rats*. FYI-OTS-0685-0415, NTIS, Springfield, VA, USA

Condé-Salazar L, Guimaraens D, Romero LV (1988) Occupational allergic contact dermatitis from anaerobic acrylic sealants. *Contact Dermatitis 18*: 129–132

Daecke C, Schaller J, Goos M (1994) Acrylates as potent allergens in occupational and domestic exposures. *Contact Dermatitis 30*: 190–191

Dempsey KJ (1982) Hypersensitivity to Sta-Lok and Loctite anaerobic sealants. *J Am Acad Dermatol 7*: 779–784

Esterbauer H, Zollner H, Scholz N (1975) Reaction of glutathione with conjugated carbonyls. *Z Naturforsch Teil C 30*: 466–473

EU (European Union) (2002) *Risk assessment report, methacrylic acid*, 1st priority list, Volume 25, Office for Official Publications of the European Communities, Luxemburg, Luxemburg

Frederick CB, Bush ML, Lomax LG, Black KA, Finch L, Kimbell JS, Morgan KT, Subramaniam RP, Morris JB, Ultman JS (1998) Application of a hybrid computational fluid dynamics and physiologically based inhalation model for interspecies dosimetry extrapolation of acidic vapors in the upper airways. *Toxicol Appl Pharmacol 152*: 211–231

Grimalt F, Romaguera C (1975) New resin allergens in shoe contact dermatitis. *Contact Dermatitis 1*: 169–174

Kanerva L, Lauerma A, Estlander T, Alanko K, Henriks-Eckerman M-L, Jolanki R (1996) Occupational allergic contact dermatitis caused by photobonded sculptured nails and a review of (meth)acrylates in nail cosmetics. *Am J Contact Dermatitis 7*: 109–115

Lovell CR, Rycroft RJG, Williams DMJ, Hamlin JW (1985) Contact dermatitis from the irritancy (immediate and delayed) and allergenicity of hydroxypropyl acrylate. *Contact Dermatitis 12*: 117–118

Morris JB, Frederick CB (1995) Upper respiratory tract uptake of acrylate ester and acidic vapors. *Inhal Toxicol 7*: 557–574

Osman R, Namboodiri K, Weinstein H, Rabinowitz JR (1988) Reactivities of acrylic and methacrylic acids in a nucleophilic addition model of their biological activity. *J Am Chem Soc 110*: 1701–1707

Parker D, Turk JL (1983) Contact sensitivity to acrylate compounds in guinea pigs. *Contact Dermatitis 9*: 55–60

Romaguera C, Grimalt F, Vilaplana J (1985) Methyl methacrylate prosthesis dermatitis. *Contact Dermatitis 12*: 172

Saillenfait AM, Bonnet P, Gallissot F, Peltier A, Fabriès JF (1999) Developmental toxicities of methacrylic acid, ethyl methacrylate, *n*-butyl methacrylate, and allyl methacrylate in rats following inhalation exposure. *Toxicol Sci 50*: 136–145

completed 09.03.2005

Methyl methacrylate

Supplement 2006

MAK value (1988)	**50 ml/m³ (ppm) ≙ 210 mg/m³**
Peak limitation (2000)	**Category I, excursion factor 2**
Absorption through the skin	–
Sensitization (1984)	**Sh**
Carcinogenicity	–
Prenatal toxicity (1985)	**Pregnancy Risk Group C**
Germ cell mutagenicity	–
BAT value	–
Synonyms	methacrylic acid methyl ester methyl α-methyl acrylate methyl 2-methylpropenoate methyl 2-methyl-2-propenoate MMA
Chemical name (CAS)	2-methyl-2-propenoic acid methyl ester
CAS number	80-62-6
1 ml/m³ (ppm) ≙ 4.2 mg/m³	**1 mg/m³ ≙ 0.238 ml/m³ (ppm)**

This supplement is based on EU Risk Assessment Report 22 (EU 2002), the IUCLID Dataset (ECB 2000), Concise International Chemical Assessment Document 4 (WHO 1998) and an IARC assessment (IARC 1994).

These reports provide detailed, comprehensive data. Therefore, only studies relevant to the evaluation are listed in the present supplement.

1 Toxic Effects and Mode of Action

Methyl methacrylate (MMA) is widely used in cement for dental and surgical prostheses. The substance is a colourless, flammable liquid with a severely irritating,

The MAK-Collection Part I: MAK Value Documentations, Vol. 26. DFG, Deutsche Forschungsgemeinschaft

ISBN: 978-3-527-32306-7

pungent and fruitlike smell and an odour threshold of 0.21 ml/m³. Since methyl methacrylate is easily polymerized or copolymerized by light, heat, ionizing radiation or chemical catalysts, it is almost always present with inhibitor additives.

Methyl methacrylate is irritating to the skin and mucous membranes. Narcotic effects, disorders of the central nervous system (CNS) and CNS depression were observed after higher doses.

In a carcinogenicity study carried out by the National Toxicology Program (NTP) with F344 rats and $B6C3F_1$ mice, no carcinogenic effects were detected after methyl methacrylate concentrations of up to 1000 ml/m^3.

On the basis of reliable case reports of the allergenic effects of methyl methacrylate on the skin and of contact sensitization in guinea pigs, methyl methacrylate is designated with an "Sh".

In vitro studies suggest that methyl methacrylate produces clastogenic effects.

In a prenatal developmental toxicity study with rats exposed by inhalation, no developmental toxicity was found up to the highest concentration of 2000 ml/m^3.

2 Mechanism of Action

The mitochondria are regarded as the main intracellular target of methyl methacrylate. If isolated rat liver mitochondria are incubated with methyl methacrylate, oxygen consumption increases. This is the result of an uncoupling of the mitochondrial respiratory chain, as seen from the expected influence on state 4 and state 3 respiration. State 4 respiration is stimulated. As has been reported for organic solvents, methyl methacrylate attacks complex I of the respiratory chain close to the rotenone binding site. This means that substrates which are oxidized in conjunction with NADH inhibit the flow of electrons and thus also ATP synthesis. Unlike classical uncouplers, methyl methacrylate stimulates the Mg^{2+}-dependent ATPase bound to the inner mitochondrial membrane. Structural changes of the inner membrane, as found with non-ionic detergents, were observed by electron microscopy. The release of enzymes indicates disintegration of the membrane (Bereznowski 1994).

3 Toxicokinetics and Metabolism

3.1 Absorption, distribution, elimination

The toxicokinetics of methyl methacrylate seem to be similar in humans and test animals (EU 2002).

3.1.1 Inhalation

After exposure of rats to methyl methacrylate concentrations of 0, 90, 437 or 2262 mg/m^3 (0, 21, 104 or 538 ml/m^3) by inhalation, 10 % to 20 % of the substance was deposited in the lower respiratory tract and metabolized there (EU 2002).

3.1.2 Ingestion

Methyl methacrylate is rapidly absorbed and distributed after ingestion. Single gavage doses of methyl methacrylate of 8 mmol/kg body weight (800 mg/kg body weight) led to maximum serum concentrations between 10 and 15 minutes after administration (Bereznowski 1995).

After administration of 5.7 and 120 mg/kg body weight of radioactively-labelled methyl methacrylate to rats, 76 % to 88 % of the radioactivity was exhaled within 10 days; 4.7 % to 7.2 % was found in the urine and 1.7 % to 3.0 % in the faeces; the remaining radioactivity was detected in the liver and adipose tissue (EU 2002).

3.1.3 Dermal absorption

Methyl methacrylate is irritating to the skin and, according to the authors, can be absorbed effectively via the skin (Rajaniemi 1986). An *in vitro* study with human (heat-separated) epidermis in a static diffusion model showed that methyl methacrylate can be absorbed through the skin and absorption is increased by occlusion. About 10 mg undiluted methyl methacrylate/cm^2 was applied for a period of 30 hours. The maximum absorption rates were measured during the first hour and were 274 µg/cm^2 and hour under occlusive conditions and 107 µg/cm^2 and hour under non-occlusive conditions. These values dropped to 152 and 3.48 µg/cm^2 and hour during the next 10 hours. Only a small amount of the dose applied penetrated the skin under non-occlusive conditions (0.56 %); this implies that methyl methacrylate evaporates from the surface of the skin (Cefic 1993).

3.1.4 Other routes of absorption

Since cement containing methyl methacrylate is used in surgery, e.g. as cement for artificial hip joints, there are numerous studies dealing, for example, with the composition of the cement (Morita *et al.* 1998), its effects on biological systems (Elmaraghy *et al.* 1998) and the determination of methyl methacrylate in blood (Gentil *et al.* 1993; Hand *et al.* 1998). These studies are not described here because they are not appropriate for deriving a MAK value.

Methyl methacrylate and methacrylic acid can be detected in the blood for a short period of time after the use of cement containing methyl methacrylate. In one study the half-life of methyl methacrylate in the blood was specified to be 47 to 55 minutes

(Svartling *et al.* 1986). In a more recent study, methyl methacrylate was no longer found in the blood after only 3 and 6 minutes. The initial and terminal half-lives were specified as being 0.3 and 3 minutes, respectively (Gentil *et al.* 1993).

3.2 Metabolism

The 1984 MAK documentation describes the metabolism of methyl methacrylate. Methyl methacrylate is hydrolyzed to methacrylic acid by carboxylesterases. Methacrylic acid is transformed via physiological metabolic pathways and enters the citric acid cycle via methylmalonyl-CoA and succinyl-CoA. Methyl methacrylate may, however, also react directly with glutathione and other sulfhydryl groups ("Methyl methacrylate", Volume 3, present series).

3.3 Species differences

In rats, hamsters and humans, methyl methacrylate is hydrolyzed to methacrylic acid by carboxylesterases, for example in the nasal mucosa. The local toxicity is attributed to the acid that is formed. Pre-treatment of rats with bis(*p*-nitrophenyl)phosphate, a carboxylesterase inhibitor, clearly reduced the severity of the nasal lesions. Investigations with rat, hamster and human nasal tissues have shown that carboxylesterases are mainly localized in goblet cells and Bowman's gland in rats, but are more generally distributed in the human olfactory epithelium. In all three species, the enzyme activity is higher in the olfactory tissue than in the respiratory tissue, by a factor of 3 in rats and humans and a factor of 12 in hamsters. The rate of metabolism of methyl methacrylate *in vitro* (V_{max}) was similar in the olfactory epithelium of rats and hamsters, but about 7 to 13 times lower in humans. In respiratory tissues, the rate of metabolism in humans was at least 6 times lower than that in the rat. The authors conclude from these findings that the level of exposure of the olfactory epithelium is lower in humans than in rats or hamsters (Mainwaring *et al.* 2001). However, it must be noted that human nasal explants were available from only 5 persons.

Mathematical models (physiologically based pharmacokinetic, PBPK) were developed to describe the dosimetry of methyl methacrylate in nasal epithelial tissue. Among other things, airflow patterns within the nose, nasal compartmentation and the distribution of the carboxylesterases in the different nasal cell types were taken into account. Assuming that the carboxylesterases are distributed in different amounts in the human nose, the models used predict 3 to 8 times lower methyl methacrylate doses in human tissue compared with those in the rat even at an increased respiration rate and the same methyl methacrylate concentrations (Andersen *et al.* 2002; EU 2002).

4 Effects in Humans

4.1 Repeated exposures

Inhalation

Irritation of the upper respiratory tract and eyes and possible CNS effects were reported in humans after exposure to methyl methacrylate concentrations of up to 250 ml/m^3. Individual publications described CNS effects and hypotension after inhalation of methyl methacrylate concentrations of about 36 to 83 ml/m^3 at the workplace for up to 11 years; functional disorders (CNS, cardiovascular system, liver and blood count) were reported after even lower occupational exposure. Since no information was provided about controls, exposure concentrations, the exposure period or smoking habits, these reports cannot be used to evaluate possible effects of methyl methacrylate ("Methyl methacrylate", Volume 3, present series).

In 91 workers exposed to person-related 8-hour mean levels of methyl methacrylate of between 4 ± 2.2 and 49 ± 26.2 ml/m^3, no significant changes in symptoms (coughing or sputum), pulmonary function, allergic reactions, blood pressure or haematology were found compared with the findings for 43 control persons. Slightly altered levels for cholesterol, albumin and total bilirubin were found particularly in the high dose group; the authors considered these findings to be clinically irrelevant. The reduced serum glucose levels that were also observed in the high exposure group were found to be reproducible; according to the authors, they must be substantiated by a properly matched control group. Changes to the skin and nervous system were also mentioned, although they were not statistically significant (NIOSH 1976; "Methyl methacrylate", Volume 3, present series). They were probably caused by dermal exposure to methyl methacrylate. The reduced serum glucose levels may be the result of the influence of shift work and circadian differences in glucose metabolism (Morgan *et al.* 2003).

A more recent study carried out between 1991 and 1993 at the Röhm factory included all 211 male workers involved in acrylic sheet production and mainly exposed to methyl methacrylate. The workers had to fill in a questionnaire and were subjected to anamnesis and rhinoscopy every six months. The workers were on average 37 years old and had spent an average of 8.8 years in acrylic sheet production. 34 % of the workers had been exposed for more than 10 years, and 16 % for more than 20 years. The current mean values for methyl methacrylate were specified to be 3 to 40 ml/m^3, earlier 8-hour means were specified to be 10 to 70 ml/m^3. The analyses were person-related. Short-term 5 to 15-minute exposure peaks were reported to be 100 to 300 ml/m^3, in one case 680 ml/m^3 (Röhm 1994). Tables giving the results of analyses of short-term exposures for 1991 and 1992 show that short-term levels were below 100 ml/m^3 in most cases. Only 3 of the 60 values reported were above 100 ml/m^3 (683, 142 and 115 ml/m^3) (Degussa 2004). The workplaces were categorized according to the median person-related exposure level (8-hour average value): Area 4: 3 to 10 ml methyl methacrylate/m^3 (7 workers); Area 3:

10 to 20 ml/m^3 (128 workers); Area 2: 20 to 30 ml/m^3 (20 workers); Area 1: 30 to 40 ml/m^3 (56 workers). Area 0 consisted of 57 newly hired workers who had presumably been exposed to methyl methacrylate several times before the examination and could therefore not be used as a control group (Degussa 2005). The workers reported no irritation, nor did rhinoscopy reveal nasal lesions. However, a few workers from Area 2 and Area 3 complained of impaired nasal breathing, dryness in the nose and burning or itching of the eyes or lacrimation in conjunction with exposure to methyl methacrylate; in these areas, there was a high rate of air exchange with low humidity caused by the exhaust air system (see Table 1). Comparable methyl methacrylate-related effects were not observed in workers of Area 1 with a higher level of exposure to methyl methacrylate caused by a lower rate of air exchange. Some workers who reported effects such as impaired nasal breathing, dryness in the nose, burning of the eyes or lacrimation attributed these mainly to peak exposures to methyl methacrylate, which were above 100 ml/m^3. There were no reports of changes to the nasal cavity epithelium or irritant effects on the upper respiratory tract that could clearly have been caused by long-term exposure to methyl methacrylate. An impaired sense of smell that could not be attributed to other causes, such as an acute cold or allergic rhinitis, was described by 2 workers of Area 2. One of these workers was later subjected to the Rhino-Test® (Muttray *et al.* 1997; see below), but his sense of smell was not impaired. The second worker could not be examined since he had apparently left the company. Workplace-related sensitization was not reported although the entire group included 27 atopics (12.8 %) with known allergies. Four workers had respiratory sensitization to enzymes, flowers or animal hair, 13 workers, 6 of them in the control group, had hay fever, 6 workers had dermal sensitization to antibiotics, animal hair, chromium and nickel and 4 workers had a food allergy. The authors point out the limitations of the study: The questionnaires were not evaluated and the data could not be evaluated statistically because there was no comparable control group (Röhm 1994).

Despite the described limitations, the above study provides evidence that no exposure-induced irritation of the eyes or respiratory tract occurs after exposure to methyl methacrylate concentrations of up to 40 ml/m^3. Irritation was observed only at higher concentrations, which, according to the authors, were more than 100 ml/m^3.

Table 1. Findings in workers from acrylic sheet production (supplemented according to Röhm 1994)

Complaints (finding No.)	8-hour average values for methyl methacrylate [ml/m^3] (current//earlier concentrations)				
		3–10//<10–30 1–6 hours/day	10–20//20–60 1–5 hours/day	20–30//40 6 hours/day	30–40//50–70 4–5 hours/day
	Area 0	Area 4	Area 3	Area 2	Area 1
No. of workers	57	7	128	20	56
Nose					
rhinitis (1)	1 (1.8 %)	1 (14 %) *not MMA-induced*	5 (3.9 %) subjective, 2 (1.6 %) of these clinically confirmed *not MMA-induced*		3 (5.4 %) subjective, 1 (1.8 %) of these clinically confirmed *not MMA-induced*
impaired nasal breathing (2)	3 (5.5 %) subjective; 2 (3.6 %) clinically confirmed		14 (11 %) subjective, 4 (3.3 %) of these clinically confirmed **7× possibly MMA-induced**	2 (10 %) subjective, 0 (–) of these clinically confirmed *not MMA-induced*	13 (23 %) subjective, 7 (12.5 %) of these clinically confirmed *not MMA-induced*
dry nose (3)		1 (14 %) subjective; 0 (–) clinical	18 (14 %) subjective, 6 (4.7 %) of these clinically confirmed **5× possibly MMA-induced**	7 (35 %) subjective, 1 (5.0 %) of these clinically confirmed **3× possibly MMA-induced** (1× confirmed), **2 of these impaired sense of smell**	4 (7.1 %) subjective, 2 (3.6 %) of these clinically confirmed *not MMA-induced*
frequent nose bleeding (6)	2 (3.6 %)		2 (1.6 %)		2 (3.6 %) *not MMA-induced*
impaired sense of smell (7)	1 (1.8 %) *not MMA-induced*		2 (1.6 %) *not MMA-induced*	2 (9.5 %) **2× possibly MMA-induced;** 1× not confirmed in secondary examination	2 (3.6 %) *not MMA-induced*
burning/itching of eyes or lacrimation (10+11)	1 (1.8 %) *not MMA-induced*		8 (6 %) **2× possibly MMA-induced**	2 (9.5 %) **1× MMA-induced (peak exposure)**	5 (9 %) *not MMA-induced*

In 1994, a Rhino-Test® with six aromas to determine hyposmia (diminished sense of smell) was carried out in 175 workers from the Röhm factory to examine whether degeneration of the nasal epithelium together with a resulting loss of the sense of smell had occurred in workers exposed to methyl methacrylate. The average exposure period was 9.6 ± 7.1 years. Up to 1988, the 8-hour mean values were between 25 and 100 ml/m^3, and from 1988 to 1994 they were 10 to 50 ml/m^3. The workers were exposed almost exclusively to methyl methacrylate; only two workers also had short-term exposures to formaldehyde up to 1990, four workers had been exposed to acrylonitrile and two other workers to formaldehyde and acrylonitrile. The proportion of smokers was higher in the exposed workers (58.3 %) than in the 88 control persons (34.1 %). In the group of exposed workers, only one worker (0.6 %), who suffered from sinusitis on the day of examination, was observed to have hyposmia. The number of workers (merely six) who from 1981 to 1994 had left the company for occupationally-related health problems was small (Muttray *et al.* 1997). This study indicates that no impairment of the sense of smell occurs, at least after exposure to methyl methacrylate concentrations of up to 50 ml/m^3.

In 8 of 40 workers who had been exposed to average methyl methacrylate concentrations of 18.5 or 21.6 ml/m^3 (range 9–21 ml/m^3 or 11.9–38.5 ml/m^3) in two factories for 5 to 10 years, or more than 10 years, an increased incidence of chronic coughing was observed compared with that in 2 of 45 control persons with the same smoking habits. Spirometric findings did not differ between exposed and control persons before the workshift, but deteriorated during the workshift in both the controls and the workers exposed to methyl methacrylate. The decline in the maximal expiratory flow at 50 % of forced vital capacity (MEF_{50}) and the ratio of MEF_{50} to maximal expiratory flow were significant in the workers exposed to methyl methacrylate compared with in the control persons. The authors attributed the increased incidence of chronic coughing and the slight respiratory obstruction to exposure to methyl methacrylate (Marez *et al.* 1993). The EU Risk Assessment Report draws attention to the small population, the lack of data about possible exposure to other irritant substances, and other inadequacies, and expresses doubts about how exposure was recorded (EU 2002). Exposure was determined only by static devices, not by personal air sampling, and the inadequate data give rise to doubts about the analytical methods used. This study is therefore not suitable for deriving a threshold value.

In a biological monitoring study, 32 male workers were examined who were exposed to methyl methacrylate concentrations of between 0.4 and 112 ml/m^3 with a geometric mean of 6.1 ml/m^3 and a median of 5.3 ml/m^3. Four workers were exposed to concentrations of more than 50 ml/m^3 and one of them was exposed to concentrations above 100 ml/m^3. No changes in haematological or biochemical parameters were found in the serum; nor was there a statistically significant difference in the prevalence of symptoms compared with those in 16 workers not exposed. Only 6 exposed workers reported frequent coughing with sputum and 4 workers reported throat irritation. Four of the 6 workers who complained of coughing and sputum and all 4 workers with throat irritation belonged to the "high" exposure group (5–112 ml/m^3; median 18 ml/m^3); however, the workers with these symptoms were not always those exposed to the highest concentrations (no other details) (Mizunuma *et al.* 1993). There are no details about the current or earlier 8-hour mean values or peak exposures to which the workers with symptoms were exposed. Nor is there any information about smoking habits. Therefore, the data cannot be used for deriving a threshold value.

In a dental laboratory, methyl methacrylate concentrations from 4.09 to 30.64 mg/m^3 (0.26–7.29 ml/m^3) were determined in the breathing zones of 8 workers. Sixteen analyses of methyl methacrylate in the air yielded concentrations between 3.68 and 38.42 mg/m^3 (0.88–9.14 ml/m^3). Persons who worked with methyl methacrylate for 20 to 30 minutes—kneading the mass containing methyl methacrylate with their bare hands—complained of the unpleasant smell and occasional eye irritation (Korczynski 1998). Since no data are available about the methyl methacrylate concentrations present when the workers complained of eye irritation, these findings cannot be used for an assessment of methyl methacrylate.

4.2 Local effects on skin and mucous membranes

Dental technicians who used liquid methyl methacrylate reported dermal lesions and finger paresthesia (Rajaniemi 1986; Seppalainen and Rajaniemi 1984).

4.3 Allergenic effects

4.3.1 Sensitizing effects on the skin

In a Belgian hospital, a total of 13833 patients were patch tested between 1978 and 1999. A reaction to (meth)acrylate was observed in 54 of a total of 7369 patients who reacted to at least one substance; 5 also reacted to 2 % methyl methacrylate in petrolatum. The number of persons tested with methyl methacrylate was not documented (Geukens and Goossens 2001). In a British hospital, 352 patients were tested with 2 % methyl methacrylate in petrolatum in the period from 1983 to 1998; 17 of them produced a reaction (Tucker and Beck 1999). Between January 1988 and October 2002, a reaction to at least one (meth)acrylate was observed in a total of 75 patients of an American hospital. A reaction to 2 % methyl methacrylate in petrolatum was also observed in 19 of 56 patients whose data could be evaluated (Sood and Taylor 2003). In another American hospital, a total of 472 patients were tested with 2 % methyl methacrylate in petrolatum between July 1994 and June 1999; 3 of 54 workers employed in a health-care related field and 3 of 418 persons not employed in health-care professions produced reactions (Shaffer and Belsito 2000). In 2001 and 2002, 4900 patients were tested with 2 % methyl methacrylate in petrolatum in a North American multicentre study; 1.4 % produced a positive result. For the periods from 1996 to 1998 and 1998 to 2000, the authors specify the proportion of reactions as being 1.6 % of 4099 persons and 1.4 % of 5812 persons tested, respectively (Pratt *et al.* 2004).

In a study of 27 patients who had contact with artificial fingernails, including 16 cosmeticians, 4 of the 21 persons tested reacted to 2 % methyl methacrylate in petrolatum and 25 of 27 to 2 % 2-hydroxyethyl methacrylate in petrolatum (Constandt *et al.* 2005). According to a Swedish study, a reaction to 2 % methyl methacrylate in petrolatum was observed in 16 of 109 dental personnel examined between 1995 and 1998. All 16 patients also reacted to 2-hydroxethyl methacrylate and 15 of them to ethylene glycol dimethacrylate as well (Wrangsjö *et al.* 2001). Of 79 dentists and 46 dental nurses who were tested in a Polish hospital between 1990 and 2000, 8 reacted to 2 % methyl methacrylate in petrolatum. All eight also produced a reaction to ethylene glycol dimethacrylate (Kiec-Swierczynska and Krecisz 2002). A reaction was observed in 20 of the 271 patients tested with 2 % methyl methacrylate in petrolatum at the Finnish Institute of Occupational Health between 1985 and 1995 (Kanerva *et al.* 1997). In a Finnish multicentre evaluation carried out in the period from 1994 to 1998, a reaction was reported in 28 of 2607 patients tested. A further 15 reactions were classified as questionable or irritant (Kanerva *et al.* 2001). Reactions to methyl methacrylate were observed in 2 of 49 Korean dental technicians, 22 of whom had contact dermatitis (Lee *et al.* 2001). Between 1992 and 1995, 143 dental technicians were tested with 2 % methyl methacrylate in petrolatum in the clinics of the Information Network of Departments of Dermatology (Informationsverbund Dermatologischer Kliniken); 18 of them (12.6 %) produced a reaction (Schnuch *et al.* 1998). However, these patients belong to the collective whose test results were already included in the 1997 documentation ("Methyl methacrylate", Volume 16, present series).

There are also individual case reports of sensitization to methyl methacrylate and reactions in patch tests in 2 dental nurses (Kanerva and Estlander 1998; Kanerva *et al.* 1998), 1 female patient with acute eczema in the contact area of a plastic catheter (Saccabusi *et al.* 2001), 1 patient with an erythematous-oedematous reaction in the contact area of an adhesive for a surgical earthing plate containing 2-hydroxyethyl acrylate and 2-hydroxyethyl methacrylate (Miranda-Romero *et al.* 1998), 1 female patient with erythematous skin changes with blisters after application of artificial nails (Mowad and Ferringer 2004), another female patient with skin reactions to components of artificial fingernails (Casse *et al.* 1998) and in several patients intolerant to dental prostheses (Bauer and Wollina 1998; Giroux and Pratt 2002; Lunder and Rogl-Butina 2000; Ruiz-Genao *et al.* 2003). The clinical relevance of the patch test reactions described in these reports is unclear in most cases as the affected patients almost always reacted to several (meth)acrylates, especially ethylene glycol dimethacrylate, and the level of methyl methacrylate in the products was only rarely specified. One female patient with anamnestic evidence of intolerance to fingernail materials containing acrylate and a temporarily inserted dental filling was subjected to a test to identify whether it would be possible to implant a knee prosthesis with bone cement containing methacrylate. For this purpose, the patient was patch tested with 2 % and 4 % mixtures of a liquid methyl methacrylate preparation (with *N,N*-dimethyl-*p*-toluidine, hydroquinone and chlorophyll); methacrylate copolymer powder (with di-benzoyl peroxide, zirconium dioxide and chlorophyll); and with a patch of the polymerized cement. After 72 hours there was an indurated erythematous reaction to the cement and the two methyl methacrylate preparations (Kaplan *et al.* 2002). In another study (Hochman and Zalkind 1997), a female patient with suspected intolerance to a dental prosthesis and a dentist who exhibited intolerance reactions while working with materials containing acrylate were apparently tested with undiluted methyl methacrylate, with the effect that the positive findings obtained cannot be evaluated. Another positive test result with methyl methacrylate in one worker with skin changes induced by an adhesive containing acrylic acid and methyl methacrylate (Bang Pedersen 1998) was not adequately documented. Four reactions to 1 % methyl methacrylate were observed in the patch testing of 520 patients with mucosal alterations which might have been caused by dental prostheses. Clinical relevance was specified for two of these reactions, but there are no details (Vilaplana and Romaguera 2000).

One female patient with a 4-year history of lesions of the palate associated with a dental prosthesis reacted to none of the acrylates or methacrylates in the patch test after 2 or 4 days. After 3 weeks, the patient went to hospital again because she had an unpleasant feeling in the test area. The repeated testing led to 3+ reactions to 2 % methyl methacrylate, 0.1 % ethyl and butyl acrylate and 2 % 2-hydroxyethyl methacrylate in petrolatum after 4 days, which the authors considered to be evidence of sensitization caused by the patch test (Vozmediano and Manrique 1998). In another case, a 1+ reaction to methyl methacrylate, which was observed only after 38 days, was reported in a female dental technician (Fowler 1999).

Methyl methacrylate was used in studies in which the effects of contact sensitizing substances on the maturation and cytokine secretion of human dendritic cells and the migration of Langerhans' cells were investigated in excised human skin to evaluate the applicability of *in vitro* test methods. In these studies, methyl methacrylate led to less pronounced effects than ethylene glycol dimethacrylate, hydroxypropyl methacrylate and 2-hydroxyethyl methacrylate (Rustemeyer *et al.* 2003).

4.3.2 Sensitizing effects on the respiratory tract

A 48-year-old female worker exposed to methyl methacrylate in the processing of an adhesive containing methyl methacrylate developed workplace-related dyspnoea and rhinorrhoea as well as other symptoms. A 30-minute provocation test, in which the patient applied "one or four ml" of the adhesive to an area of 10 × 10 cm in an exposure chamber, led to a 24 % decrease in the peak expiratory flow (PEF) only after 8 hours. Up to the seventh hour, the reduction in PEF was about 16 % at most and was also lower in subsequent determinations with the exception of the analysis after 12 hours (about –18 %). A 46-year-old dental technician who had been in the profession for 20 years reported tiredness and respiratory symptoms such as a cough and chest tightness which disappeared when she was ill or on holiday, but recurred within one week after she had returned to her job. A workplace-related provocation test, during which the patient processed a prosthesis made of 10 ml methacrylate powder (probably polymethyl methacrylate) and 10 ml methacrylate liquid, led to a dual reaction with a decrease in the PEF of a maximum 26 % (no other details). The grinding of a "piece of methacrylate" (probably polymethyl methacrylate) led to a delayed 15 % decrease in the PEF in another female worker, who developed asthmatic symptoms after exposure for one month to (poly)methyl methacrylate in the manufacture of hearing aids. The authors determined an increase in non-specific respiratory tract reactivity in the histamine test after the provocation (Savonius *et al.* 1993a, 1993b). A recent communication reported occupationally induced hypersensitivity pneumonitis in a 20-year-old and a 24-year-old female dental technician. Dyspnoea and coughing occurred in both employees a few weeks or about half a year after beginning training. The blood gas analysis revealed hypoxaemia (65 and 55 mm Hg), and pulmonary function tests showed the diffusion capacity to be reduced and in one case the total lung capacity to be reduced (67 % of the expected level). The 20-year-old patient was exposed to a methyl methacrylate aerosol two months later. There was a 30 % increase in lymphocytes in the bronchioalveolar lavage (BAL) fluid, and pulmonary function tests revealed a 20 % decrease in the diffusion capacity for carbon monoxide (TLCO). No other findings were communicated. About four weeks after the symptoms had subsided in the second patient as a result of oral corticosteroid therapy, she returned to her laboratory workplace, and one day later a pronounced cough recurred. After another two days, she was re-examined in hospital, and hypoxaemia (58 mm Hg), an increase in the lymphocyte count in the BAL fluid, reduced forced vital capacity (2.0 l; 50 %) and reduced forced expiratory volume in the first second (FEV_1; 0.85 l; 24 %) were again found (Scherpereel *et al.* 2004).

A 44-year-old secretary reported experiencing respiratory symptoms (rhinorrhoea, dyspnoea and coughing attacks) for the past two years 15 to 20 minutes after beginning copying on a black and white copier with a toner containing a polymer of styrene and *n*-butyl methacrylate. The histamine test revealed non-specific bronchial hyperreactivity ($PC_{20(histamine)}$: 2.16 mg/ml). In a workplace-related provocation test, respiratory symptoms (dyspnoea) occurred after 18 minutes. The FEV_1 was reduced at this time by 21 %, after 1 hour by 24 % and after 4 hours by 19 %. In a provocation test with methyl methacrylate, which was heated to 80°C, a FEV_1 decrease of 30 % occurred after 1 hour and a 24 % decrease after 5 hours. Provocation tests with polystyrene and potato flour induced no respiratory symptoms or spirometric alterations. The nasal lavage fluid contained an increased amount of eosinophils only after the workplace-related provocation and after provocation with methyl methacrylate. In addition, the two provocation tests led to an increase in the permeability index (from 6.5 % to 16.1 % and from 9.1 % to 19.7 %) after 24 hours (Wittczak *et al.* 2003).

Hoarseness and sore throat, as well as nasal symptoms and dyspnoea, occurred in a 48-year-old dental nurse after almost 27 years of exposure to acrylates. Prick tests with ubiquitous allergens, acrylates (no other details), chloramine-T and latex and a provocation test with a mixture of 2-hydroxyethyl methacrylate and bisphenol A diglycidyl methacrylate yielded negative results. Pulmonary function tests did not reveal abnormal findings. Provocation tests with liquid methyl methacrylate (10 drops; trade product with stabilizers and other additives) and with polymethyl methacrylate powder (10 ml; trade product) led to mucosal alterations not described in detail and to symptoms of the upper respiratory tract, indicating rhinitis and pharyngitis. An increase in airway resistance (70 %) was determined rhinomanometrically. However, the FEV_1 decrease was only 6 % at most, and the PEF was reduced by 20 % 16 hours after provocation. PEF monitoring revealed no workplace-related changes. After about two years of continued work with reduced exposure to acrylates and local corticosteroid therapy, further PEF monitoring revealed fluctuations of 350–470 l/min on working days and

420–460 l/min on work-free days; the authors concluded from this that the patient had occupational asthma (Piirilä *et al.* 1998).

A female dental technician developed dyspnoea, wheezing, coughing and rhinorrhoea 6 to 8 months after her first contact with preparations containing methyl methacrylate. At the time of examination, the patient had already been exposed to methyl methacrylate for 13 years. Prick tests with ubiquitous allergens yielded negative results. Pronounced stridor and dyspnoea and a decrease in the FEV_1 and PEF occurred in the provocation test with methyl methacrylate. In the nasal lavage fluid there was an increase in leukocytes, eosinophils and basophils and an increase in the eosinophilic cationic protein and mast cell tryptase (Wittczak *et al.* 1996).

In a cross-sectional study, the pulmonary function parameters of 19 male dental technicians exposed to methyl methacrylate concentrations of 0.16 to 4.38 ml/m^3 (TWA; arithmetic mean: 1.40 ml/m^3; geometric mean: 0.91 ml/m^3) were compared with those of 9 male workers not exposed. Peak exposures of a maximum 37.71 ml/m^3 occurred during the processing of thermosetting resins. In the group of exposed persons, the parameters FVC and FEV1.0 (determined as %FVC/Ht and %FEV1.0/Ht) were reduced and the workers specified a higher incidence of respiratory symptoms (coughing) (Nishiwaki *et al.* 2001). Since no immunological investigations were carried out, the study cannot be used for assessing the sensitizing effects of methyl methacrylate on the respiratory tract. The same applies to other studies in which respiratory functions were monitored in workers exposed to methyl methacrylate or possible respiratory symptoms were recorded but no provocation tests or other immunological investigations were carried out among the specific workers (see also EU 2002).

4.4 Reproductive toxicity

Male and female workers exposed to methyl methacrylate and vinyl chloride reported sexual disorders that were not specified in detail. Since these studies are only available as abstracts and there are no further details, they cannot be used for assessment (EU 2002).

The EU Risk Assessment Report describes a study of a cohort of women who had been occupationally exposed to methyl methacrylate from 1976 to 1985. The evaluation of a total of 502 pregnancies, for example, revealed an increased incidence of spontaneous abortions among women exposed to more than 20 mg/m^3 (about 5 ml/m^3) and asphyxia, malformations (no other details) and stillbirths among the newborn babies of women exposed to below 10 mg/m^3 (about 2.5 ml/m^3). Since the study is based only on retrospective data and there are no details of controls or workplace and exposure conditions, it is not possible to attribute the described effects to an exposure to methyl methacrylate. Therefore, this study cannot be used for assessment (EU 2002).

4.5 Genotoxicity

The number of SCEs (sister chromatid exchanges) was increased in 31 workers exposed to methyl methacrylate (7.85 ± 2.66) compared with the number in 31 controls adjusted for age and smoking behaviour (7.49 ± 2.33). The incidence of SCEs was significantly increased only in the group of workers exposed to peak methyl methacrylate concentrations of 114 to 400 ml/m^3. Responsible for this increase were some cells with a large number of SCEs (Marez *et al.* 1991). As a result of methodological inadequacies, the specified exposure concentrations are questionable (see Section 4.1: Marez *et al.* 1993).

4.6 Carcinogenicity

Large mortality studies were carried out in two US companies involved in acrylic sheet production (Collins *et al.* 1989; Walker *et al.* 1991). The cohorts were exposed mainly to ethyl acrylate and methyl methacrylate. There was co-exposure to ethylene dichloride, methylene chloride and acrylonitrile. Increased mortality resulting from colon cancer was significant in one factory and not significant in the other. A non-significant increase in rectal cancer was identified in the first factory. The increases were most obvious in the workers from the earliest production period and in workers with the highest exposure levels. There was, however, no relationship between the tumour risk and increasing methyl methacrylate exposure (IARC 1994).

5 Animal Experiments and *in vitro* Studies

5.1 Acute toxicity

5.1.1 Inhalation

The 4-hour LC_{50} for rats is 29800 mg/m^3 (7092 ml/m^3) (EU 2002).

Mice were exposed to methyl methacrylate concentrations of 740 to 33000 ml/m^3 for 30 minutes to determine the RD_{50}. The respiratory rates were not consistently reduced by more than 25 % at any exposure concentration (ACGIH 2001). Therefore, no RD_{50} could be determined.

Female F344 rats exposed to methyl methacrylate concentrations of 200 ml/m^3 for 6 hours were found to have lesions in the nasal olfactory epithelium characterised by degeneration and atrophy (Mainwaring *et al.* 2001).

5.1.2 Ingestion

Oral LD_{50} values of 8 to 10 ml/kg body weight (7552–9440 mg/kg body weight) were obtained in rats (EU 2002).

5.1.3 Dermal absorption

The dermal LD_{50} was greater than 5000 mg/kg body weight in rabbits under occlusive conditions (EU 2002).

5.2 Subacute, subchronic and chronic toxicity

Inhalation

Most of the available studies with repeated administration have already been described in the 1984 MAK documentation ("Methyl methacrylate", Volume 3, present series). What follows is a description of relevant studies published since that time; for comparison, the earlier studies relevant to the evaluation are described below.

Groups of 5 female F344 rats were exposed to methyl methacrylate concentrations of 0, 110 or 400 ml/m^3 in whole-body exposure chambers for 6 hours per day, for 1, 2, 5, 10 or 28 days. Animals were examined 4, 13, 24 or 36 weeks after the end of exposure to assess the reversibility of the findings. The only finding in both exposure groups was damage to the olfactory epithelium. The lesions induced by methyl methacrylate concentrations of 110 ml/m^3 were reversible during the exposure period. The lesions caused by 400 ml/m^3 were repaired after 13 weeks, but minimal respiratory metaplasia was observed, and there were focal adhesions between the septum and turbinates and between the turbinates themselves (Hext *et al.* 2001).

Degeneration of the olfactory epithelium, bronchopneumonia, interstitial pneumonia, haemorrhages, atelectasis, oedema, emphysema and bronchial epithelial hyperplasia occurred after exposure of at least 10 rats per group to methyl methacrylate concentrations of 0 or 1000 ml/m^3 for 6 hours per day, on 5 days per week for 4 weeks, under poor and normal ventilation conditions. Bronchopneumonia with abscesses was observed only in rats under poor ventilation conditions; in addition, glutathione levels were significantly decreased and malondialdehyde levels were significantly increased in rats of this group. No difference was observed in superoxide dismutase activity. According to the authors, the poor air exchange rate led to higher concentrations; they point out that adequate protection systems should be in place in operating theatres, which is rarely the case in Turkey (Aydin *et al.* 2002).

In a carcinogenicity study of the NTP ("Methyl methacrylate", Volume 3, present series), groups of 50 male F344 rats were exposed to methyl methacrylate concentrations of 0, 500 or 1000 ml/m^3, female rats were exposed to methyl methacrylate concentrations of 0, 250 or 500 ml/m^3 and male and female B6C3F$_1$ mice were exposed to 0, 500 or 1000 ml/m³ over 102 weeks, for 6 hours per day, on 5 days per week. Body weight gains were reduced in animals of all exposure groups, and there were non-neoplastic lesions in the nasal cavity. Rats and mice were found to have inflammation of the nasal cavity and degeneration of the olfactory epithelium; mice also had hyperplasia of the nasal cavity epithelium (Chan *et al.* 1988). No NOAEL was obtained in this study.

In a combined chronic toxicity and carcinogenicity study carried out by Rohm and Haas in 1979, groups of 70 male and 70 female F344 rats were exposed to methyl methacrylate concentrations of 0, 25, 100 or 400 ml/m³ for 2 years. Ten animals from each group were investigated after 13 and 52 weeks. Body weight gains were reduced only in the females of the high exposure group from the 52nd week. Haematological and clinicochemical parameters and urinalyses were unchanged. Histopathological changes were observed only in the nasal cavity. The nasal tissues—three to four sections per animal—were re-evaluated in 1992 and 1997 (Lomax *et al.* 1997). Sections of trachea, pharynx and larynx were no longer preserved. Degeneration, atrophy, hyperplasia, inflammation and metaplasia in the olfactory epithelium and hyperplasia and inflammation in the respiratory epithelium were observed in the animals of the two high exposure groups (Table 2). The NOAEL in this study was 25 ml/m³ (EU 2002).

Table 2. Incidence of nasal lesions in F344 rats after exposure for two years to methyl methacrylate (EU 2002; Lomax *et al.* 1997)

Findings	Methyl methacrylate concentration (ml/m³)							
	males				females			
	0	25	100	400	0	25	100	400
olfactory epithelium								
number of animals examined (n)	39	47	48	38	44	45	41	41
basal cell hyperplasia (%)	13	6	69	87	0	2	44	76
degeneration/ atrophy (%)	0	0	86	100	0	0	59	95
chronic mucosal and submucosal inflammation (%)	0	0	35	76	0	0	12	61
metaplasia (%)	0	0	2	39	0	0	17	51
respiratory epithelium								
number of animals examined (n)	44	47	48	42	45	45	41	42
Bowman's gland and goblet cell hyperplasia (%)	2	0	2	60	0	0	2	21
chronic mucosal and submucosal inflammation (%)	9	0	4	60	4	0	0	21

statistical significance of the findings not specified in the publication

In a carcinogenicity study carried out by Rohm and Haas in 1979, golden hamsters were exposed to methyl methacrylate concentrations of 0, 25, 100 or 400 ml/m^3 over 78 weeks, for 6 hours per day, on 5 days per week. The re-evaluation of the findings in 1997, in which two to four sections were evaluated per animal, revealed no nasal lesions (EU 2002; Lomax *et al.* 1997).

5.3 Local effects on skin and mucous membranes

5.3.1 Skin

In a range-finding study with 2 rabbits, occlusive application of 0.5 ml undiluted methyl methacrylate for 4 hours was found to be weakly irritating to the skin. In addition, blanching, eschar formation and desiccation of the skin were reported (Rohm & Haas 1982).

Groups of 2 male rabbits were treated dermally for 24 hours with methyl methacrylate doses of 0, 200, 2000 or 5000 mg/kg body weight under occlusive conditions. Well-defined to severe erythema with blanching and moderate to severe oedema with pocketing were observed after 24 hours. After 14 days, skin irritation was still present in animals treated with methyl methacrylate doses of 2000 and 5000 mg/kg body weight. After 3 days, no irritation was observed in animals treated with methyl methacrylate doses of 200 mg/kg body weight. After 2 days, eschar formation was found in the animals of the 2000 and 5000 mg/kg groups. On day 12, eschar was observed to be sloughing off with new hair growth. Desiccation of the skin was observed in animals of all exposure groups (EU 2002).

5.3.2 Eyes

In a range-finding study with 2 rabbits, the instillation of 0.1 ml undiluted methyl methacrylate into the eyes led to conjunctival redness in both rabbits after 24 hours, which was no longer present after 72 hours (Rohm & Haas 1982). No effects on the iris or cornea were observed in another study with 6 rabbits (EU 2002).

5.4 Allergenic effects

5.4.1 Sensitizing effects on the skin

All animals produced a reaction in a modified maximization test with two groups of 5 animals (intradermal induction with 10 % methyl methacrylate in corn oil/physiological saline, epicutaneous induction with 25 % methyl methacrylate in sunflower oil; challenge occlusively on day 14 with 25 % methyl methacrylate in corn oil, occlusively with 25 % methyl methacrylate in DMSO/corn oil and on day 28 non-occlusively with 25 % methyl methacrylate in DMSO/ethanol or non-occlusively with 50 % methyl methacrylate in DMSO/ethanol) (Rustemeyer *et al.* 1998). Two of 14 female Hartley guinea pigs reacted in another modified maximization test (intradermal and epicutaneous inductions with 10 % methyl methacrylate in olive oil; challenge occlusively with 1 % methyl methacrylate in acetone). After challenge with 1 % methyl methacrylate in acetone, no reaction was observed in 10 animals which had been treated with 1 % or 0.1 % methyl methacrylate preparations for induction (Kanazawa *et al.* 1999). Further positive findings were obtained in a modified Freund's Complete Adjuvant (FCA) test. In this test, a total of 300 µl 10 % methyl methacrylate in FCA/water (1:1) was injected intradermally into both flanks (2 × 50 µl) and the ears (2 × 50 µl) and necks (100 µl) of groups of 5 animals for induction. Challenge was carried out on day 14 by occlusive application of 25 µl of a preparation of 25 % methyl methacrylate in corn oil or in DMSO/corn oil and on day 28 by non-occlusive application of 25 % methyl methacrylate in DMSO/ethanol (4:1) or a preparation of 50 % methyl methacrylate in DMSO/ethanol. Ten of 10 and 9 of 10 animals reacted after occlusive and non-occlusive challenge treatment, respectively. Almost all of the animals sensitized with methyl methacrylate also reacted to 10 % ethylene glycol dimethacrylate. There were markedly fewer cross-reactions with 2-hydroxyethyl methacrylate and 2-hydroxypropyl methacrylate (Rustemeyer *et al.* 1998). In later studies with the modified FCA test, male and female inbred guinea pigs (no other details) were also sensitized with 10 % methyl methacrylate in FCA/water (1:1) and reacted to 50 % methyl methacrylate in DMSO/ethanol (4:1) after non-occlusive challenge treatment. However, if 175 µl undiluted methyl methacrylate was administered orally to the animals 26, 20 and 14 days before the beginning of sensitization, it was possible to induce tolerance, which was manifest in the clearly less pronounced reaction to the challenge treatment (Rustemeyer *et al.* 2001).

5.4.2 Sensitizing effects on the respiratory tract

There are no data available for sensitization of the respiratory tract induced by methyl methacrylate.

5.5 Reproductive toxicity

5.5.1 Fertility

There are no valid studies of fertility available.

Long-term inhalation studies with rats and mice and an oral long-term study with the administration of methyl methacrylate in drinking water revealed no histopathological changes in the male or female sex organs (see Section 5.2).

5.5.2 Developmental toxicity

Rats

In a prenatal developmental toxicity study carried out according to OECD Test Guideline 414, groups of 27 pregnant CD rats (Sprague-Dawley) were exposed to methyl methacrylate concentrations of 0, 99, 304, 1178 or 2028 ml/m^3 for 6 hours a day, on days 6 to 15 of gestation. Reduced feed consumption and reduced maternal body weight gains were recorded in all exposure groups throughout the exposure period. The minimal reductions in body weight gain observed on days 6 to 8 of gestation after 99 and 304 ml/m^3 were transient. The incidences of foetuses with variations and retardations per litter were somewhat higher in all exposed groups than in the control group, but there was no clear relationship to the concentration. A significant increase was observed only for the incidence of variations (particularly of rudimentary 14th ribs) at the second highest concentration of 1178 ml/m^3. Here, 3 foetuses from two litters were found to have malformations (1× omphalocoele of the abdomen and 2× enlarged adrenal glands) compared with 1 foetus in the control group (duplication of the hypothalamus). The authors do not regard the findings as substance-induced as there was no dose-response relationship, and they conclude that there was no substance-related embryotoxicity or foetotoxicity even at concentrations that resulted in maternal toxicity (Solomon *et al.* 1993).

In two independent tests carried out by Imperial Chemical Industries Limited (ICI) in 1977, rats were exposed to methyl methacrylate concentrations of 0, 100 or 1000 ml/m^3 and to 0, 25, 100 or 1000 ml/m^3 on days 6 to 15 of gestation. The no observed adverse effect concentration (NOAEC) for maternal toxicity was specified to be 1000 ml/m^3. The authors reported an increase in the number of early resorptions in the high exposure group in both tests and of late resorptions in only one test. The authors derived a NOAEC of 100 ml/m^3 from their results. Because of this study's limitations (insufficient randomization of test animals, inadequate test protocol and poor documentation), the authors' interpretations could not be followed in the Risk Assessment Report (EU 2002).

Another inhalation study with rats (Nicholas *et al.* 1979) is not useful because of the high, toxic concentration of methyl methacrylate administered of 110000 mg/m^3 (26180 ml/m^3) for 17 or 54 minutes per day. Exposure led to deaths, loss of body weight and reduced feed consumption in the dams. Early resorptions were observed. Foetal body weights were reduced, crown–rump lengths were shorter and there were haematomas and retarded ossification (EU 2002).

The intraperitoneal injection of methyl methacrylate in doses of 0, 0.133, 0.266 or 0.443 ml/kg body weight (0, 126, 251 or 418 mg/kg body weight) on days 5, 10 and 15 of gestation revealed no maternal effects in rats. Compared with in untreated controls, there was a higher incidence of resorptions, and foetal body weights were slightly reduced, but all values were in the range of those of the control animals with intraperitoneal injection of water, saline or oil. The foetuses were found to have a higher, dose-dependent incidence of anomalies (haemangiomas) (2.3 %, 8.0 % and 16.7 %; controls 0–2 %), but no malformations or other effects of developmental toxicity (Singh *et al.* 1972). The haemangiomas were presumably the result of the irritant effects of the substance after intraperitoneal injection and are therefore not relevant for the assessment of the developmental toxicity of methyl methacrylate under inhalation conditions.

Mice

The exposure of groups of pregnant CD-1 mice (n = 38, 32, 18) to methyl methacrylate concentrations of 0, 100 or 400 ml/m^3 for 6 hours per day, on days 4 to 13 of gestation, led only to slight, but statistically significant differences in the body weights of the foetuses (no other details); no teratogenic effects were induced (ICI 1980).

Exposure of 18 pregnant ICR mice to methyl methacrylate concentrations of 1330 ml/m^3 for 2 hours twice daily on days 6 to 15 of gestation revealed only slightly increased foetal weights, but no evidence of developmental toxicity. There are no data for maternal toxicity (McLaughlin *et al.* 1978).

Rabbits

In a study carried out by ICI in 1977, the intraperitoneal injection of 0.004, 0.04 or 0.4 ml/kg body weight (3.8, 38 or 376 mg/kg body weight) on days 6 to 18 of gestation led in rabbits to a high incidence of peritonitis (probably the result of the irritant effects of methyl methacrylate) and an increase in the respiration rate in the high dose group. In this dose group, also the foetal weights were significantly reduced and the number of resorptions was increased. There were no malformations (EU 2002).

5.6 Genotoxicity

5.6.1 *In vitro*

Methyl methacrylate yielded negative results in bacterial gene mutation tests (EU 2002; IARC 1994).

In vitro genotoxicity studies in mammalian test systems are shown in Table 3.

At concentrations with moderate toxicity, methyl methacrylate induced small colonies in the mouse lymphoma test without, but especially with S9 mix (Dearfield *et al.* 1991; Doerr *et al.* 1989; Myhr *et al.* 1990), which are evidence of a clastogenic effect of the substance. Similar evidence was obtained in the micronucleus test (Doerr *et al.* 1989) and in chromosomal aberration tests (Anderson *et al.* 1990; Doerr *et al.* 1989). However, in the studies by Doerr *et al.* (1989), positive effects were not related to the concentration tested but linked with severe toxicity. In contrast, "authentic clastogens" demonstrate very steep dose–response relationships. The SCE data showing more pronounced effects at a later time of preparation confirm these data. If the SCE data are considered separately, they have little relevance, but in the context of the other data they substantiate the presence of a genotoxic potential *in vitro*, which is detected only when there are also toxic effects.

Table 3. *In vitro* genotoxicity studies with methyl methacrylate in mammalian test systems (according to IARC 1994)

Test system		Concentration	Results without MA	Results with MA	References
SCE	CHO cell line	5 µg/ml	–		Anderson *et al.* 1990
		16–1250 µg/ml	+		
		50–500 µg/ml		–	
		1600–5000 µg/ml		+	
gene mutation, $TK^{+/-}$ locus	mouse lymphoma cell line L5178Y	250–3000 µg/ml	(+)[1]		Dearfield *et al.* 1991
		500–1000 µg/ml		+[1]	
gene mutation, $TK^{+/-}$ locus	mouse lymphoma cell line L5178Y	250–3000 µg/ml	+[1]		Doerr *et al.* 1989; Moore *et al.* 1988
gene mutation, $TK^{+/-}$ locus	mouse lymphoma cell line L5178Y	125–250 nl/ml (118–235 µg/ml)	–		Myhr *et al.* 1990
		500–1000 nl/ml (470–940 µg/ml); toxic at 1500 nl/ml (1410 µg/ml)	+		
		125 nl/ml (118 µg/ml)		–	
		250–1500 nl/ml (235–1410 µg/ml)		+	
gene mutation, $TK^{+/-}$ locus	mouse lymphoma cell line L5178Y	up to 100 nl/ml (94 µg/ml)	–		EU 2002
		100–250 nl/ml (94–235 µg/ml)		(+)	
MN	mouse lymphoma cell line L5178Y	1000–3000 µg/ml	(+)[2]	not investigated	Doerr *et al.* 1989
CA	CHO cell line	up to 500 µg/ml	–		Anderson *et al.* 1990
		1600, 3000 µg/ml	+		
		160–1600 µg/ml		–	
		5000 µg/ml		+	
CA	mouse lymphoma cell line L5178Y	1000–3000 µg/ml	(+)[3]	not investigated	Doerr *et al.* 1989; Moore *et al.* 1988

CA: chromosomal aberrations; MA: metabolic activation; MN: micronuclei; SCE: sister chromatid exchange; +: positive; (+): weakly positive; –: negative

[1] small colonies

[2] a maximum of 25 % of cells with aberrations; negative controls 9 %

[3] a maximum of 39 % of cells with aberrations; negative controls 15 % and positive controls 47 %; "not all cultures yielded positive results" (no other details)

5.6.2 *In vivo*

A dominant lethal test carried out by ICI in 1976 yielded negative results after inhalation exposure of mice to methyl methacrylate concentrations of 100, 1000 or 9000 ml/m^3. The animals were exposed for 6 hours daily, on 5 days per week, for 8 weeks and mated weekly (EU 2002; "Methyl methacrylate", Volume 3, present series).

A micronucleus test with bone marrow cells of mice yielded negative results after single intraperitoneal methyl methacrylate doses of 4500 mg/kg body weight and after four intraperitoneal doses of 1100 mg/kg body weight (Hachitani *et al.* 1981).

A bone marrow chromosomal aberration test in rats yielded negative results after single intraperitoneal methyl methacrylate doses of 650 or 900 mg/kg body weight. Increased aberrations were reported after single intraperitoneal doses of 1300 mg/kg body weight (17 % with aberrations; controls 1.8 %), but it was not specified whether the evaluation was carried out including gaps. In another test with repeated intraperitoneal methyl methacrylate doses of 650 mg/kg body weight, positive findings were described after treatment for 2 and 4 weeks, and negative findings after 6 and 8 weeks (Fedyukovich and Egorova 1991). There is no plausible explanation for this unusual time–effect relationship; therefore, the findings are of questionable reliability (EU 2002).

Two bone marrow chromosomal aberration tests in rats were carried out by ICI in 1976 and 1979 (EU 2002). In the first test, positive findings were obtained after a single 2-hour exposure or after five 5-hour exposures to 9000 ml/m^3 by inhalation. No aberrations were observed after exposure to 100 or 1000 ml/m^3, although the 2-hour exposure, rather than the five exposures, led to a questionably positive result. In the second test, weakly positive findings were reported after exposure to 400 and 700 ml/m^3. These results cannot be assessed because positive findings were only obtained including gaps, and 11 % hydroquinone, which is genotoxic itself, was used in the tests.

5.7 Carcinogenicity

No carcinogenic effects were observed in a carcinogenicity study carried out by the NTP with F344 rats and B6C3F$_1$ mice with methyl methacrylate concentrations of up to 1000 ml/m^3 ("Methyl methacrylate", Volume 3, present series; Chan *et al.* 1988) or in carcinogenicity studies carried out by Rohm and Haas in F344 rats and golden hamsters with methyl methacrylate concentrations of up to 400 ml/m^3 (EU 2002).

5 Manifesto (MAK value, classification)

Clastogenic effects were observed in genotoxicity studies *in vitro* at toxic doses. To evaluate this finding, the results from *in vivo* studies are very important. Because of methodological limitations, these are problematical and of little use to the evaluation. Therefore, they cannot counter the suspicion resulting from the *in vitro* studies that the substance has clastogenic effects. The carcinogenicity studies revealed no evidence of carcinogenic effects in either rats, mice or hamsters. On the basis of the available data, classification as a carcinogen is not required. Since there are only limited data, classification in one of the categories for germ cell mutagens is not possible.

In several studies with repeated exposures carried out in rats, distinct nasal lesions in the olfactory epithelium were observed at methyl methacrylate concentrations of 100 ml/m^3. A NOAEC of 25 ml/m^3 was obtained in a 2-year study. However, *in vitro* studies of the carboxylesterase level in the nose and PBPK models revealed higher exposure of the olfactory epithelium in rats than in humans. Therefore, only the results from studies of exposed persons are used for deriving the MAK value.

Studies of workers involved in acrylic sheet production who were almost exclusively exposed to methyl methacrylate revealed no rhinologically detectable irritant effects (Röhm 1994) or impairment of the sense of smell after an average 8.8 years of employment with 8-hour mean exposure values for methyl methacrylate of up to 40 ml/m^3 (Muttray *et al.* 1997). Sensory irritation was reported only after short-term exposure peaks of more than 100 ml/m^3 (Röhm 1994). On the basis of these results, the MAK value of 50 ml/m^3 has been retained.

Since local irritation occurred in workers only after short-term exposure peaks of more than 100 ml/m^3, Peak Limitation Category I with an excursion factor of 2 can be retained.

An *in vitro* study, in which an absorption rate of 107 µg/cm^2 and hour was determined, is available for the assessment of dermal absorption. This rate would correspond to the absorption of 214 mg methyl methacrylate after one hour of exposure of the hands and forearms (2000 cm^2). The systemic NOAEC is about 100 ml/m^3 (420 mg/m^3); reduced body weight gains were observed at 400 ml/m^3 in female rats in the carcinogenicity study. Assuming an inhaled volume of 10 m^3, the calculated amount absorbed by the skin is only 1/20 of the systemic NOEC (no observed effect concentration; 4200 mg), and dermal absorption thus makes no relevant contribution to systemic toxicity. Methyl methacrylate is, therefore, still not designated with an "H".

Both the findings in humans which have been published since the 1997 MAK documentation and the results from animal studies demonstrate that methyl methacrylate has contact sensitizing potential. Some supplementary findings regarding effects on the respiratory tract in humans are also available. Nevertheless, these findings are not sufficient to establish that methyl methacrylate can induce sensitization of the respiratory tract. Methyl methacrylate is therefore still designated with an "Sh", but not with an "Sa".

Methyl methacrylate has to date been classified in Pregnancy Risk Group C. A prenatal developmental toxicity study in rats with exposure by inhalation that was carried out according to valid guidelines revealed no developmental toxicity up to the highest concentrations (> 2000 ml/m^3). Other prenatal developmental toxicity studies carried out in rats and rabbits are not of use to the evaluation because of their methodological inadequacies or unphysiological administration (intraperitoneal). Only early studies are available for mice. Although slight, but statistically significant differences in foetal weights were observed in a study in which mice were exposed to methyl methacrylate concentrations of 100 or 400 ml/m^3, another study revealed no developmental toxicity other than increased foetal body weights at the only methyl methacrylatc concentration tested of 1330 ml/m^3. In view of the findings obtained in inhalation studies with rats and mice, methyl methacrylate remains in Pregnancy Risk Group C.

7 References

ACGIH (American Conference of Governmental Industrial Hygienists) (2001) Methyl methacrylate. in: *Documentation of TLVs and BEIs*, ACGIH, Cincinnati, OH, USA

Andersen ME, Green T, Frederick CB, Bogdanffy MS (2002) Physiologically based pharmacokinetic (PBPK) models for nasal tissue dosimetry of organic esters: assessing the state-of-knowledge and risk assessment applications with methyl methacrylate and vinyl acetate. *Regul Toxicol Pharmacol 36*: 234–245

Anderson BE, Zeiger E, Shelby MD, Resnick MA, Gulati DK, Ivett JL, Loveday KS (1990) Chromosome aberration and sister chromatid exchange test results with 42 chemicals. *Environ Mol Mutagen 16, Suppl 18*: 55–137

Aydin O, Attila G, Dogan A, Aydin MV, Canacankatan N, Kanik A (2002) The effects of methyl methacrylate on nasal cavity, lung, and antioxidant system (an experimental inhalation study). *Toxicol Pathol 30*: 350–356

Bang Pedersen N (1998) Allergic contact dermatitis from acrylic resin repair of windscreens. *Contact Dermatitis 39*: 99

Bauer A, Wollina U (1998) Denture-induced local and systemic reactions to acrylate. *Allergy 53*: 722–723

Bereznowski Z (1994) Effect of methyl methacrylate on mitochondrial function and structure. *Int J Biochem 26*: 1119–1127

Bereznowski Z (1995) *In vivo* assessment of methyl methacrylate metabolism and toxicity. *Int J Biochem Cell Biol 27*: 1311–1316

Casse V, Salmon-Ehr V, Mohn C, Kalis B (1998) Dépigmentation durable secondaire à des tests positifs aux dérivés des méthacrylates (Chronic depigmentation due to positive patch tests for methacrylate derivatives) (French). *Ann Dermatol Venereol 125*: 56–57

Cefic (European Chemical Industry Council) (1993) *Methyl methacrylate: in vitro absorption through human epidermis*. Zeneca Central Toxicology Lab., 14.07.1993, Cefic Methylacrylate Toxicology Committee, Brussels, Belgium

Chan PC, Eustis SL, Huff JE, Haseman JK, Ragan H (1988) Two-year inhalation carcinogenesis studies of methyl methacrylate in rats and mice: inflammation and degeneration of nasal epithelium. *Toxicology 52*: 237–252

Collins JJ, Page LC, Caporossi JC, Utidjian HM, Saipher JN (1989) Mortality patterns among men exposed to methyl methacrylate. *J Occup Med 31*: 41–46

Constandt L, Hecke EV, Naeyaert JM, Goossens A (2005) Screening for contact allergy to artificial nails. *Contact Dermatitis 52*: 73–77

Dearfield KL, Harrington-Brock K, Doerr CL, Rabinowitz JR, Moore MM (1991) Genotoxicity in mouse lymphoma cells of chemicals capable of Michael addition. *Mutagenesis 6*: 519–525

Degussa (2004) Communication from Dr. Müllerschön to the Commission Secretariat, dated 25.11.2004

Degussa (2005) Communication from Dr. Müllerschön to the Commission Secretariat, dated 20.02.2005

Doerr CL, Harrington-Brock K, Moore MM (1989) Micronucleus, chromosome aberration, and small-colony TK mutant analysis to quantitate chromosomal damage in L5178Y mouse lymphoma cells. *Mutat Res 222*: 191–203

ECB (European Chemicals Bureau) (2000) *Methyl methacrylate*. IUCLID dataset, 19.02.2000, ECB, Ispra, Italy

Elmaraghy AW, Humeniuk B, Anderson GI, Schemitsch EH, Richards RR (1998) The role of methylmethacrylate monomer in the formation and haemodynamic outcome of pulmonary fat emboli. *J Bone Joint Surg Br 80*: 156–561

EU (European Union) (2002) *Methyl methacrylate*. Risk assessment report, 1st priority list, volume 22, Office for Official Publications of the European Communities, Luxemburg, Luxemburg

Fedyukovich LV, Egorova AB (1991) [Genotoxic effects of acrylates] (Article in Russion). *Gig Sanit 12*: 62–64

Fowler Jr JF (1999) Late patch test reaction to acrylates in a dental worker. *Am J Contact Dermatitis 10*: 224–225

Gentil B, Paugam C, Wolf C, Lienhart A, Augereau B (1993) Methylmethacrylate plasma levels during total hip arthroplasty. *Clin Orthop Relat Res 287*: 112–116

Geukens S, Goossens A (2001) Occupational contact allergy to (meth)acrylates. *Contact Dermatitis 44*: 153–159

Giroux L, Pratt MD (2002) Contact dermatitis to incontinency pads in a (meth)acrylate allergic patient. *Am J Contact Dermatitis 13*: 143–145

Hachitani N, Taketani A, Takizawa Y (1981) Studies on mutagenicity of life-related environmental agents. III Ames and mouse bone marrow micronucleus assay of acryl resin monomers and major additives. *Nippon Koshu Eisei Zasshi 29*: 236–239

Hand GC, Henderson M, Mace P, Sherif N, Newman JH, Goldie DJ (1998) Methyl methacrylate levels in unwashed salvage blood following unilateral total knee arthroplasty. *J Arthroplasty 13*: 576–579

Hext PM, Pinto PJ, Gaskell BA (2001) Methyl methacrylate toxicity in rat nasal epithelium: investigation of the time course of lesion development and recovery from short term vapour inhalation. *Toxicology 156*: 119–128

Hochman N, Zalkind M (1997) Hypersensitivity to methyl methacrylate: mode of treatment. *J Prosthet Dent 77*: 93–96

IARC (International Agency for Research on Cancer) (1994) *Methyl methacrylate*. IARC monographs on the evaluation of carcinogenic risk to humans, Volume 60, IARC, Lyon, France, 445–474

ICI (Imperial Chemical Industries Limited) (1980) *Toxicology of methylmethacrylate – sponsors communication document*. ICI, Plastic Division 11.01.1980, London, England

Kanazawa Y, Yoshida T, Kojima K (1999) Structure-activity relationships in allergic contact dermatitis induced by methacrylates. Studies of the influence of side-chain length of methacrylates. *Contact Dermatitis 40*: 19–23

Kanerva L, Estlander T (1998) Contact leukoderma caused by patch testing with dental acrylics. *Am J Contact Dermatitis 9*: 196–198

Kanerva L, Jolanki R, Estlander T (1997) 10 years of patch testing with the (meth)acrylate series. *Contact Dermatitis 37*: 255–258

Kanerva L, Mkola H, Henriks-Eckerman ML, Jolanki R, Estlander T (1998) Fingertip paresthesia and occupational allergic contact dermatitis caused by acrylics in a dental nurse. *Contact Dermatitis 38*: 114–116

Kanerva L, Rantanen T, Aalto-Korte K, Estlander T, Hannuksela M, Harvima RJ, Hasan T, Horsmanheimo M, Jolanki R, Kalimo K, Lahti A, Lammintausta K, Lauerma A, Niinimäki A, Turjanmaa K, Vuorela AM (2001) A multicenter study of patch test reactions with dental screening series. *Am J Contact Dermatitis 12*: 83–87

Kaplan K, Della Valle CJ, Haines K, Zuckerman JD (2002) Preoperative identification of a bone-cement allergy in a patient undergoing total knee arthroplasty. *J Arthroplasty 17*: 788–791

Kiec-Swierczynska M, Krecisz B (2002) Allergic contact dermatitis in dentists and dental nurses. *Exog Dermatol 1*: 27–31

Korczynski RE (1998) Occupational health concerns in the denture industry. *Appl Occup Environ Hyg 13*: 299–303

Lee JY, Yoo JM, Cho BK, Kim HO (2001) Contact dermatitis in Korean dental technicians. *Contact Dermatitis 45*: 13–16

Lomax LG, Krivanek ND, Frame SR (1997) Chronic inhalation toxicity and oncogenicity of methyl methacrylate in rats and hamsters. *Food Chem Toxicol 35*: 393–407

Lunder T, Rogl-Butina M (2000) Chronic urticaria from an acrylic dental prosthesis. *Contact Dermatitis 43*: 232–233

Mainwaring G, Foster JR, Lund V, Green T (2001) Methyl methacrylate toxicity in rat nasal epithelium: studies of the mechanism of action and comparisons between species. *Toxicology 158*: 109–118

Marez T, Shirali P, Hildebrand HF, Haguenoer JM (1991) Increased frequency of sister chromatid exchange in workers exposed to high doses of methylmethacrylate. *Mutagenesis 6*: 127–129

Marez T, Shirali P, Haguenoer JM (1992) Continuous ambulatory electrocardiography among workers exposed to methylmethacrylate. *Int Arch Occup Environ Health 64*: 373–375

Marez T, Edme JL, Boulenguez C, Shirali P, Haguenoer JM (1993) Bronchial symptoms and respiratory function in workers exposed to methylmethacrylate. *Br J Ind Med 50*: 894–897

McLaughlin RE, Reger SI, Barkalow JA, Allen MS, Dafazio CA (1978) Methylmethacrylate: a study of teratogenicity and fetal toxicity of the vapor in the mouse. *J Bone Joint Surg Am 60*: 355–358

Miranda-Romero A, Martinez M, Sanchez-Sambucety P, Aragoneses H, Garcia Munoz CM (1998) Allergic contact dermatitis from the acrylic adhesive of a surgical earthing plate. *Contact Dermatitis 38*: 279–280

Mizunuma K, Kawai T, Yasugi T, Horiguchi S, Takeda S, Miyashita K, Taniuchi T, Moon CS, Ikeda M (1993) Biological monitoring and possible health effects in workers occupationally exposed to methyl methacrylate. *Int Arch Occup Environ Health 65*: 227–232

Moore MM, Amtower A, Doerr CL, Brock KH, Dearfield KL (1988) Genotoxicity of acrylic acid, methyl acrylate, ethyl acrylate, methyl methacrylate, and ethyl methacrylate in L5178Y mouse lymphoma cells. *Environ Mol Mutagen 11*: 49–63

Morgan L, Hampton S, Gibbs M, Arendt J (2003) Circadian aspects of postprandial metabolism. *Chronobiol Int 20*: 795–808

Morita S, Furuya K, Ishihara K, Nakabayashi N (1998) Performance of adhesive bone cement containing hydroxyapatite particles. *Biomaterials 19*: 1601–1606

Mowad CM, Ferringer T (2004) Allergic contact dermatitis from acrylates in artificial nails. *Dermatitis 15*: 51–53

Muttray A, Schmitt B, Klimek L (1997) Effects of methyl methacrylate on the sense of smell. *Cent Eur J Occup Environ Med 3*: 58–66

Myhr B, McGregor D, Bowers L, Riach C, Brown AG, Edwards I, McBride D, Martin R, Caspary WJ (1990) L5178Y mouse lymphoma cell mutation assay results with 41 compounds. *Environ Mol Mutagen 16, Suppl 18*: 138–167

Nicholas CA, Lawrence WH, Autian J (1979) Embryotoxicity and fetotoxicity from maternal inhalation of methyl methacrylate monomer in rats. *Toxicol Appl Pharmacol 50*: 451–458

NIOSH (US National Institute for Occupational Safety and Health) (1976) *A study of methyl methacrylate exposure and employee health*. Cromer J, Kronoveter K, US Department of Health, Education, and Welfare, Public Health Service, Center for Disease Control, NIOSH, Cincinnati, OH, USA

Nishiwaki Y, Saitoh T, Takebayashi T, Tanaka S, Etoh N, Eitaki Y, Omae K (2001) Cross-sectional study of health effects of methyl methacrylate monomer among dental laboratory technicians. *J Occup Health 43*: 375–378

Piirilä P, Kanerva L, Keskinen H, Estlander T, Hytönen M, Tuppurainen M, Nordman H (1998) Occupational respiratory hypersensitivity by preparations containing acrylates in dental personnel. *Clin Exp Allergy 28*: 1404–1411

Pratt MD, Belsito DV, de Leo VA, Fowler Jr JF, Fransway AF, Maibach HI, Marks JG, Toby Mathias CG, Rietschel RL, Sasseville D, Sherertz EF, Storrs FJ, Taylor JS, Zug K (2004) North American Contact Dermatitis Group patch-test results, 2001–2002 study period. *Dermatitis 15*: 176–183

Rajaniemi R (1986) Clinical evaluation of occupational toxicity of methylmethacrylate monomer to dental technicians. *J Soc Occup Med 36*: 56–59

Röhm (1994) *Medical examination of workers in acrylic sheet production exposed to methyl methacrylate*. Röhm GmbH, Darmstadt

Rohm & Haas (1982) *Acute range finding toxicity studies with methyl methacrylate in rats and rabbits with cover letter dated 071789*. Rohm & Haas Toxicology Department, Spring House, PA, USA, OTS0544282, New Doc ID 86-890001378S, NTIS, Springfield, VA, USA

Ruiz-Genao DP, de Vega MJM, Perez JS, Garcia-Díez A (2003) Labial edema due to an acrylic dental prosthesis. *Contact Dermatitis 48*: 273–274

Rustemeyer T, de Groot J, von Blomberg BME, Frosch PJ, Scheper RJ (1998) Cross-reactivity patterns of contact-sensitizing methacrylates. *Toxicol Appl Pharmacol 148*: 83–90

Rustemeyer T, de Groot J, von Blomberg BME, Frosch PJ, Scheper RJ (2001) Induction of tolerance and cross-tolerance to methacrylate contact sensitizers. *Toxicol Appl Pharmacol 176*: 195–202

Rustemeyer T, Preuss M, von Blomberg BME, Das PK, Scheper RJ (2003) Comparison of two *in vitro* dendritic cell maturation models for screening contact sensitizers using a panel of methacrylates. *Exp Dermatol 12*: 682–691

Saccabusi S, Boatto G, Asproni B, Pau A (2001) Sensitization to methyl methacrylate in the plastic catheter of an insulin pump infusion set. *Contact Dermatitis 45*: 47–48

Savonius B, Keskinen H, Tuppurainen M, Kanerva L (1993a) Occupational respiratory disease caused by acrylates. *Clin Exp Allergy 23*: 416–424

Savonius B, Keskinen H, Tuppurainen M, Kanerva L (1993b) Erratum: occupational respiratory disease caused by acrylates. *Clin Exp Allergy 23*: 712

Scherpereel A, Tillie-Leblond I, Pommier de Santi P, Tonnel AB (2004) Exposure to methyl methacrylate and hypersensitivity pneumonitis in dental technicians. *Allergy 59*: 890–892

Schnuch A, Uter W, Geier J, Frosch PJ, Rustemeyer T (1998) Contact allergies in healthcare workers. Results from the IVDK. *Acta Derm Venereol (Stockh) 78*: 358–363

Seppalainen AM, Rajaniemi R (1984) Local neurotoxicity of methyl methacrylate among dental technicians. *Am J Ind Med 5*: 471–477

Shaffer MP, Belsito DV (2000) Allergic contact dermatitis from glutaraldehyde in health care workers. *Contact Dermatitis 43*: 150–156

Singh AR, Lawrence WH, Autian J (1972) Embryonic-fetal toxicity and teratogenic effects of a group of methacrylate esters in rats. *J Dent Res 51*: 1632–1638

Solomon HM, McLaughlin JE, Swenson RE, Hagan JV, Wanner FJ, O'Hara GP, Krivanek ND (1993) Methyl methacrylate: inhalation developmental toxicity study in rats. *Teratology 48*: 115–125

Sood A, Taylor JS (2003) Acrylic reactions: a review of 56 cases. *Contact Dermatitis 48*: 346–347

Svartling N, Pfäffli P, Tarkkanen L (1986) Blood levels and half-life of methyl methacrylate after tourniquet release during knee arthoplasty. *Arch Orthop Trauma Surg 105*: 36–39

Tucker SC, Beck MH (1999) A 15-year study of patch testing to (meth)acrylates. *Contact Dermatitis 40*: 278–279

Vilaplana J, Romaguera C (2000) Contact dermatitis and adverse oral mucous membrane reactions related to the use of dental prostheses. *Contact Dermatitis 43*: 183–185

Vozmediano J, Manrique A (1998) Active sensitization to (meth)acrylates. *Contact Dermatitis 39*: 314

Walker AM, Cohen AJ, Loughlin JE, Rothman KJ, DeFonso LR (1991) Mortality from cancer of the colon or rectum among workers exposed to ethyl acrylate and methyl acrylate. *Scand J Work Environ Health 17*: 7–19

WHO (World Health Organisation) (1998) *Methyl methacrylate*. Concise International Chemical Assessment Document 4, WHO, Geneva, Switzerland

Wittczak T, Palczynski C, Szulc B, Gorski P (1996) Bronchial asthma with inflammation of the nose mucous membrane induced by occupational exposure to methyl methacrylate in a dental technician (Polish). *Med Pr 47*: 259–266

Wittczak T, Walusiak J, Ruta U, Palczynski C (2003) Occupational asthma and allergic rhinitis due to xerographic toner. *Allergy 58*: 957
Wrangsjö K, Swartling C, Meding B (2001) Occupational dermatitis in dental personnel: contact dermatitis with special reference to (meth)acrylates in 174 patients. *Contact Dermatitis 45*: 158–163

completed 09.03.2005

N-Methyl-2-pyrrolidone (vapour)[1]

Supplement 2002

MAK value (1994)	**19 ml/m³ (ppm) ≙ 80 mg/m³**
Peak limitation (2002)	**Category II, excursion factor 2**
Absorption through the skin (1992)	**H**
Sensitization	**–**
Carcinogenicity	**–**
Prenatal toxicity (1994)	**Pregnancy Risk Group C**
Germ cell mutagenicity	**–**
BAT value	**–**
Synonyms	1-methylazacyclopentan-2-one *N*-methyl-2-ketopyrrolidine *N*-methyl-2-oxypyrrolidine *N*-methylpyrrolidinone *N*-methyl-2-pyrrolidinone 1-methyl-5-pyrrolidinone *N*-methylpyrrolidone 1-methyl-2-pyrrolidone
Chemical name (CAS)	1-methyl-2-pyrrolidinone
CAS number	872-50-4

Peak Limitation Category

In a multi-generation study in rats, the body weight development of the pups was delayed after exposure to *N*-methyl-2-pyrrolidinone concentrations of 116 ml/m³. The NOEC (no observed effect concentration) was 206 mg/m³ (50 ml/m³). The MAK value was therefore set at 80 mg/m³ (19 ml/m³) ("*N*-methyl-2-pyrrolidone", Volume 10, present series). In a 2-year study, no histopathological damage was found in the noses of rats after concentrations of 400 mg/m³ (about 100 ml/m³) (Lee *et al.* 1987). Eye irritation and headaches were reported after exposure at the workplace to 0.7 ml/m³.

[1] A more recent supplement follows.

The MAK-Collection Part I: MAK Value Documentations, Vol. 26. DFG, Deutsche Forschungsgemeinschaft

ISBN: 978-3-527-32306-7

Concentrations of 16 ml/m³ were described as "immediately uncomfortable". Concentrations of 49 to 83 ml/m³ were reported as being "unbearable". These were 8-hour mean values, and the occurrence of peak concentrations can be assumed, which could have been responsible for the effects (Beaulieu and Schmerber 1991).

In a study with volunteers, however, no symptoms were observed after exposure for 8 hours to 12 ml/m³ (Åkesson and Paulsson 1997). These findings contradict the results of Beaulieu and Schmerber (1991) with exposure to 16 ml/m³. The results of the study by Åkesson and Paulsson (1997) are considered to be more significant as the level of exposure was well monitored. It is unclear whether irritation is the critical effect. The classification of the substance in Peak Limitation Category II has therefore been provisionally retained, and an excursion factor of 2 has been set.

7 References

Åkesson B, Paulsson K (1997) Experimental exposure of male volunteers to *N*-methyl-2-pyrrolidone (NMP): acute effects and pharmacokinetics of NMP in plasma and urine. *Occup Environ Med 54*: 236–240

Beaulieu HJ, Schmerber KR (1991) M-Pyrol™ (NMP) use in the microelectronics industry. *Appl Occup Environ Hyg 6*: 874–880

Lee KP, Chromey NC, Culik R, Barnes JR, Schneider PW (1987) Toxicity of *N*-methyl-2-pyrrolidone (NMP): teratogenic, subchronic, and two-year inhalation studies. *Fundam Appl Toxicol 9*: 222–235

completed 24.04.2002

N-Methyl-2-pyrrolidone (vapour)

Supplement 2006

MAK value (1994)[1]	**20 ml/m³ (ppm) ≙ 82 mg/m³**
Peak limitation (2002)	**Category II, excursion factor 2**
Absorption through the skin (1992)	**H**
Sensitization	–
Carcinogenicity	–
Prenatal toxicity (1994)	**Pregnancy Risk Group C**
Germ cell mutagenicity	–
BAT value[2]	**150 mg 5-hydroxy-*N*-methyl-2-pyrrolidone/l urine**
Synonyms	1-methylazacyclopentan-2-one *N*-methyl-2-ketopyrrolidine *N*-methyl-2-oxypyrrolidine *N*-methylpyrrolidinone *N*-methyl-2-pyrrolidinone 1-methyl-5-pyrrolidinone *N*-methylpyrrolidone 1-methyl-2-pyrrolidone
Chemical name (CAS)	1-methyl-2-pyrrolidinone
CAS number	872-50-4
1 ml/m³ (ppm) ≙ 4.120 mg/m³	**1 mg/m³ ≙ 0.243 ml/m³ (ppm)**

Since publication of the documentation for *N*-methyl-2-pyrrolidone (NMP) in 1994 ("*N*-Methyl-2-pyrrolidone", Volume 10, present series), studies have been carried out that make it necessary to reassess the MAK value, the ability of *N*-methyl-2-pyrrolidone to penetrate the skin, the pregnancy risk group and the carcinogenic effects.

---

[1] adjusted to preferred value in 1998
[2] BAT value established in 2007

The MAK-Collection Part I: MAK Value Documentations, Vol. 26. DFG, Deutsche Forschungsgemeinschaft

ISBN: 978-3-527-32306-7

1 Toxic Effects and Mode of Action

N-Methyl-2-pyrrolidone is an organic solvent with low vapour pressure, which is readily absorbed through the skin. During inhalation exposure, aerosol formation occurs with increasing concentration, temperature and humidity; the aerosol can condense onto the skin and be absorbed through the skin. However, at workplace exposure levels not exceeding the MAK value, aerosol formation is not to be expected at normal humidity ("*N*-Methyl-2-pyrrolidone", Volume 10, present series).

Undiluted *N*-methyl-2-pyrrolidone is a skin irritant and causes skin damage, which leads to its increased uptake. *N*-Methyl-2-pyrrolidone solutions of 50% were found to be irritating to rabbit skin, while solutions of 5% were non-irritant. The uptake of *N*-methyl-2-pyrrolidone through the skin is increased in the presence of organic solvents such as *d*-limonene. The irritation threshold for *N*-methyl-2-pyrrolidone in the human respiratory tract is not known. Concentrations of 20 ml/m^3 were not found to have irritative effects in volunteers.

In most animal studies there is little conformity to the profile of effects of *N*-methyl-2-pyrrolidone after repeated exposure. Effects are observed only at high concentrations or doses. In Wistar rats, the aerosol caused irritation after concentrations of 1000 mg/m^3 (243 ml/m^3) and above and changes in the haematopoietic system and testes after concentrations of 3000 mg/m^3 (730 ml/m^3) and above ("*N*-Methyl-2-pyrrolidone", Volume 10, present series). In studies with oral administration of *N*-methyl-2-pyrrolidone, toxic effects were found after 3 months at concentrations of 7500 mg/kg diet (433 and 565 mg/kg body weight and day for male and female animals) and above, and effects on the testes after administration of 15000 mg/kg diet (678 mg/kg body weight and day) for 2 years. In carcinogenicity studies with administration of *N*-methyl-2-pyrrolidone in the diet, an increase in liver tumours was observed in mice in the high concentration group after *N*-methyl-2-pyrrolidone doses of more than 1000 mg/kg body weight and day. *N*-Methyl-2-pyrrolidone was not, however, found to be genotoxic *in vitro* and *in vivo* in the relevant studies.

Recent developmental toxicity studies with inhalation exposure and oral administration of *N*-methyl-2-pyrrolidone confirm the findings of the earlier investigations. Inhalation exposure to concentrations of 120 ml/m^3 and oral doses of 400 mg/kg body weight and day led to reduced maternal body weight gains; reduced body weight gains were found also in the progeny. Variations and malformations were observed after oral exposure to doses of 500 mg/kg body weight and day and above. *N*-Methyl-2-pyrrolidone was not found to cause sensitization.

2 Mechanism of Action

Administration with the diet of *N*-methyl-2-pyrrolidone doses of more than 1000 mg/kg body weight and day for 18 months led to the formation of liver tumours in mice (Malley *et al.* 2001; NMP Producers Group 1999a). Studies were carried out specifically to clarify the mechanism of tumour formation.

Enzyme induction and peroxisome proliferation were investigated in male and female B6C3F$_1$ mice given *N*-methyl-2-pyrrolidone with the diet for 2 weeks. The dose used of 7200 mg/kg diet, corresponding to 1364 mg/kg body weight and day for male animals and 1945 mg/kg body weight and day for female animals, led to liver tumours in a carcinogenicity study performed with the same strain of mice. The levels of cytochrome P450, ethoxyresorufin *O*-deethylase (EROD) and pentoxyresorufin *O*-depentylase (PROD) were determined in 10 animals of each sex. Another 5 animals of each sex were tested for cyanidine-sensitive palmitoyl-CoA (PALCoA) oxidation. In addition, the liver was examined by means of light and electron microscopy for changes in structure or amount of peroxisomes, endoplasmic reticulum and mitochondria. *N*-Methyl-2-pyrrolidone led to a slight increase in PALCoA in male animals compared with the levels determined in untreated controls. A slight increase in the amount of peroxisomes was observed by electron microscope in 2 of the 5 male animals. There was no evidence of enzyme induction (NMP Producers Group 2002a).

In another study, cell proliferation (S-phase response) was investigated in 10 male and 10 female B6C3F$_1$ mice after administration with the diet for 1 and 4 weeks of the same *N*-methyl-2-pyrrolidone doses used in the experiment described above (NMP Producers Group 2002a). In addition to general parameters, such as body weight and food consumption, also the uptake of bromodeoxyuridine (BrdU) into the DNA of the liver was evaluated; this was carried out microscopically following immunohistochemical staining. One week before the animals were killed, BrdU was administered subcutaneously by means of a mini pump. After one week of exposure, the incidence of cell proliferation in the liver increased by a factor of 6.9 in the male mice and a factor of 3.3 in the female mice relative to that in the untreated control group. The rate of mitosis was increased in the male animals. Minimal or slight centrilobular hypertrophy of the liver was observed, particularly in the male animals (9/10). The incidence in the female mice was 1/10 animals. Fat deposition was decreased in the males; in the female animals a loss of deposited fat was observed and liver weights were reduced. After treatment for 4 weeks with *N*-methyl-2-pyrrolidone, the average body weight of the male animals was 4.6% below that of the controls. The incidence of cell proliferation was increased by a factor of 2.1 in the male animals and by a factor of 1.7 in the female animals. The number of apoptotic cells was increased in the male mice. Minimal or slight centrilobular hypertrophy was observed in 7/10 male and 2/10 female mice. The effects on the fat content of the liver were comparable with those found in the one week experiment. All in all, the study produced clear evidence that *N*-methyl-2-pyrrolidone administered with the diet for 1 or 4 weeks causes cell proliferation in the liver (NMP Producers Group 2002b).

3 Toxicokinetics and Metabolism

3.1 Absorption, distribution and elimination

The available studies of absorption show that *N*-methyl-2-pyrrolidone is readily absorbed after inhalation and ingestion, and by the skin.

3.1.1 Inhalation

Rats exposed to ^{14}C-NMP concentrations of 10 ml/m^3 for 6 hours absorbed 7.3% of the radioactivity and after exposure to 100 ml/m^3, 9.5% of the radioactivity. 15% to 20% of the radioactivity absorbed was not recovered; according to the authors, it was probably eliminated with lost urine. The elimination half-life of the radioactivity in the plasma was 2 to 3 hours. Most of the radioactivity was eliminated with the urine within 12 hours and made up about 70% to 80% of the amount absorbed. 5% to 12% was eliminated with the faeces. The main urinary metabolite was identified as 5-hydroxy-*N*-methyl-2-pyrrolidone (NMP Producers Group 1995a).

3.1.2 Ingestion

The elimination of the radioactivity from the plasma of rats given single oral ^{14}C-NMP doses of 5 and 50 mg took approximately 3 and 3.8 hours, respectively. After single ^{14}C-NMP doses of 50 mg/kg body weight followed by 7 non-labelled doses of NMP, *N*-methyl-2-pyrrolidone was eliminated from the plasma faster than after single doses. Regardless of the dose, 69% to 86% of the radioactivity was found in the urine and 5% to 9% in the faeces (NMP Producers Group 1995a).

3.1.3 Dermal absorption

Particularly absorption through the skin has been studied extensively, as this often represents the main form of exposure in humans. The details of the studies carried out to determine the rate of absorption of radioactively labelled *N*-methyl-2-pyrrolidone *in vitro* are found in Table 1, the details of the studies to determine the rate of absorption *in vivo* are found in Table 2.

Table 1. *In vitro* and *in vivo* studies of the rate of dermal absorption of radioactively labelled *N*-methyl-2-pyrrolidone

Skin sample	NMP concentration	NMP dose (volume)	Exposure period (hours)	Maximum rate of absorption (µg/cm² and hour)	References
in vitro					
human, thorax, dermatomed, 1–2 mm	no other details (presumably 100%)	800 mg/cm² (500 µl/0.64 cm²)	6	17100	Ursin *et al.* 1995
human, breast, back or abdomen, stripped, 0.2–0.4 mm	0.1% in H_2O	0.4 mg/cm² (250 µl/0.64 cm²)	0–3 3–7	0.14 0.32	NMP Producers Group 2003b
	0.3% in H_2O	1.2 mg/cm² (250 µl/0.64 cm²)	0–3 3–7	0.36 1.0	
	1.0% in H_2O	4.0 mg/cm² (250 µl/0.64 cm²)	0–3 3–7	1.33 2.94	
	3.0% in H_2O	12 mg/cm² (250 µl/0.64 cm²)	0–3 3–7	3.51 8.57	
	10% in H_2O	40 mg/cm² (250 µl/0.64 cm²)	0–3 3–7	16.9 35.7	
	30% in H_2O	120 mg/cm² (250 µl/0.64 cm²)	0–3 3–7	36.4 80.4	
human, breast, abdomen, stripped, 0.2–0.4 mm	30% in H_2O	2.8 mg/cm² (6 µl/0.64 cm²)	3	579	NMP Producers Group 2002c
		110 mg/cm² (250 µl/0.64 cm²)	8	114	
	100%	9.7 mg/cm² (6 µl/0.64 cm²)	3	1650	
		400 mg/cm² (250 µl/0.64 cm²)	6.5	10057	
	65% in *d*-limonene	7.3 mg/cm² (6 µl/0.64 cm²)	3	6331	
		258 mg/cm² (250 µl/0.64 cm²)	5	64957	
rat, dorsum, stripped, 0.2–0.4 mm	30% in H_2O	2.8 mg/cm² (6 µl/0.64 cm²)	3	950	NMP Producers Group 2002c
	100%	9.7 mg/cm² (6 µl/0.64 cm²)	3	3113	
	65% in *d*-limonene	7.3 mg/cm² (6 µl/0.64 cm²)	3	12905	

Table 1. continued

Skin sample	NMP concentration	NMP dose (volume)	Exposure period (hours)	Maximum rate of absorption (µg/cm^2 and hour)	References
rat, dorsum, intact (full-thickness), 0.9–1.5 mm	3% in H_2O	13 mg/cm^2 (400 µl/cm^2)	52	4700 [T_{max} 32 hours]	Payan *et al.* 2003
	6% in H_2O	26 mg/cm^2 (400 µl/cm^2)	52	5700 [T_{max} 34 hours]	
	12.5% in H_2O	52 mg/cm^2 (400 µl/cm^2)	52	7600 [T_{max} 35 hours]	
	50% in H_2O	206 mg/cm^2 (400 µl/cm^2)	52	5100 [T_{max} 45 hours]	
	100%	412 mg/cm^2 (400 µl/cm^2)	24	7700 [T_{max} 6.1 hours]	
	100%	103 mg/cm^2 (100 µl/cm^2)	24	5300 [T_{max} 3.4 hours]	
	100%	26 mg/cm^2 (25 µl/cm^2)	24	2100 [T_{max} 1.9 hours]	
in vivo					
rat, dorsum, occlusive	100%	21 mg/cm^2 (20 µl/cm^2)	2	9700 [T_{max} 30 min]	Payan *et al.* 2003
		42 mg/cm^2 (40 µl/cm^2)	2	23400 [T_{max} 45 min]	

T_{max}: time until the maximum rate of absorption is achieved

Absorption rates

After application of a limited volume (finite application) of *N*-methyl-2-pyrrolidone of 6 µl/0.64 cm^2 to human and rat skin, the rates of absorption were, as to be expected after application of undiluted *N*-methyl-2-pyrrolidone, about 3 times higher than after application of a 30% aqueous *N*-methyl-2-pyrrolidone solution. After application of an excess (infinite application) of undiluted *N*-methyl-2-pyrrolidone of 250 µl/0.64 cm^2 to human skin, the absorption rate of approximately 10000 µg/cm^2 and hour was almost 100 times higher than after infinite application of a 30% aqueous *N*-methyl-2-pyrrolidone solution (114 µg/cm^2 and hour). For undiluted *N*-methyl-2-pyrrolidone the rate of absorption after infinite application was almost 8 times higher than after finite application (NMP Producers Group 2002c). A possible explanation could be the skin damage caused by the great excess of undiluted *N*-methyl-2-pyrrolidone used in the infinite application; undiluted *N*-methyl-2-pyrrolidone was shown in this study to cause skin damage (see below) (NMP Producers Group 2002c). In another *in vitro* study with infinite application (250 µl/0.64 cm^2) of aqueous *N*-methyl-2-pyrrolidone solutions of various concentrations to human skin, the rate of absorption was found to increase proportionally with increasing *N*-methyl-2-pyrrolidone concentration. For the 30% aqueous *N*-methyl-2-pyrrolidone solution, the rate of absorption was 80 µg/cm^2 and day after application for 3 to 7 hours (NMP Producers Group 2003b) and thus was similar to the rate of absorption found in the study described above with infinite application of a 30% aqueous *N*-methyl-2-pyrrolidone solution for 8 hours (114 µg/cm^2 and hour) (NMP Producers Group 2002c).

Intact (full thickness) rat skin yielded an absorption rate of 7700 µg/cm^2 and hour (Payan *et al.* 2003) after infinite application of undiluted *N*-methyl-2-pyrrolidone *in vitro*; this was slightly lower than the rate of absorption yielded with the markedly thinner stripped human skin of about 10000 µg/cm^2 and hour (NMP Producers Group 2002c). After finite applications (6 µl/0.64 cm^2) of undiluted or 30% *N*-methyl-2-pyrrolidone, the rates of absorption for stripped rat skin were about twice that for stripped human skin of the same thickness (NMP Producers Group 2002c). This demonstrates that rat skin and human skin are of similar permeability for *N*-methyl-2-pyrrolidone.

In vitro application of *N*-methyl-2-pyrrolidone (65%) in *d*-limonene to rat and human skin markedly increased the rate of absorption compared with that after undiluted *N*-methyl-2-pyrrolidone: after finite application to human and rat skin by about 4 times and after infinite application to human skin by about 6 times (NMP Producers Group 2002c).

Undiluted *N*-methyl-2-pyrrolidone applied to human skin in infinite volumes (250 µl/0.64 cm^2) for 0.5 to 4 hours led to increasing skin damage, the most severe effects within the first hour. The integrity of the skin was analyzed by determining the permeation of tritium-labelled water before and after the application of *N*-methyl-2-pyrrolidone. The authors concluded that the skin damage during the first hour facilitates the absorption of *N*-methyl-2-pyrrolidone (NMP Producers Group 2002c). Corresponding investigations of the integrity of the skin with aqueous *N*-methyl-2-pyrrolidone solutions are not available.

Absorption *in vivo*

After dermal application of 20 µl undiluted ^{14}C-NMP/cm^2 for 2 hours to 10 cm^2 of the shaved dorsal skin of rats and occlusion with a plastic cap, 60% of the substance was absorbed. The maximum rate of absorption of 9.7 µg/cm^2 and hour was reached after 30 minutes of exposure and thereafter decreased; the authors attributed this to the hygroscopic properties of *N*-methyl-2-pyrrolidone and the resulting dilution with water. After the application of 40 µl/cm^2 the maximum rate of absorption was achieved after 45 minutes and amounted to 23.4 µg/cm^2 and hour. After occlusive application of 20 µl/cm^2 for 24 hours, 80% was absorbed (Payan *et al.* 2003).

Table 2. *In vivo* studies of the dermal absorption of radioactively labelled *N*-methyl-2-pyrrolidone in rats

Occlusion	NMP concentration	NMP dose (applied volume)	Exposure period (hours)	Absorption (%)	References
plastic cap	100%	206 mg/animal (20 µl/cm², 10 cm²)	2 24	60 85	Payan *et al.* 2003
carbon filter (the fraction not absorbed was found mainly in the filter)	3% in H_2O	0.4 mg/animal (120 µl/animal)	24	0.6	NMP Producers Group 1998
	10% in H_2O	12 mg/animal	24	1.3	
	30% in H_2O	37 mg/ animal	24	3.5	
	65% in H_2O	80 mg/animal	24	14.2	
	100%	123 mg/animal	24	30	
	3% in *d*-limonene	0.4 mg/animal	24	16	
	10% in *d*-limonene	12 mg/animal	24	55	
	30% in *d*-limonene	37 mg/animal	24	70	
	65% in *d*-limonene	46 mg/animal	24	37	
aluminium foil	in H_2O (no other details)	about 50 mg/ animal (100 µl/ animal; 10 mg/kg body weight)	6	43 (♀) 44 (♂)	NMP Producers Group 1995a
non-occlusive	100%	103 mg/animal (100 µl/animal, 10 cm²)	1	69	NMP Producers Group 2003a
carbon filter	100%	103 mg/animal (100 µl/animal, 10 cm²)	1	57	
aluminium foil	100%	103 mg/animal (100 µl/animal, 10 cm²)	1	50	

In an *in vivo* study of absorption, undiluted ^{14}C-NMP (10 µl/cm²; 120 µl/animal) was applied to the shaved dorsal skin of rats and a carbon filter was used for occlusion (NMP Producers Group 1998). This study is, however, only of limited relevance as *N*-methyl-2-pyrrolidone was adsorbed onto the carbon filter and thus could not be absorbed dermally to full extent. Within 24 hours 30% of the radioactivity administered had been absorbed. The highest levels of activity were found in the blood, liver and kidneys 3 to 6 hours after the application. After the application of aqueous *N*-methyl-2-pyrrolidone solutions, the rates of absorption were lower: 0.6% (3% NMP), 1.3% (10% NMP), 3.5% (30% NMP) and 14.2% (65% NMP). The application of *N*-methyl-2-pyrrolidone in limonene enhanced the absorption of *N*-methyl-2-pyrrolidone: 16% (3% NMP), 54% (10% NMP), 70% (30% NMP), 65% (37% NMP) (NMP Producers Group 1998).

^{14}C-*N*-methyl-2-pyrrolidone doses of 10 mg/kg body weight in aqueous solution (100 µl/animal) were applied to the shaved dorsal skin of rats for 6 hours, and the area was covered with aluminium foil and the foil taped down; 43% to 44% of the applied radioactivity was absorbed. The elimination half-life for the radioactivity from the plasma was about 6 hours. 36% to 57% of the radioactivity absorbed was eliminated with the urine, most of it within 36 hours. 3% to 5.3% was found in the faeces and 26% to 54% was found in the treated areas of skin (NMP Producers Group 1995a). One study examined the effects of different occlusion conditions on the absorption of undiluted ^{14}C-labelled *N*-methyl-2-pyrrolidone (100 µl/animal) through the shaved dorsal skin of rats over a period of one hour. The highest level of absorption of almost 69% was achieved with non-occlusive application. Under semi-occlusive conditions (carbon filter 0.5 and 2 cm above the skin to determine the evaporating *N*-methyl-2-pyrrolidone), 57% was absorbed and under occlusive conditions (aluminium foil) 50% was absorbed (NMP Producers Group 2003a). The lower level of absorption with complete occlusion may be explained by the mixing of *N*-methyl-2-pyrrolidone with transdermally secreted water. This endogenous water cannot escape and thus, as a result of the dilution of the substance, leads to a lower level of absorption through the skin.

3.2 Metabolism

N-Methyl-2-pyrrolidone is hydroxylated to 5-hydroxy-*N*-methyl-2-pyrrolidone (5-HNMP), which oxidizes to *N*-methylsuccinimide (MSI) and is further hydroxylated to 2-hydroxy-*N*-methylsuccinimide (2-HMSI) (Åkesson and Jönsson 1997; Jönsson and Åkesson 2001). In humans and rats, 5-HNMP is the main urinary metabolite after inhalation, ingestion and dermal absorption. Particularly 5-HNMP, but also MSI and 2-HMSI, are used for biological monitoring investigations (Åkesson and Jönsson 2000; Åkesson and Paulsson 1997; Akrill *et al.* 2002; Anundi *et al.* 2000; Bader *et al.* 2003; BASF 2003; Jönsson and Åkesson 2001, 2003; Ligocka *et al.* 2002; NMP Producers Group 2003c; Penney 2003). Also *N*-methyl-2-pyrrolidone itself has been suggested as a marker for biological monitoring in plasma and urine (Xiaofei *et al.* 2000).

In comparative studies with rats and human microsomes *in vitro* and with volunteers, it was demonstrated that a specific cytochrome P450 isoenzyme (CYP2E1) is substantially involved in the metabolism of *N*-methyl-2-pyrrolidone in rats, but only to a lesser extent in humans (Ligocka *et al.* 2003).

4 Effects in Humans

To determine chemosensory effects, 15 healthy male non-smokers were exposed to constant *N*-methyl-2-pyrrolidone concentrations of 10, 40 or 80 mg/m^3 (2.4, 9.6, 19.2 ml/m^3) for 8 hours, or on one day to four increasing *N*-methyl-2-pyrrolidone concentrations of 25 mg/m^3 to 160 mg/m^3 (15-minute exposure peaks). The volunteers were exposed in an exposure chamber. In an additional experiment to determine purely dermal uptake of *N*-methyl-2-pyrrolidone, volunteers were exposed to constant *N*-methyl-2-pyrrolidone concentrations of 80 mg/m^3 in an exposure chamber while the inhalation of *N*-methyl-2-pyrrolidone was prevented by a face mask that provided filtered fresh air. The exposure was carried out both without and with temporary physical activity on a bicycle ergometer (6 × 10 minutes at 75 watts). The following end points were investigated: palpebral closure frequency, nasal airflow, neuropsychological attention function tests, subjective acute symptoms, intensity of chemosensory sensations, breathing frequency and changes in the odour threshold. In a preliminary data analysis, no evidence of irritative effects was found under the exposure conditions tested. The analyses showed that physical activity did not influence the variables for the chemosensory effects tested. After *N*-methyl-2-pyrrolidone concentrations of 40 mg/m^3, however, an increase in odour intensity and odour nuisance was observed. After 80 mg/m^3 and increasing concentrations up to 160 mg/m^3, the odour intensity was moderate and the odour nuisance mild to moderate. The parameter of trigeminal sensation "pungent" was rated as weak only at exposure peaks of 160 mg/m^3. The authors concluded from the findings that, because of the very weak trigeminal sensations and the lack of trigeminally mediated symptoms or physiological indicators of sensory irritation, concentrations of up to 80 mg/m^3 lead only to olfactory effects, but not to irritation (NMP Producers Group 2005).

Six male volunteers were exposed on 4 different days to *N*-methyl-2-pyrrolidone concentrations of 0, 10, 25 or 50 mg/m^3 (2.4; 6.1; 12.2 ml/m^3) for 8 hours in an exposure chamber. At the end of the exposure, *N*-methyl-2-pyrrolidone concentrations of 0.33, 0.99 and 1.6 mg/l were determined in the plasma. The average elimination rate of *N*-methyl-2-pyrrolidone in the urine from the beginning of exposure until 2 hours after the end of exposure was 27, 111 and 275 µg/hour. Blood tests were carried out to determine the number of leukocytes, neutrophils, eosinophils, lymphocytes, basophils, monocytes and thrombocytes, and the serum concentrations of IgE, bilirubin, alkaline phosphatase, γ-glutamyl transferase, aspartate aminotransferase and alanine aminotransferase. Rhinometric examinations and lung function tests were performed in addition. In a questionnaire the volunteers were asked about odour perception, irritation of the eyes, nose and respiratory tract, discomfort and other irritative effects. The exposure to *N*-methyl-2-pyrrolidone was not found to influence any of the end points investigated (Åkesson and Paulsson 1997).

Headaches and chronic eye irritation were reported even at *N*-methyl-2-pyrrolidone concentrations as low as 0.7 ml/m^3 in a publication from the microelectronics industry, in which, according to the authors, the *N*-methyl-2-pyrrolidone concentrations determined using stationary air sampling devices are usually between 0.02 and

1.5 ml/m^3. *N*-Methyl-2-pyrrolidone concentrations of 67 ml/m^3 were described as "immediately unbearable" (Beaulieu and Schmerber 1991). A conclusive evaluation of the reported findings cannot be made in view of the inadequate description of the methods and findings, possible additional effects resulting from the warming of *N*-methyl-2-pyrrolidone to 85°C or to over 140°C, and physical impairment from ventilation systems or other chemicals. The findings have, therefore, not been used to derive a MAK value.

5 Animal Experiments and *in vitro* Studies

5.1 Subacute, subchronic and chronic toxicity

5.1.1 Ingestion

Details of the studies are given in Table 3.

Rats

Groups of 5 male and 5 female Sprague-Dawley rats were given *N*-methyl-2-pyrrolidone concentrations of 0, 2000, 6000, 18000 or 30000 mg/kg with the diet for 28 days. The first effects to be seen were reduced body weight gains and reduced food consumption in the male animals after concentrations of 18000 mg/kg diet and above, and in the female animals after 30000 mg/kg diet. Changes in clinicochemical parameters appeared after concentrations of 6000 mg/kg diet and above (reduced serum albumin concentration) in the male animals and after 18000 mg/kg diet and above (reduced serum albumin and total protein concentrations) in the females. In addition, centrilobular hypertrophy was observed in the liver of male and female animals after concentrations of 18000 mg/kg diet and above. Atrophy and degeneration of the seminiferous tubules in the testes were observed in 1/5 male animals after concentrations of 18000 mg/kg diet and above, and in all the animals of the high dose group. After 30000 mg/kg diet also haematological changes, such as a slightly reduced number of lymphocytes, and histopathological changes, such as reduced cell density in the bone marrow and thymus atrophy, were found in male and female animals. Yellow discoloration of the urine was observed after concentrations of 18000 mg/kg diet and above, which is considered to be evidence of the bioavailability of the substance and not an adverse effect. The NOAEL (no observed adverse effect level) in this study was determined by the authors to be 6000 mg/kg diet (429 mg/kg body weight and day) for the male animals and 18000 mg/kg diet (1548 mg/kg body weight and day) for the female animals (Malek *et al.* 1997). As a result of the liver cell hypertrophy observed in animals of both sexes at 18000 mg/kg diet, the NOAEL was given as 6000 mg/kg diet (429 and 493 mg/kg body weight and day) for both sexes.

Table 3. Studies with repeated oral administration of *N*-methyl-2-pyrrolidone

Species, strain, number of animals per sex and group	Exposure	Findings	References
rat Sprague-Dawley (Crl:CD®BR), 5 ♂ and 5 ♀	**4 weeks**, 0, 2000, 6000, 18000, 30000 mg/kg diet (about 0, 149/161, 429/493, 1234/1548, 2019/2269 (♂/♀) mg/kg body weight and day)	**6000 mg/kg diet**: NOAEL **≥6000 mg/kg diet**: serum albumin (♂) decreased **≥18000 mg/kg diet:** body weight gains (♂) decreased, food consumption (♂) decreased, yellow discoloration of the urine, serum glucose (♂) decreased, total protein (♀) decreased, serum albumin (♂+♀) decreased, liver cell hypertrophy (♂+♀), testicular atrophy and degeneration **30000 mg/kg diet:** body weight gains (♀) decreased, food consumption (♀) decreased, lymphocytes (♂+♀) decreased, cholesterol (♂+♀) increased, alkaline phosphatase (♂) decreased, thymus atrophy (♂+♀)	Malek *et al.* 1997
rat Sprague-Dawley (Crl:CD®BR), 20–26 ♂ and 20–26 ♀	**3 months** 0, 3000, 7500, 18 000 mg/kg diet (about 0, 169/217, 433/565, 1057/1344 (♂/♀) mg/kg body weight and day)	**3000 mg/kg diet**: NOAEL **≥3000 mg/kg diet**: yellow discoloration of the urine **≥7500 mg/kg diet**: body weight gains decreased, food consumption decreased, number of leukocytes and lymphocytes (♂) decreased, foot splay (♂) **18000 mg/kg diet**: relative liver weights (♀) increased, centrilobular liver cell hypertrophy (♀), haemosiderin in the spleen (♂+♀), relative kidney weights (♂+♀) increased, number of leukocytes and lymphocytes (♀) decreased, "low arousal" (♂), "slight palpebral closure" (♂)	Malley *et al.* 1999; NMP Producers Group 1995b
rat Sprague-Dawley (Crl:CD®BR), 62 ♂ and 62 ♀	**2 years** 0, 1600, 5000, 15000 mg/kg diet (about 0, 66/88, 207/283, 678/939 (♂/♀) mg/kg body weight and day)	**5000 mg/kg diet**: NOAEL (♂+♀) **≥5000 mg/kg diet**: yellow discoloration of the urine **15 000 mg/kg diet**: body weight gains decreased, food consumption decreased, survival (♂) decreased, severe chronic progressive nephropathy (♂) increased, accumulation of macrophages containing pigment in the spleen (♂+♀), depletion of mesenteric lymph nodes (♂), bilateral degeneration and atrophy of the seminiferous tubules in the testes (♂), oligospermia (♂), osteodystrophy (♂)	Malley *et al.* 2001

Table 3. continued

Species, strain, number of animals per sex and group	Exposure	Findings	References
mouse B6C3F$_1$, 5 ♂ and 5 ♀	**4 weeks**, 0, 500, 2500, 7500, 10000 mg/kg diet (about 0, 130/180, 720/920, 2130/2970, 2670/4060 (♂/♀) mg/kg body weight and day)	**2500 mg/kg diet**: NOAEL (♂) **≥2500 mg/kg diet**: yellow discoloration of the urine **7500 mg/kg diet**: NOAEL (♀) **≥7500 mg/kg diet**: swelling of the epithelium of the distal kidney tubules (♂) **10000 mg/kg diet**: intercurrent death (1/5 ♂), swelling of the epithelium of the distal kidney tubules (♂+♀), alkaline phosphatase decreased	Malek *et al.* 1997; NMP Producers Group 1994
mouse B6C3F$_1$, 10 ♂ and 10 ♀	**4 weeks,** **3 months**, 0, 1000, 2500, 7500 mg/kg diet (about 0, 277, 619, 1931 mg/kg body weight and day)	<u>**4 weeks:**</u> **1000 mg/kg diet**: NOAEL **≥2500 mg/kg diet**: yellow discoloration of the urine, cholesterol (♀) increased, triglycerides (♂) decreased **7500 mg/kg diet**: alkaline phosphatase (♂) decreased, calcium (♂) decreased <u>**3 months**:</u> **1000 mg/kg diet**: NOAEL **≥2500 mg/kg diet**: yellow discoloration of the urine, relative liver weights increased (♂), centrilobular liver cell hypertrophy (♂+♀)	Malley *et al.* 1999; NMP Producers Group 1995b
mouse B6C3F$_1$, 50 ♂ and 50 ♀	**18 months,** 0, 600, 1200, 7200 mg/kg diet (about 0, 89/115, 173/221, 1089/1399 (♂/♀) mg/kg body weight and day)	**600 mg/kg diet**: NOAEL (♂) **1200 mg/kg diet:** NOAEL (♀) **≥1200 mg/kg diet**: yellow discoloration of the urine, body weight gains (♂) slightly decreased, relative liver weights (♂) increased, centrilobular liver cell hypertrophy (♂) **7200 mg/kg diet**: relative liver weights (♀) increased, liver foci increased, liver adenomas and liver carcinomas increased	Malley *et al.* 2001; NMP Producers Group 1999a

NOAEL: no observed adverse effect level

In the subsequent study, groups of 20 to 26 male and female Sprague-Dawley rats were given *N*-methyl-2-pyrrolidone in concentrations of 0, 3000, 7500 or 18000 mg/kg diet over a period of 3 months. Afterwards 10 male and 10 female rats of the control group and high dose group were observed for one month without treatment. Functional and morphological neurotoxicity tests were also part of the study. The main effects in this study were also on body weight, food consumption and efficiency and were observed after concentrations of 7500 mg/kg diet and above. After 18000 mg/kg diet, increased relative liver, lung and kidney weights, and reduced relative brain weights were observed in male and female animals; in addition, increased relative testis weights were observed in male animals. The increased relative liver weights in female animals result from adaptive liver cell hypertrophy. Slight irregularities were detected in the male animals for only 3 of the 36 parameters in the neurotoxicity tests. Foot splay was found after concentrations of 7500 mg/kg diet and above. As a result of a slight sedation, "low arousal" was observed in the group given 18000 mg/kg diet and "slight palpebral closure" was found more often. Yellow discoloration of the urine observed after concentrations of 3000 mg/kg diet and above is not considered an adverse effect. The NOAEL in this study was found to be 3000 mg/kg diet, corresponding to 169 mg/kg body weight and day for male animals and 217 mg/kg body weight and day for female animals (Malley *et al.* 1999; NMP Producers Group 1995c).

In a carcinogenicity study, *N*-methyl-2-pyrrolidone concentrations of 0, 1600, 5000 or 15000 mg/kg were administered with the diet to groups of 62 male and 62 female Sprague-Dawley rats for 2 years. Survival of the female rats was not reduced, while that of the male rats of the high dose group was. Responsible for this is the increased incidence of the severe, chronic progressive nephropathy found in the male animals. This is a typical age-related finding, particularly in male rats. Apart from impairment of body weight development with correspondingly reduced body weights, reduced food consumption and poor efficiency in rats of both sexes, no other adverse substance-related findings were observed after 15000 mg/kg diet. After concentrations of 5000 mg/kg diet, yellow discoloration of the urine was observed, which is not considered to be an adverse effect. The NOAEL in this study was 5000 mg/kg diet, corresponding to 207 mg/kg body weight and day for male animals and 283 mg/kg body weight and day for female animals (Malley *et al.* 2001).

Mice

In B6C3F$_1$ mice, the administration of *N*-methyl-2-pyrrolidone in concentrations of 0, 500, 2500, 7500 or 10000 mg/kg diet for 4 weeks to 5 male and 5 female animals per group did not result in impairment of body weight development or food consumption. Swelling of the epithelium in the area of the distal kidney tubules was found in the male animals after concentrations of 7500 mg/kg diet and above, and in the female animals after 10000 mg/kg diet. One male animal of the high dose group died prematurely as a result of the kidney toxicity. The urine became yellowish in colour after concentrations of 2500 mg/kg diet and above. The NOAEL in this study was 2500 mg/kg diet, corresponding to 720 mg/kg body weight and day for male animals (Malek *et al.* 1997; NMP Producers Group 1994). As effects were not observed in female animals in this study until concentrations of 10000 mg/kg diet, the NOAEL for female animals is 7500 mg/kg diet, corresponding to 2970 mg/kg body weight and day.

In the subsequent study, groups of 10 male and 10 female B6C3F$_1$ mice were given *N*-methyl-2-pyrrolidone concentrations of 0, 1000, 2500 or 7500 mg/kg diet for 4 weeks (satellite group) or for 3 months (main group). After 4 weeks, slight changes in some clinicochemical parameters were observed after concentrations of 2500 mg/kg diet and above. These findings were, however, found to be transient, as they could no longer be detected 3 months later. After administration of *N*-methyl-2-pyrrolidone with the diet for 3 months, increased liver weights were found in the male mice given concentrations of 2500 mg/kg diet and above. Histological liver changes were seen in animals of both sexes in the form of centrilobular hypertrophy after concentrations of 2500 mg/kg diet (2/10 ♂, 3/10 ♀) and 7500 mg/kg diet (9/10 ♂, 10/10 ♀) (controls 1/10 ♂ and 1/10 ♀). Yellow discoloration of the urine was also found in this study after concentrations of 2500 mg/kg diet and above. The NOAEL in this study for male and female animals was 1000 mg/kg diet, corresponding to 277 mg/kg body weight and day (Malley *et al.* 1999; NMP Producers Group 1995b).

In a carcinogenicity study, groups of 50 male and 50 female $B6C3F_1$ mice were given *N*-methyl-2-pyrrolidone in concentrations of 0, 600, 1200 or 7200 mg/kg diet. Survival was not affected in either the male or the female animals. The absolute kidney weights were significantly reduced in the male animals of all exposure groups, as were the relative kidney weights in the male animals of the two low dose groups. As no clear dose–response relationship was evident and histopathological changes were not observed, the authors regarded these changes as not substance-related. The authors also regarded the increased relative testis weights in the male animals after concentrations of 600 mg/kg diet and above and increased absolute brain weights in the female animals after 600 mg/kg diet and above as not substance-related, as no histopathological changes occurred. Increased relative liver weights and centrilobular liver cell hypertrophy were found in male animals after 1200 mg/kg diet and above. After 7200 mg/kg diet, the relative liver weights were increased also in female animals, and increased incidences of preneoplastic foci of the liver, liver adenomas and liver carcinomas were found in animals of both sexes. The NOAEL in this study was 600 mg/kg diet (89 mg/kg body weight and day) for the male animals and 1200 mg/kg diet (221 mg/kg body weight and day) for the female animals (Malley *et al.* 2001; NMP Producers Group 1999a).

5.2 Local effects on skin and mucous membranes

The application of undiluted *N*-methyl-2-pyrrolidone to the shaved dorsal skin of rabbits for 5 to 15 minutes caused severe redness of the skin and subsequent scaling. Oedema developed after contact with the substance for 20 hours ("*N*-Methyl-2-pyrrolidone", Volume 10, present series). In a study of sensitizing effects (see below), a 5% aqueous *N*-methyl-2-pyrrolidone solution did not produce irritative effects, while a 50% solution led to slight erythema at the site of application after 24 hours (WHO 2001).

5.3 Allergenic effects

In a modified Draize test, a 5% *N*-methyl-2-pyrrolidone solution applied repeatedly did not produce sensitization in guinea pigs (no other details) ("*N*-Methyl-2-pyrrolidone", Volume 10, present series).

In an unpublished test by E. I. Du Pont de Nemours and Company from 1976, 10 male albino guinea pigs received intradermal injections of 0.1 ml of a 1% *N*-methyl-2-pyrrolidone solution in 0.9% physiological saline once a week for 4 weeks. Two weeks later, 0.05 ml of a 5% or 50% aqueous *N*-methyl-2-pyrrolidone solution (vol/vol) was applied to the shaved intact shoulder skin and rubbed in lightly. There was no mention of occlusion in the IPCS report (WHO 2001). Nine animals without intradermal injection served as controls. Evidence of sensitization was not found after either 24 or 48 hours. After 24 hours slight erythema was observed at the site of application after exposure to the 50% solution in 6/10 intradermally injected animals and in 4/9 control animals. No effects were seen after 48 hours. The 5% solution did not produce irritation (WHO 2001).

5.4 Reproductive toxicity

5.4.1 Developmental toxicity

Details of the studies are given in Table 4.

5.4.2 Inhalation

Sprague-Dawley rats were exposed on days 6 to 20 of gestation to *N*-methyl-2-pyrrolidone concentrations of 0, 30, 60 or 120 ml/m^3 for 6 hours a day in whole body exposure chambers. The highest concentration corresponded to the maximum vapour pressure at room temperature. Groups of 25 to 26 pregnant animals were used. After concentrations of 60 and 120 ml/m^3, the treatment led to reduced body weight development in the dams between days 6 and 13 of gestation. In addition, exposure to 120 ml/m^3 caused a reduction in food consumption between days 13 and 21 of gestation. Maternal toxicity was accompanied by reduced foetal weights only after concentrations of 120 ml/m^3. Otherwise, no signs of foetotoxicity were found at any other concentration. Particularly the number and form of malformations were similar in all groups. The NOAEC (no observed adverse effect concentration) in this study for maternal toxicity was 30 ml/m^3, the NOAEC for developmental toxicity 60 ml/m^3 (Saillenfait *et al.* 2003).

Table 4. Studies of the developmental toxicity of *N*-methyl-2-pyrrolidone in rats

Strain, number of animals per group	Exposure	Findings	References
inhalation			
Sprague-Dawley, 25–26 pregnant ♀	**GD 6–20**, 0, 30, 60, 120 ml/m^3, 6 hours/day, whole body exposure	**30 ml/m^3:** NOAEC maternal toxicity **60 ml/m^3:** NOAEC developmental toxicity; dams: body weight gains (GD 6–13) decreased **120 ml/m^3:** dams: food consumption (GD 13–21) decreased; foetuses: body weights decreased	Saillenfait *et al.* 2003
ingestion			
Sprague-Dawley (Crl:CD®BR), 25 pregnant ♀	**GD 6–15**, 0, 40, 125, 400 mg/kg body weight and day, gavage	**125 mg/kg body weight and day:** NOAEL maternal and developmental toxicity **400 mg/kg body weight and day:** dams: body weight gains (GD 6–15) decreased; foetuses: body weights decreased, stunted foetuses increased	GAF 1992
Sprague-Dawley, 25–27 pregnant ♀	**GD 6–20**, 0, 125, 250, 500, 750 mg/kg body weight and day, gavage	**125 mg/kg body weight and day:** NOAEL maternal and developmental toxicity **≥250 mg/kg body weight and day:** dams: body weight gains (GD 6–21) decreased, corrected body weights decreased; foetuses: body weights decreased **≥500 mg/kg body weight and day:** dams: food consumption (GD 9–21) decreased, post-implantation losses increased, resorptions increased; foetuses: malformations increased (external, skeletal, visceral), skeletal variations increased, delayed ossification of skull and sternum **750 mg/kg body weight and day:** foetuses: mortality increased, visceral variations increased	Saillenfait *et al.* 2002

GD: gestation day; NOAEC: no observed adverse effect concentration; NOAEL: no observed adverse effect level

5.4.3 Ingestion

The study of the kinetics of orally administered ^{14}C-labelled *N*-methyl-2-pyrrolidone in pregnant rats, only available as an abstract, confirms earlier reports that *N*-methyl-2-pyrrolidone is transferred to the foetuses and concentrations equal to those in the dams are found in the foetal organs (Sitarek 2003).

Groups of 25 pregnant Sprague-Dawley rats were given gavage doses of an aqueous *N*-methyl-2-pyrrolidone solution of 0, 40, 125 or 400 mg/kg body weight and day from days 6 to 15 of gestation. The treatment caused reduced body weight gains only in the dams of the high dose group between days 6 and 15 of gestation. The sole symptom of maternal toxicity was accompanied by reduced foetal weights and an increased number of stunted foetuses. Otherwise, no effects on the number of implantations, live foetuses per litter, resorptions per litter or the distribution of the sex of the foetuses were seen in any dose group. Also the number and form of malformations were almost the same in each group. The NOAEL in this study was 125 mg/kg body weight and day for maternal and developmental toxicity (GAF 1992).

In a study with a prolonged treatment interval carried out according to OECD Test Guideline 414, Sprague-Dawley rats were given gavage doses of an aqueous *N*-methyl-2-pyrrolidone solution of 0, 125, 250, 500 or 750 mg/kg body weight and day on days 6 to 20 of gestation. After doses of 500 mg/kg body weight and day and above, body weight gains were significantly reduced in the dams between days 6 and 21 of treatment and food consumption was reduced. Also the corrected body weight gains (body weight on days 6–21 minus the uterus weight) were reduced. These changes were statistically significant. Body weight gains were also markedly reduced after doses of 250 mg/kg body weight, but these changes were not statistically significant. The values were about 9% below those of the controls and thus in the same range as the statistically significant reduced foetal weights in this dose group. The post-implantation losses and number of resorptions were increased after 500 mg/kg body weight and day and above, with a steep dose–response curve. A marked increase in malformations (external, visceral, skeletal) was observed in the foetuses after 500 mg/kg body weight and day and above. Among the malformations were anal atresia, absent tails, cardiovascular malformations and various abnormalities of the spine. Further findings of developmental toxicity were reduced foetal weights after 250 mg/kg body weight and day and above, delayed ossification (skull and sternum) and an increase in skeletal variations after 500 mg/kg body weight and day and above; in addition, there were a very small proportion of live foetuses and an increase in visceral variations after 750 mg/kg body weight and day (Saillenfait *et al.* 2002). Although the authors state the NOAEL for maternal findings to be 250 mg/kg body weight and day, the NOAEL for maternal and developmental toxicity is, in fact, 125 mg/kg body weight and day considering the reduced body weight gains observed at this dose.

5.4.4 Multi-generation studies

The details of the generation studies are summarized in Table 5.

Table 5. Multi-generation studies with rats administered *N*-methyl-2-pyrrolidone with the diet

Species, strain, number of animals per sex and group	Exposure	Findings	References
Wistar, 25 ♂ and 25 ♀	**Two-generation study**, 0, 50, 160, 500/350 mg/kg body weight and day $\mathbf{F_0}$: from 10 weeks before mating, ♂: until the end of the 2nd mating period (F_{1b}), ♀: until weaning of the 2nd litter (F_{1b}) $\mathbf{F_{1b}}$: after weaning, throughout mating, ♂: until the end of the 2nd mating period (F_{2b}), ♀: during gestation until the end of lactation of the 2nd litter (F_{2b})	**160 mg/kg body weight and day**: NOAEL parental systemic and developmental toxicity **≥160 mg/kg body weight and day**: yellow discoloration of the urine **350 mg/kg body weight and day**: NOAEL reproductive ability/fertility **500/350 mg/kg body weight and day:** $\mathbf{F_0}$ ♂: kidney weights increased, kidney tubule dilation; ♀: body weight gains decreased, food consumption decreased; $\mathbf{F_{1a}}$: number of live births decreased, mortality up to day 21 *post partum* increased, body weight gains decreased; $\mathbf{F_{1b}}$: after reduction to 350 mg/kg body weight and day, mortality up to postnatal day 21 increased, body weight gains decreased **350 mg/kg body weight and day:** $\mathbf{F_1}$ ♂: kidney weights increased, kidney tubule dilation; ♀: body weight gains decreased, kidney weights increased, calcification of the kidney papilla; $\mathbf{F_{2a}}$: mortality up to postnatal day 4 increased, body weight gains decreased; $\mathbf{F_{2b}}$: mortality up to postnatal day 21 increased, body weight gains decreased	NMP Producers Group 1999b
Sprague-Dawley (Crl:CD®BR), 30 ♂ and 30 ♀	**Two-generation study**, 0, 50, 160, 500/350 mg/kg body weight and day $\mathbf{F_0}$: from 10 weeks before mating, ♂: until the end of the 2nd mating period (F_{1b}), ♀: until weaning of the 2nd litter (F_{1b}) $\mathbf{F_{1b}}$: after weaning, throughout mating, ♂: until the end of the 2nd mating period (F_{2b}), ♀: during gestation until the end of lactation of the 2nd litter (F_{2b})	**160 mg/kg body weight and day**: NOAEL parental systemic and developmental toxicity **350 mg/kg body weight and day** NOAEL reproductive ability/fertility **500/350 mg/kg body weight and day:** $\mathbf{F_0}$ ♂: kidney and liver weights increased; $\mathbf{F_0}$ ♀: body weight gains and food consumption decreased, spleen weights increased; $\mathbf{F_{1a}}$: size of litter decreased, live births decreased, mortality up to postnatal day 21 increased, birth weights and body weight gains decreased; $\mathbf{F_{1b}}$: no effects after reduction to 350 mg/kg body weight and day **350 mg/kg body weight and day:** $\mathbf{F_1}$ **♂:** kidney weights increased; ♀: calcification in the kidneys; $\mathbf{F_{2a}}$: birth weights decreased; $\mathbf{F_{2b}}$: mortality up to postnatal day 21 increased, birth weights and body weight gains decreased	NMP Producers Group 1999c

In the MAK documentation from 1994 ("*N*-Methyl-2-pyrrolidone", Volume 10, present series), a two-generation study with oral administration to Sprague-Dawley rats is described, in which clear effects were observed after doses of 500 mg/kg body weight and day, and the NOAEL was given as 160 mg/kg body weight and day (GAF 1991). After doses of 50 and 160 mg/kg body weight and day, reduced fertility indices were seen in the F_1 and F_2 generations (the reasons for which were unclear) and resulted in inconsistent evaluations of the data. Because of this, further two-generation studies were carried out by two independent institutes with Sprague-Dawley rats (Huntingdon Life Science, England) and Wistar rats (BASF AG, Ludwigshafen). In these studies, 30 Sprague-Dawley and 25 Wistar rats per sex and group were given *N*-methyl-2-pyrrolidone doses of initially 0, 50, 160 or 500 mg/kg body weight and day with the diet for 10 weeks before and during mating, and during gestation, lactation and the mating intervals. The concentrations in the diet were regularly adapted to body weight development. Because of the excessive mortality among the pups of the first litter (F_{1a}) in both rat strains, the high dose was reduced to 350 mg/kg body weight and day for the rest of the study. There were two litters per generation. After weaning, the pups of the second litter (F_{1b}) were subjected to the same pattern of treatment, and became the parent animals of the next generation. Fertility parameters, including cycle determination and sperm parameters, and toxicity and developmental parameters, including the sexual maturity of the pups, were assessed as well as the general clinical parameters. In addition, the parent animals were subjected to a comprehensive histological examination with particular consideration of the reproductive organs.

The changes observed were very similar in both rat strains. An impairment in reproductive ability, including fertility, was not detected clinically or histopathologically. While signs of systemic toxicity (reduced body weight gains, reduced food consumption, increased kidney weights and histological findings in the kidneys) were seen in the Wistar rats in the parent animals of the high dose group after doses of 500 mg/kg body weight and day and also after reduction to 350 mg/kg body weight and day, hardly any signs of systemic toxicity were observed in the Sprague-Dawley rats after reduction to 350 mg/kg body weight and day. Developmental toxicity manifested itself in increased mortality of the pups and delayed body weight development, with corresponding changes in organ weights, after doses of 500 and 350 mg/kg body weight and day in both rat strains and in the pups from both litters of one generation. Thus, the NOAEL for fertility was 350 mg/kg body weight and day. The NOAEL for systemic and developmental toxicity was 160 mg/kg body weight and day (NMP Producers Group 1999b, 1999c).

5.5 Genotoxicity

5.5.1 *In vitro*

N-Methyl-2-pyrrolidone was not found to be genotoxic in various *Salmonella* mutagenicity tests, a mouse lymphoma test, an HPRT (hypoxanthine guanine phosporibosyl transferase) test with CHO cells (a cell line derived from Chinese hamster ovary) and an UDS test (test for DNA repair synthesis) with primary rat hepatocytes ("*N*-Methyl-2-pyrrolidone", Volume 10, present series). Evidence of possible aneugenic effects of *N*-methyl-2-pyrrolidone in *Saccharomyces cerevisiae* ("*N*-Methyl-2-pyrrolidone", Volume 10, present series) is neither relevant for the evaluation nor supported by the *in vivo* findings.

5.5.2 *In vivo*

The results of a dominant lethal test ("*N*-Methyl-2-pyrrolidone", Volume 10, present series) are not considered evidence that the substance has genotoxic effects as the findings are within the range of biological variability.

A micronucleus test with polychromatic erythrocytes from the bone marrow of mice and a chromosomal aberration test with bone marrow cells of the Chinese hamster did not yield any evidence of clastogenic effects of *N*-methyl-2-pyrrolidone ("*N*-Methyl-2-pyrrolidone", Volume 10, present series).

5.6 Carcinogenicity

5.6.1 Inhalation

A carcinogenicity study with inhalation exposure of groups of 90 male and 90 female CD-1 rats is described in the documentation from 1994. In this study, exposure to *N*-methyl-2-pyrrolidone vapour concentrations of 0, 41.2 or 412 mg/m^3 (0, 10 or 100 ml/m^3) 5 times a week for 6 hours a day did not lead to a dose-dependent increase in tumour incidences (Lee *et al.* 1987).

5.6.2 Ingestion

Details of the results of the carcinogenicity studies with oral administration are given in Table 6.

In a two-year carcinogenicity study, 62 male and 62 female Sprague-Dawley rats were given food containing *N*-methyl-2-pyrrolidone in concentrations of 0, 1600, 5000 or 15000 mg/kg. The tumour incidences were not increased in either the male or the female rats (Malley *et al.* 2001; see Section 5.1).

Table 6. Oral studies of the carcinogenicity of *N*-methyl-2-pyrrolidone

Authors:	Malley *et al.* 2001
Substance:	*N*-methyl-2-pyrrolidone (99.8% pure)
Species/strain:	rat/Sprague-Dawley (Crl:CD®BR), groups of 62 ♂ and 62 ♀
Administration route:	diet
Concentration:	0, 1600, 5000, 15000 mg/kg diet (♂ about 0, 66, 207, 678 mg/kg body weight and day; ♀ about 0, 88, 283, 939 mg/kg body weight and day)
Duration:	2 years
Findings:	5000 mg/kg diet: NOAEL (♂ + ♀) ≥5000 mg/kg diet: yellow discoloration of the urine 15000 mg/kg diet: body weight gains decreased, food consumption decreased, severe chronic progressive nephropathy (♂) increased
Tumours:	no increase in the incidence of tumours
Authors:	BASF 2004; Malley *et al.* 2001; NMP Producers Group 1999a
Substance:	*N*-methyl-2-pyrrolidone (99.8% pure)
Species/strain	mouse/B6C3F$_1$, groups of 50 ♂ and 50 ♀
Administration route:	diet
Concentration:	0, 600, 1200, 7200 mg/kg diet (♂ about 0, 89, 173, 1089 mg/kg body weight and day; ♀ about 0, 115, 221, 1399 mg/kg body weight and day)
Duration:	18 months
Findings:	600 mg/kg diet: NOAEL (♂) 1200 mg/kg diet: NOAEL (♀) ≥1200 mg/kg diet: liver weights (♂) increased, centrilobular liver cell hypertrophy (♂) 7200 mg/kg diet: liver weights (♀) increased

Tumours:	Concentration (mg/kg diet)			
	0	600	1200	7200
male animals				
survival (days 0–561)	49/50 (98%)	48/50 (96%)	50/50 (100%)	49/50 (98%)
liver:				
hypertrophy	0/50 (0%)	0/50 (0%)	3/50 (6%)	43/50 (86%)#**
foci of phenotypically changed hepatocytes	5/50 (10%)	5/50 (10%)	6/50 (12%)	25/50 (50%)#**
clear cell foci	0/50 (0%)	1/50 (2%)	1/50 (2%)	6/50 (12%)#*
basophilic foci	4/50 (8%)	2/50 (4%)	3/50 (6%)	5/50 (10%)
eosinophilic foci	1/50 (2%)	2/50 (4%)	2/50 (4%)	19/50 (38%)#**
adenomas	5/50 (10%)	2/50 (4%)	4/50 (8%)	12/50 (24%)#
carcinomas	4/50 (8%)	1/50 (2%)	3/50 (6%)	13/50 (26%)#*
adenomas and carcinomas (tumour hosts)	9/50 (18%)	3/50 (6%)	7/50 (14%)	19/50 (38%)#*

Table 6. continued

Tumours:	Concentration (mg/kg diet)			
	0	600	1200	7200
male animals				
lungs:				
adenomas	1/50 (2%)[1]	2/50 (4%)	3/50 (6%)	5/50 (10%)
adenocarcinomas	1/50 (2%)[2]	4/50 (8%)	6/50 (12%)	2/50 (4%)
adenomas and adenocarcinomas	2/50 (4%)[3]	6/50 (12%)	9/50 (18%)	7/50 (14%)
female animals				
survival (days 0–561)	49/50 (98%)	48/50 (96%)	50/50 (100%)	48/50 (96%)
liver:				
hypertrophy	0/50 (0%)	1/50 (2%)	0/50 (0%)	0/50 (0%)
foci of phenotypically changed hepatocytes	3/50 (6%)	2/50 (4%)	1/50 (2%)	17/50 (34%)#**
clear cell foci	0/50 (0%)	0/50 (0%)	0/50 (0%)	1/50 (2%)
basophilic foci	2/50 (4%)	2/50 (4%)	1/50 (2%)	10/50 (20%)#*
eosinophilic foci	1/50 (2%)	0/50 (0%)	0/50 (0%)	6/50 (12%)#
adenomas	2/50 (4%)	2/50 (4%)	1/50 (2%)	7/50 (14%)#
carcinomas	0/50 (0%)	0/50 (0%)	0/50 (0%)	3/50 (6%)#
adenomas and carcinomas (tumour hosts)	2/50 (4%)	2/50 (4%)	1/50 (2%)	10/50 (20%)#*
lungs:				
adenomas	1/50 (2%)	1/50 (2%)	0/50 (0%)	0/50 (0%)
adenocarcinomas	1/50 (2%)	0/50 (0%)	2/50 (4%)	2/50 (4%)
adenomas and adenocarcinomas	2/50 (4%)	1/50 (2%)	2/50 (4%)	2/50 (4%)

$p \leq 0.05$ Cochran-Armitage trend test
* $p<0.05$ and ** $p<0.01$ Fisher's exact test
1 historical controls: 40/700 (5.7%); range: 0/50–8/50 (0–16%)
2 historical controls: 17/700 (2.4%); range: 0/50–6/50 (0–12%)
3 historical controls: 57/700 (8.1%); range: 0/50–10/50 (0–20%)

The carcinogenic potential of *N*-methyl-2-pyrrolidone in mice was investigated in an 18-month feeding study. Groups of 50 male and 50 female B6C3F$_1$ mice were given *N*-methyl-2-pyrrolidone with the diet in concentrations of 0, 600, 1200 or 7200 mg/kg. Survival was not impaired in either the male or the female animals. Increased liver weights, an increase in the incidence of foci of phenotypically changed hepatocytes and an increase in the incidence of liver adenomas and liver carcinomas were seen in the mice of both sexes after doses of 7200 mg/kg diet. Although trend test analyses indicate a trend, the data show that relevant and, according to Fisher's exact test, statistically significant increases in the incidence of liver adenomas and liver carcinomas occur only in the high dose group. In addition, an increased incidence of adenomas and adenocarcinomas of the lung was observed (6/50 (12%), 9/50 (18%) and 7/50 (14%)) in the male mice of all three exposure groups compared with that in the control group (2/50 (4%)). The pulmonary tumours hardly affected survival. Only one of 50 animals of the high dose group died from an adenocarcinoma before the end of the study (Malley *et al.* 2001; NMP Producers Group 1999a). The incidence of these tumours, which occur spontaneously in the B6C3F$_1$ mouse strain, varies considerably. The historical control values for pulmonary tumours in male B6C3F$_1$ mice, found in 14 studies (including the NMP study) carried out by the laboratory (BASF AG, Ludwigshafen) between the years 1983 and 2000, amounted to an average of 57/700 (8.1%) with a range of 0/50 to 10/50 (20%) for adenomas and adenocarcinomas of the lung combined (BASF 2004). An average value of 40/350 (11.4%) is obtained if only the values of the most recent 7 studies (including the NMP study) carried out between 1993 and 2000 are considered. In accordance with the evaluation of the authors and particularly because the findings were not dose-dependent, the lung tumours are regarded as not being substance-related.

6 Manifesto (MAK value/classification)

N-Methyl-2-pyrrolidone was not found to be genotoxic in various *in vitro* and *in vivo* studies. Carcinogenic effects were not observed in long-term studies with rats after either inhalation (Lee *et al.* 1987) or oral exposure (Malley *et al.* 2001). However, the incidences of preneoplastic lesions and adenomas and carcinomas of the liver were increased in mice of both sexes in the high dose group of a feeding study (Malley *et al.* 2001). Despite these findings, *N*-methyl-2-pyrrolidone has not been classified in a category for carcinogenic substances as the increased incidence of liver tumours was observed only in mice, but not in rats, and solely in the high exposure group with doses of more than 1000 mg/kg body weight and day. This finding is not, therefore, considered to be relevant for humans.

The MAK value for *N*-methyl-2-pyrrolidone was set in 1994 at 80 mg/m^3 (19 ml/m^3), as no effects were detected in a multi-generation study with rats after concentrations of 206 mg/m^3 (50 ml/m^3). Exposure to 478 mg/m^3 (116 ml/m^3) caused delayed body weight gains in the pups (DuPont 1990). No other inhalation studies have been reported in the meantime, and the MAK value has therefore been retained. In accordance with the preferred value approach, it has merely been adjusted to 20 ml/m^3, which corresponds to 82 mg/m^3. If the MAK value is observed, local irritative effects are not to be expected. In a study with volunteers, whole body exposure to 80 mg/m^3 for 8 hours with and without physical activity caused a moderate odour nuisance, but no irritative effects. Even at exposure peaks of 160 mg/m^3, weak eye irritation or irritation of the nasal mucosa was reported only in individual cases. No irritative effects were found for the physiological variables (palpebral closure frequency, nasal respiratory flow) even at exposure peaks.

Peak Limitation Category II with an excursion factor of 2 has been retained.

As *N*-methyl-2-pyrrolidone is readily absorbed through the skin, designation with an "H" is still necessary.

The studies of allergenic effects cited in the 1994 MAK documentation and another study of the sensitizing effects in guinea pigs showed that *N*-methyl-2-pyrrolidone causes skin irritation, but not sensitizing effects. The substance is therefore not designated with an "Sh". There are no data available regarding designation with an "Sa" (for substances that cause sensitization of the airways).

In 1994 *N*-methyl-2-pyrrolidone was classified in Pregnancy Risk Group C because prenatal toxicity was not observed until concentrations of 1000 mg/m^3 (243 ml/m^3) in rabbits and 680 mg/m^3 (165 ml/m^3) in rats. The NOAEC for rabbits was given as 500 mg/m^3 (122 ml/m^3) and that for rats as 360 mg/m^3 (87 ml/m^3) ("*N*-Methyl-2-pyrrolidone", Volume 10, present series). A recent developmental toxicity study with rats (Saillenfait *et al.* 2003) yielded a NOAEC of 60 ml/m^3 for developmental toxicity. Toxic effects on development, in the form of slightly reduced foetal body weights, were found at 120 ml/m^3; this concentration also resulted in reduced maternal food consumption and reduced maternal body weight gains. Pregnancy Risk Group C can therefore be retained; dermal absorption of *N*-methyl-2-pyrrolidone should continue to be excluded.

No data are available that would justify classification of the substance in a category for germ cell mutagenicity.

7 References

Åkesson B, Jönsson BAG (1997) Major metabolic pathway for *N*-methyl-2-pyrrolidone in humans. *Drug Metab Dispos 25*: 267–269

Åkesson B, Jönsson BAG (2000) Biological monitoring of *N*-methyl-2-pyrrolidone using 5-hydoxy-*N*-methyl-2-pyrrolidone in plasma and urine as the biomarker. *Scand J Work Environ Health 26*: 213–218

Åkesson B, Paulsson K (1997) Experimental exposure of male volunteers to *N*-methyl-2-pyrrolidone (NMP): acute effects and pharmacokinetics of NMP in plasma and urine. *Occup Environ Med 54*: 236–240

Akrill P, Cocker J, Dixon S (2002) Dermal exposure to aqueous solutions of *N*-methyl pyrrolidone. *Toxicol Lett 134*: 265–269

Anundi H, Langworth S, Johanson G, Lind ML, Åkesson B, Friis L, Itkes N, Söderman E, Jönsson BAG, Edling C (2000) Air and biological monitoring of solvent exposure during graffiti removal. *Int Arch Occup Environ Health 73*: 561–569

Bader M, Rosenberger W, Rebe T, Wrbitzky R (2003) V40: Feldstudie zur äußeren und inneren Belastung mit *N*-Methyl-2-pyrrolidon bei der Anwendung von Harzlösern zur Kessel- und Werkzeugreinigung (V40: Field study of external and internal exposure to *N*-methyl-2-pyrrolidone after the use of resin solvents for cleaning boilers and tools) (German). *Arbeitsmed Sozialmed Umweltmed 38*: 135

BASF (2003) *Dermal and inhalative uptake of NMP during paint stripping of furniture using end use products containing NMP*. 15.11.2003, BASF AG, Ludwigshafen, unpublished report

BASF (2004) *Stellungnahme zur Frage der Inzidenz von Lungentumoren nach oraler Gabe von NMP über 18 Monate bei männlichen B6C3F1-Mäusen* (The incidence of lung tumours in male B6C3F1 mice after oral administration of NMP for 18 months) (German) 08.12.2004, BASF AG, Ludwigshafen, unpublished report

Beaulieu HJ, Schmerber KR (1991) M-Pyrol™ (NMP) use in the microelectronics industry. *Appl Occup Environ Hyg 6*: 874–880

DuPont (1990) *1-Methyl-2-pyrrolidone (NMP): reproductive and developmental toxicity in the rat.* Haskell Laboratory, No. 294–90, E. I. du Pont de Nemours and Company Newark, Delaware, USA, unpublished report

GAF (1991) *Multigeneration rat reproduction study with N-methylpyrrolidone.* Exxon Biomedical Sciences, East Millestone, NJ, USA, Nr. 236535, GAF Corporation, New York, NY, USA, unpublished report

GAF (1992) *Developmental toxicity study in rats with N-methylpyrrolidone.* Exxon Biomedical Sciences, East Millestone, NJ, USA, No. 136534, GAF Corporation, New York, NY, USA, unpublished report

Jönsson BAG, Åkesson B (2001) *N*-Methylsuccinimide in plasma and urine as a biomarker of exposure to *N*-methyl-2-pyrrolidone. *Int Arch Occup Environ Health 74*: 289–294

Jönsson BAG, Åkesson B (2003) Human experimental exposure to *N*-methyl-2-pyrrolidone (NMP): toxicokinetics of NMP, 5-hydroxy-*N*-methyl-2-pyrrolidone, *N*-methylsuccinimide and 2-hydroxy-*N*-methylsuccinimide (2-HMSI), and biological monitoring using 2-HMSI as a biomarker. *Int Arch Occup Environ Health 76*: 267–274

Lee KP, Chromey NC, Culik R, Barnes JR, Schneider PW (1987) Toxicity of *N*-methyl-2-pyrrolidone (NMP): teratogenic, subchronic, and two-year inhalation studies. *Fundam Appl Toxicol 9*: 222–235

Ligocka D, Lison D, Haufroid V (2002) Quantitative determination of 5-hydroxy-*N*-methylpyrrolidone in urine for biological monitoring of *N*-methylpyrrolidone exposure. *J Chromatogr B 778*: 223–230

Ligocka D, Lison D, Haufroid V (2003) Contribution of CYP2E1 to *N*-methyl-2-pyrrolidone metabolism. *Arch Toxicol 77*: 261–266

Malek DE, Malley LA, Slone TW, Elliott GS, Kennedy GL, Mellert W, Deckardt K, Gembardt C, Hildebrand B, Murphy SR, Bower DB, Wright GA (1997) Repeated dose toxicity study (28 days) in rats and mice with *N*-methylpyrrolidone (NMP). *Drug Chem Toxicol 20*: 63–77

Malley LA, Kennedy GL, Elliott GS, Slone TW, Mellert W, Deckardt K, Gembardt C, Hildebrand B, Parod RJ, McCarthy TJ, Griffiths JC (1999) 90-Day subchronic toxicity studies in rats and mice fed *N*-methylpyrrolidone (NMP) including neurotoxicity evaluation in rats. *Drug Chem Toxicol 22*: 455–480

Malley LA, Kennedy GL, Elliott GS, Slone TW, Mellert W, Deckardt K, Kuttler K, Hildebrand B, Banton MI, Parod RJ, Griffiths JC (2001) Chronic toxicity and oncogenicity of *N*-methylpyrrolidone (NMP) in rats and mice by dietary administration. *Drug Chem Toxicol 24*: 315–338

NMP Producers Group (1994) *Repeated dose toxicity study with N-methylpyrrolidone in B6C3F1 mice. Administration in the diet for 4 weeks (range-finding study).* BASF AG, No. 40C0225/93030, NMP Producers Group INC, Bergeson & Campbell, PC, Washington DC, USA, unpublished report

NMP Producers Group (1995a) *Oral, dermal, and inhalation pharmacokinetics and disposition of [2-^{14}C]NMP in the rat.* E. I. DuPont de Nemours and Company, Haskell Laboratory for Toxicology and Industiral Medicine, No. 630–95, NMP Producers Group INC, Bergeson & Campbell, PC, Washington DC, USA, unpublished report

NMP Producers Group (1995b) *N-Methylpyrrolidone – subchronic oral toxicity study in B6C3F1 mice. Administration in the diet for 3 months.* BASF AG, No. 60C0225/93053, NMP Producers Group INC, Bergeson & Campbell, PC, Washington DC, USA, unpublished report

NMP Producers Group (1995c) *Subchronic oral toxicity*: 90-day feeding and neurotoxicity study in rats with *N*-methylpyrrolidone (NMP). E. I. du Pont de Nemours and Company, Haskell Laboratory for Toxicology and Industrial Medicine, No. 9737–001, NMP Producers Group INC, Bergeson & Campbell, PC, Washington DC, USA, unpublished report

NMP Producers Group (1998) *[^{14}C]-N-Methylpyrrolidone*: topical application: dermal absorption study in the rat. Huntingdon Life Sciences Ltd, No. 982974, NMP Producers Group INC, Bergeson & Campbell, PC, Washington DC, USA, unpublished report

NMP Producers Group (1999a) *N-Methylpyrrolidone – carcinogenicity study in B6C3F1 mice. Administration in the diet for 18 months.* BASF AG, No. 76C0225/93065, NMP Producers Group INC, Bergeson & Campbell, PC, Washington DC, USA, unpublished report

NMP Producers Group (1999b) *N-Methylpyrrolidone (NMP) – two generation reproduction toxicity study in Wistar rats. Administration in the diet.* BASF AG, No. 70R0056/97008, NMP Producers Group INC, Bergeson & Campbell, PC, Washington DC, USA, unpublished report

NMP Producers Group (1999c) *Two-generation reproduction toxicity study with N-methylpyrrolidone (NMP) in Sprague Dawley rats. Administration in the diet.* Huntingdon Life Science, East Millestone, No. 97–4106 NMP Producers Group INC, Bergeson & Campbell, PC, Washington DC, USA, unpublished report

NMP Producers Group (2002a) *N-Methylpyrrolidone – liver enzyme induction study in B6C3F1 mice – administration in the diet for 2 weeks.* BASF AG, No. 99C0225/93071, NMP Producers Group INC, Bergeson & Campbell, PC, Washington DC, USA, unpublished report

NMP Producers Group (2002b) *N-Methylpyrrolidone – S-phase response study in the liver of B6C3F1 mice; administration in the diet for 1 and 4 weeks.* BASF AG, No. 99C0225/93070, NMP Producers Group INC, Bergeson & Campbell, PC, Washington DC, USA, unpublished report

NMP Producers Group (2002c) *[^{14}C]-N-methylpyrrolidone: in vitro dermal penetration study using human and rat skin.* Huntingdon Life Sciences Ltd., No. NPG 002/004698, NMP Producers Group INC, Bergeson & Campbell, PC, Washington DC, USA, unpublished report

NMP Producers Group (2003a) *The in vivo percutaneous absorption of [^{14}C]-N-methylpyrrolidone in the rat.* Inveresk Research, No. 22650, NMP Producers Group INC, Bergeson & Campbell, PC, Washington DC, USA, unpublished report

NMP Producers Group (2003b) *N-Methylpyrrolidone: in vitro dermal penetration through human skin.* Safepharm Laboratories, No. 1695/001, NMP Producers Group INC, Bergeson & Campbell, PC, Washington DC, USA, unpublished report

NMP Producers Group (2003c) *Workplace respiratory, dermal and systemic exposure to N-methyl pyrrolidone in the UK.* Health and Safety Laboratory, No. HSL/ECO/2003/04, NMP Producers Group INC, Bergeson & Campbell, PC, Washington DC, USA, unpublished report

NMP Producers Group (2005) *Human volunteer study on chemosensory effects and evaluation of a threshold limit value in biological material of N-methyl-2-pyrrolidone (NMP) after inhalation and dermal exposure. Chemosensory effects.* Institute for Occupational Physiology at the University of Dortmund (IfADo), Draft report 12.09.2005, NMP Producers Group INC, Bergeson & Campbell, PC, Washington DC, USA, unpublished

Payan JP, Boudry I, Beydon D, Fabry JP, Grandclaude MC, Ferrari E, Andre CJ (2003) Toxicokinetics and metabolism of *N*-[^{14}C]methyl-2-pyrrolidone in male Sprague Dawley rats: *in vivo* and *in vitro* percutaneous absorption. *Drug Metab Dispos 31*: 659–669

Penney MS (2003) Physiologically-based pharmacokinetic modelling of the skin absorption of *N*-methyl-pyrrolidone *in vivo*. *Toxicology 192*: 79–80

Saillenfait AM, Gallissot F, Langonne I, Sabate JP (2002) Developmental toxicity of *N*-methyl-2-pyrrolidone administered orally to rats. *Food Chem Toxicol 40*: 1705–1712

Saillenfait AM, Gallissot F, Morel G (2003) Developmental toxicity of *N*-methyl-2-pyrrolidone in rats following inhalation exposure. *Food Chem Toxicol 41*: 583–588

Sitarek K (2003) Excretion and maternal-fetal distribution of *N*-methyl-2-pyrrolidone in rats. *Reprod Toxicol 17*: 505

Ursin C, Hansen CM, van Dyk JW, Jensen PO, Christensen IJ, Ebbehoej J (1995) Permeability of commercial solvents through living human skin. *Am Ind Hyg Assoc J 56*: 651–660

WHO (World Health Organisation) (2001) *N-Methyl-2-pyrrolidone.* IPCS – Concise International Chemical Assessment Document 35, WHO, Genf, Schweiz

Xiaofei E, Wada Y, Nozaki JI, Miyauchi H, Tanaka S, Seki Y, Koizumi A (2000) A linear pharmacokinetic model predicts usefulness of *N*-methyl-2-pyrrolidone (NMP) in plasma or urine as a biomarker for biological monitoring for NMP exposure. *J Occup Health 42*: 321–327

completed 02.02.2006

Sulfuric acid

Supplement 2007

MAK value (1999)	**0.1 mg/m³ I (inhalable aerosol fraction)**
Peak limitation (2000)	**Category I, excursion factor 1 momentary value: 0.2 mg/m³**
Absorption through the skin	–
Sensitization	–
Carcinogenicity (1999)	**Category 4**
Prenatal toxicity (1999)	**Pregnancy Risk Group C**
Germ cell mutagenicity	–
BAT value	–
Synonyms	battery acid hydrogen sulfate oil of vitriol
Chemical name (CAS)	sulfuric acid
CAS number	7664-93-9

Deep inhalation of sulfuric acid concentrations of 0.38 mg/m³ while performing strenuous physical exercise led in healthy volunteers to coughing (Avol *et al.* 1988a). Concentrations of 0.45 mg/m³ (Utell *et al.* 1983b), 1.0 mg/m³ (Frampton *et al.* 1992) and 2 mg/m³ (Linn *et al.* 1989) caused irritation of the throat, but no change in lung function parameters was observed (Linn *et al.* 1989). After lower concentrations, however, changes in mucociliary clearance were detected. These findings were inconsistent. In studies of one research group a consistent reduction in mucociliary clearance was observed after concentrations of 0.3 mg/m³ and above. Also in animal studies, which, however, were not sufficiently well-founded, slight effects were seen after concentrations of 0.3 mg/m³ in some cases (see 1999 MAK documentation "Sulfuric acid", Volume 15, present series).

In the meantime, an inhalation study with rats, carried out according to current test guidelines, has been published. Groups of 10 female Wistar rats were exposed via the nose to sulfuric acid aerosols (generated from a 49.7% solution) in concentrations of 0, 0.3, 1.4 and 5.5 mg/m³ for 6 hours a day, on 5 days a week, for 4 weeks. A satellite group was exposed for 5 days. Two other satellite groups comprised animals from the

The MAK-Collection Part I: MAK Value Documentations, Vol. 26. DFG, Deutsche Forschungsgemeinschaft

ISBN: 978-3-527-32306-7

high exposure group with 4-week and 8-week observation phases. Changes in body weight gains or organ weights were not observed in any of the groups. Changes were found only in the larynx, but not in the lungs or nose. A significant increase in cell proliferation in the larynx was observed after exposure for 5 days and 4 weeks in the medium and high exposure groups, but not in the low exposure group. Squamous cell metaplasia in level 1 of the larynx was found in all exposure groups (see Table 1). The degree of severity was described as minimal in the low exposure group, and as severe in the high group. Squamous cell metaplasia in level 2 was observed only in the high exposure group. The authors regard the findings in the low exposure group as adaptive and give a NOAEC (no observed adverse effect concentration) of 0.3 mg/m^3 (European Sulphuric Acid Association, CEFIC 2000). The Commission does not agree with this, as the findings in the low exposure group are regarded as substance-related. Therefore 0.3 mg/m^3 is regarded as the LOAEC (lowest observed adverse effect concentration). In view of the minimal severity of the findings, this LOAEC is probably close to the NOAEC, however.

Table 1. Incidence of microscopic findings in the larynx of rats exposed to sulfuric acid aerosols for 4 weeks (European Sulphuric Acid Association, CEFIC 2000)

Microscopic findings	Sulfuric acid aerosol concentration (mg/m^3)			
	0	0.3	1.4	5.5
Larynx				
number of animals investigated	10	10	9	10
squamous cell metaplasia, level 1 (number of animals affected)	0	6	9	10
minimal	0	6	0	0
slight	0	0	6	0
moderate	0	0	3	3
severe	0	0	0	7
squamous cell metaplasia, lateral, level 2 (number of animals affected)	0	0	0	6
parakeratosis, level 2 (number of animals affected)	0	0	0	1
ventral pouch (number of animals affected)	0	1	1	3
foreign body granulomas	0	0	0	2
foreign bodies	0	1	1	0
feed components	0	0	0	1

Manifesto (MAK value/classification)

In 1999 the MAK value was set at 0.1 mg/m^3, as in studies of one research group a consistent reduction in mucociliary clearance was observed in volunteers after concentrations of 0.3 mg/m^3 and above. In view of the paucity of the findings, the new 28-day inhalation study with rats with a LOAEC of 0.3 mg/m^3 does not justify lowering the MAK value. This has been provisionally retained; to verify the effects in this dose range and determine a clear NOAEC, further, longer-term investigations are necessary.

References

European Sulphuric Acid Association, CEFIC (2000) *Sulphuric acid aerosol: 28 day sub-acute inhalation study in the rat.* Central Toxicology Laboratory, Alderley Park Macclesfield, Cheshire, UK, Report No CTL/P/6278, CEFIC, Brussels, Belgium

completed 29.03.2006

Tetrahydrothiophene

Supplement 2006

MAK value (2005)	**50 ml/m^3 (ppm) ≙ 180 mg/m^3**
Peak limitation (2005)	**Category I, excursion factor 1**
Absorption through the skin	–
Sensitization	–
Carcinogenicity	–
Prenatal toxicity (2005)	**Pregnancy Risk Group D**
Germ cell mutagenicity	–
BAT value	–
Synonyms	tetramethylenesulfide thiacyclopentane thiolane thiophane
Chemical name	tetrahydrothiophene
CAS number	110-01-0
1 ml/m^3 (ppm) ≙ 3.66 mg/m^3	**1 mg/m^3 ≙ 0.27 ml/m^3 (ppm)**

Toxic Effects and Mode of Action

Tetrahydrothiophene is a sulfurous, heterocyclic hydrocarbon. It is colourless, has an intensive and unpleasant odour, is insoluble in water and chemically stable. It is used as an industrial solvent, but above all as a gas odorization agent (in concentrations of about 1 ml/m^3).

The vapour is strongly irritative to the skin and mucous membranes. If inhaled, it causes irritation of the respiratory passages and congestion in the lungs. Headaches, palpitation, dizziness, nausea and general discomfort occur even after short-term exposures. High concentrations of tetrahydrothiophene induce hyperactivity, accompanied by characteristic motor hyperreflexia. Tetrahydrothiophene is anaesthetizing in

The MAK-Collection Part I: MAK Value Documentations, Vol. 26. DFG, Deutsche Forschungsgemeinschaft

ISBN: 978-3-527-32306-7

higher doses. Death occurs with the symptoms of narcosis. Long-term exposure to non-lethal concentrations causes in addition to behavioural disturbances also liver damage (Henschler 1979).

Toxicokinetics and Metabolism

Studies with mice showed that tetrahydrothiophene is eliminated as a methylated sulfonium ion (Mozier and Hoffman 1990).

Effects in Humans

The odour threshold is given as 1 ml/m^3 (HSDB 2004).

Animal Experiments and *in vitro* Studies

Acute toxicity

Inhalation

The results of studies of the acute inhalation toxicity of tetrahydrothiophene are shown in Table 1.

Groups of 5 male and 5 female Wistar rats were exposed in a glass chamber for 30 minutes or one hour to air saturated with technical-grade tetrahydrothiophene at 20°C. The symptoms of intoxication observed were non-specific behavioural disturbances, breathing difficulties, irritation of the visible mucous membranes of the eyes and nose, and increased lacrimation and salivation; the rats were strongly sedated. The lungs of the animals killed at the end of the recovery period were greenish in colour and inflated (Bayer AG 1981).

Table 1. Studies of the acute inhalation toxicity of tetrahydrothiophene

Species	Concentration		Exposure (hours)	End point	References
	mg/m³	ml/m³			
rat					
♂	28086–32176		4	LC_{50}	Bayer AG 1981
♀	about 32000		4	LC_{50}	Bayer AG 1981
♂		5900	4	LC_{50}	Pennwalt 1986
♀		6700	4	LC_{50}	Pennwalt 1986
♂/♀		6270	4	LC_{50}	Pennwalt 1986
		42160	1	LC_{50}	Pennwalt 1975

In another study, groups of 10 male and 10 female Wistar rats were exposed for 4 hours to technical-grade tetrahydrothiophene in concentrations of 4205, 6707, 20860, 28086 or 32176 mg/m³ (1136, 1811, 5632, 7583 and 8688 ml/m³); an additional group of 20 female animals was exposed to 36028 mg/m³ (9728 ml/m³). During the first day of the recovery period, the visible mucous membranes of the eyes and nose were seen to be irritated. For up to two days after exposure, the animals were strongly sedated and lay in latero-abdominal position, for up to seven days they had breathing difficulties, and for up to nine days persistent non-specific behavioural disturbances. In a large number of the animals killed at the end of the recovery period, the lungs were grey in colour and inflated in some regions. The LC_{50} after 4-hour exposure to tetrahydrothiophene was given for male animals as 28086 to 32176 mg/m³ (7583 to 8688 ml/m³), and for female animals as about 32000 mg/m³ (8640 ml/m³) (Bayer AG 1981).

Six groups of 5 Sprague-Dawley rats were exposed for 4 hours via inhalation in whole animal exposure chambers to nominal tetrahydrothiophene concentrations of 0 to 9480 ml/m³ (analysed values of 0 to 6300 ml/m³). Lacrimation, dyspnoea, reduced activity, matted fur, exhaustion and general poor condition were observed in the animals. The LC_{50} for both sexes was calculated to be 6270 ml/m³ (22948 mg/m³), that for male animals to be 5900 ml/m³ (21594 mg/m³) and for female animals to be 6700 ml/m³ (24522 mg/m³) (Pennwalt 1986).

Five rats per group were exposed to tetrahydrothiophene concentrations of 100000 to 225000 mg/m³ (27000 to 60750 ml/m³) for one hour in an inhalation chamber. The symptoms observed were peripheral vasodilatation, slight lacrimation, dilation of the pupils, ataxia, dyspnoea and loss of the righting reflex. The LC_{50} after exposure for one hour was given as 155000 mg/m³ (according to the report, corresponding to 42160 ml/m³) (Pennwalt 1975).

Ingestion

The LD_{50} values after oral administration of tetrahydrothiophene are shown in Table 2.

Table 2. Studies of the acute toxicity of tetrahydrothiophene after oral administration and dermal application

Species	Dose (mg/kg body weight)	End point	References
rat (oral)			
♂	2080	LD_{50}	Bayer AG 1981
♀	2275	LD_{50}	Bayer AG 1981
♂	2000	LD_{50}	Pennwalt 1985b
♀	1750	LD_{50}	Pennwalt 1985b
♂/♀	1850	LD_{50}	Pennwalt 1985b
rat (dermal)			
♂	3335	LD_{50}	Bayer AG 1981
♀	3700	LD_{50}	Bayer AG 1981
rabbit (dermal)			
♂/♀	> 2000	LD_{50}	Pennwalt 1985a

Groups of ten male or ten female Wistar rats were given gavage doses of technical-grade tetrahydrothiophene of 50 to 3500 mg/mg body weight. Central nervous symptoms of intoxication (staggering gait, reduced motility, cramped posture, latero-abdominal position) and breathing difficulties were observed within a few minutes after administration of the dose. Autopsy of the animals still alive at the end of the observation period did not reveal any organ damage. The LD_{50} was given as 2080 mg/kg body weight for male rats, and 2275 mg/kg body weight for female rats (Bayer AG 1981).

Groups of 5 male or 5 female Sprague-Dawley rats were given single oral doses of tetrahydrothiophene of 800 to 2000 mg/kg body weight, 6 male and 4 female animals 3125 mg/kg body weight. Within 24 hours after administration of the dose, ataxia, tremor, hypoactivity and sometimes exhaustion, the formation of foam on the snout, breathing anomalies and reduced food consumption were observed in all animals. No differences were found between the animals killed after 14 days and the animals of the control group. The LD_{50} for male and female animals was calculated to be 1850 mg/kg body weight, that for male animals to be 2000 mg/kg body weight and that for female animals to be 1750 mg/kg body weight (Pennwalt 1985b).

Dermal application

The LD_{50} values after dermal application of tetrahydrothiophene are shown in Table 2.

Doses of technical grade tetrahydrothiophene of 500 to 5000 µl/kg body weight (corresponding to 499 to 4994 mg/kg body weight) were applied occlusively to the skin of groups of 5 male Wistar rats, and doses of 100 to 5000 µl/kg body weight (corresponding to 100 to 4994 mg/kg body weight) to groups of 5 females. The exposure time was 24 hours. The animals were observed for 14 days. Within a few minutes after application of the dose, central nervous symptoms of intoxication (staggering gait, reduced motility, cramped posture, latero-abdominal position) and breathing difficulties were observed. The urine was bloody and discoloured, the faeces were black. Apart from healed scars on the dorsal skin, no organ changes were observed in animals killed after the observation period. The LD_{50} was given as 3335 mg/kg body weight for male rats, and 3700 mg/kg body weight for female rats (Bayer AG 1981).

Single doses of tetrahydrothiophene of 2000 mg/kg body weight were applied occlusively for 24 hours to the shaved skin of groups of 5 male and 5 female rabbits. There are no details available of the purity of the test substance. All animals survived the 14-day recovery period. Marked skin lesions (necrosis with scabbing and chapped skin; flaking of the scabs), which persisted over the duration of the study, were observed in most animals. Most animals suffered from hyperpnoea. Tremor, dyspnoea, irregular breathing, the formation of pus in the nasal region and reddening of the eyes were observed in one or two animals. Unusual findings were a changed surface of the spleen in five animals, discoloration of the kidneys in three animals and ruptures in the livers of two animals. The authors gave an LD_{50} after dermal application of tetrahydrothiophene of more than 2000 mg/kg body weight (Pennwalt 1985a).

Subacute, subchronic and chronic toxicity

Inhalation

Groups of 10 male Wistar rats were exposed in a chamber to nominal tetrahydrothiophene concentrations of 0, 100, 200 or 500 ml/m^3 (366, 732 and 1830 mg/m^3) for 1 hour a day, on 5 days a week, for 4 weeks. The purity of the test substance and analytically determined concentrations were not given. Unusual behaviour in the animals was not observed during the exposure period. All animals survived the treatment. Reduced body weight gains were observed in the animals exposed to 100 ml/m^3 in the second and third weeks and in the animals exposed to 200 ml/m^3 in the third week; this was not observed at the high concentration. The organ weights (liver), haematology, urinalysis and microscopic investigations (heart, liver, eyes, nasal cavity, trachea, lungs and bronchi, kidneys, testes, adrenal glands, spleen) did not reveal any adverse effects of the test substance. The urea levels of the animals of all concentration groups and the lipids of the animals of the high concentration group were reduced (CREP 1974).

Groups of 5 male and 5 female Wistar rats were exposed to nominal tetrahydrothiophene concentrations (purity 99.1%) of 0, 10, 100 or 1000 ml/m^3 (actual concentrations 10.4 ± 2.0, 96 ± 0.9 and 961 ± 24 ml/m^3, corresponding to 38.1 ± 7.3, 351 ± 3 and 3517 ± 88 mg/m^3) in whole animal exposure chambers. The animals were exposed daily for 6 hours, on 5 days a week, for 28 days. The test criteria corresponded to those of OECD Test Guideline 412. Haematological, clinico-biochemical and pathological investigations were carried out on the day of autopsy (day 29 of the experiment). No clinical symptoms that, in the opinion of the authors, could be connected with the exposure to tetrahydrothiophene were observed in the exposed animals compared with the symptoms in the control group exposed to filtered air. In the male rats of the low and medium exposure groups, the mean body weights were significantly increased, compared with the mean body weight of the control group, only at the end of the study.

In the female animals, a significant decrease in body weights was determined only in the high exposure group after the first and third week of the experiment. The mean food consumption of the male and female animals of the high exposure group in the second week of the study was significantly lower than that in the control group. Also the female rats of the two other exposure groups consumed less food than the female animals of the control group (10 ml/m^3 group: weeks 1 and 2; 96 ml/m^3 group: week 1). Increased Hepato-Quick® values (this corresponds to the thromboplastin time) relative to those in the control group were determined in the animals exposed to concentrations of 10 or 96 ml/m^3, while the Hepato-Quick® values were decreased relative to those in the control group in the rats exposed to 961 ml/m^3. The authors note that it is not certain that this is a substance-related effect. The lack of a dose–response relationship suggests it is not. Histopathological examination revealed no evidence of tetrahydrothiophene-induced damage. The NOAEC (no observed adverse effect concentration) is thus 961 ml/m^3 (BG Chemie 1988).

Groups of 10 male and 10 female Sprague-Dawley rats were exposed to tetrahydrothiophene concentrations of 0, 50, 275 or 1500 ml/m^3 (180, 990 and 5410 mg/m^3, purity at least 99%) for 6 hours a day, on 5 days a week, in an inhalation chamber. The analysed concentrations were 51, 236 and 1442 ml/m^3 (184, 852 and 5201 mg/m^3). The study was carried out in accordance with the OECD Guideline for Good Laboratory Practice and OECD Test Guideline 413 (the satellite group of animals recommended in the guideline was, however, not included). An increase in the symptoms with increasing concentration is visible in Table 3, showing the clinical symptoms of irritation (lacrimation, salivation, closing of the eyes). Increased irritation (closing of the eyes) was first seen at concentrations of 236 ml/m^3. The meaning of the symptoms "licking the inside of the mouth" and "head swaying from side to side" is unclear and it is questionable whether this was substance-related. The animals of the high exposure group and the male animals of the 236 ml/m^3 group drank more water than did the animals of the control group. Ophthalmological investigations did not yield any unusual findings. There were slight differences between the values for haematological parameters in the female animals of the medium and high exposure groups and those in the control animals; these were not regarded by the authors as toxicologically relevant, with the exception of an increase in white blood cells and lymphocytes in the female animals exposed to 1442 ml/m^3. In these animals also the pH of the urine was slightly increased. Gross pathological examination and the determination of organ weights (lungs, testes, thyroid gland, epididymis, liver, prostate, spleen, ovaries, adrenal gland, brain, heart, pituitary, kidneys) did not yield any unusual findings. An increase in liver weights was observed in the female animals of the high exposure group; according to the authors, this is not toxicologically relevant. In the histopathological examinations, the lungs were investigated in all animals, while in the animals of the control group and those exposed to 1442 ml/m^3 also a series of other tissues was examined. This included tissue from the nasal region of the animals (rostral and caudal nasal cavity). Traces of blood were found in the nasal cavity in one male animal of the high concentration group, sinusoidal blood in the nasal region in one female animal of the control group. The changes were regarded as coincidental and not unusual in this strain of rat. The NOAEC for local irritation is therefore 51 ml/m^3, the NOAEC for systemic effects 1442 ml/m^3 (Pennwalt 1988b).

For clarity, the time at which clinical symptoms occurred in the inhalation studies (see also the section "Developmental toxicity") is shown in Table 3. No clinical symptoms occurred in the CREP (1974) and BG Chemie (1988) studies.

Local effects on skin and mucous membranes

Skin

Cellulose pads soaked with about 0.5 ml (499 mg) technical-grade tetrahydrothiophene were placed on the hairless skin of the inner ear of constrained rabbits and weighted with a glass stopper. The exposure period was one hour (two animals), two hours (one animal) and eight hours (one animal). The animals were followed up for 14 days. After one hour, slight reddening and slight corrosion were observed, after two hours slight reddening. After eight hours, moderate reddening and swelling, and deeper corrosion were seen. The effects were observed over the whole recovery period. In the report, tetrahydrothiophene is described as strongly irritative to the the skin of rabbits (Bayer AG 1981).

0.5 ml (499 mg) undiluted tetrahydrothiophene (no details of purity) was applied occlusively to the shaved dorsal skin of 4 male and 2 female New Zealand White rabbits for around 4 hours. Examinations were carried out 4.5, 24, 48 and 72 hours, and 7, 10 and 14 days after the application. The evaluation was carried out according to Draize. Barely perceptible to well-defined erythema and oedema (score 1 to 2 on a scale with a maximum score of 4) were observed after 4.5 hours; within 24 hours to 7 days, however, damage to the skin (necrosis and eschar formation, score 3 to 4) was seen in 4 of 6 animals. The scabs began to heal within 10 to 14 days in 3 of the 4 animals. In the fourth animal, the scabbing lasted until the end of the experiment (Pennwalt 1985d).

Eyes

0.1 ml (100 mg) technical-grade tetrahydrothiophene was instilled into the conjunctival sac of one eye in two rabbits. The recovery period was 14 days. Lacrimation developed after a few minutes and lasted for up to six hours. The moderate reddening and severe swelling of the conjunctiva seen during the first 24 hours after the instillation was still present in mild form up to the third day of observation. As a result of the severe lacrimation and thus rapid washing out of the substance, tetrahydrothiophene is described in the report as causing only moderate irritation of the mucous membranes (Bayer AG 1981).

Table 3. Clinical effects in various strains of rat after inhalation exposure to tetrahydrothiophene. Shown are the number of weeks or days over which the effects occurred. Details of the number of affected animals in the individual groups are not given in the studies

Rat strain/concentration	Effects											References
	Licking inside of mouth	Head swaying/ extreme agitation	Head swaying	Salivation	Lacrimation	Eyes closed/ partially closed	Reduced response to knocking	Hunched posture	Rubbing of snout, paws on cage bottom	Excessive salivation	Pilo erection	
Sprague-Dawley												
51 ml/m^3	10 weeks	–	2 weeks	1 week	1 week	–	–	–	–	–	–	Pennwalt 1988b
236 ml/m^3	13 weeks	–	–	3 weeks	2 weeks	10 weeks	–	–	–	–	–	
1442 ml/m^3	13 weeks	–	13 weeks	13 weeks	13 weeks	12 weeks	13 weeks	4 weeks	13 weeks	13 weeks	3 weeks	
COBS CD BR												
234 ml/m^3	observed (no further details)				observed (no further details)							Pennwalt 1988a
782 ml/m^3	observed (no further details)	–	–	–	observed (no further details)	observed (no further details)	–	–	observed (no further details)	–	–	
1910 ml/m	observed (no further details)	observed (no further details) (end of treatment)	–	observed (no further details)	observed (no further details)	observed (no further details)	–	–	observed (no further details)	–	–	
125 ml/m^3	–	–	–	–	–	–	–	–	–	–	–	Pennwalt 1988a
320 ml/m^3	3 days	–	–	–	–	–	–	–	–	–		
800 ml/m^3	8 days	–	–	2 days	–	2 days	–	–	3 days	–	–	
2000 ml/m^3	9 days	–	–	7 days	7 days	10 days	–	9 days	6 days	–	–	

– Effect not observed

In 6 rabbits (3 males, 3 females), 0.1 ml (100 mg) tetrahydrothiophene (no details of purity) was instilled into one of both eyes; the other eye served as control. The first examination was carried out one hour after instillation. The eyes were rinsed and examined around 24 hours after instillation of the substance; further examinations were carried out after 48 and 72 hours, and 7 and 10 days. Evaluation was carried out according to Draize (total score depending on the animal from 19 to 39 after one hour, from 4 to 55 after 24 hours, from 4 to 57 after 48 hours and from 4 to 46 after 72 hours; maximum score 110). Severe irritation of the conjunctiva (redness, chemosis, purulent discharge, necrosis) and inflammation of the iris were observed in all animals, corneal clouding and ulceration in 3 animals. Hair loss around the eye was observed in 3 animals. The effects on the conjunctiva and iris were most pronounced one hour after instillation of the substance, those on the cornea 24 hours after the treatment. The animals were without eye irritation within seven to ten days after the treatment (Pennwalt 1985c).

In 6 albino rabbits, 0.1 ml (100 mg) tetrahydrothiophene (no other details) was instilled into the conjunctival sac of one of both eyes. The instillation caused pain and blepharospasms. Severe inflammation, chemosis and slightly deepened folds of the iris were observed during the first four hours. Twenty-four hours after the treatment, only mild inflammation of the conjunctiva was still visible. This report considered tetrahydrothiophene not to be an eye irritant (Pennwalt 1975).

Reproductive and developmental toxicity

Developmental toxicity

Groups of 8 female rats (Crl: COBS CD BR) were exposed in a range-finding study to tetrahydrothiophene concentrations of 125, 320, 800 or 2000 ml/m^3 (450, 1152, 2880 and 7200 mg/m^3) for 6 hours a day, for 10 days (gestation days 6 to 15). Clinical symptoms were not observed in the animals exposed to 125 ml/m^3. Licking the inside of the mouth was the only symptom, seen on 3 days, in the animals exposed to 320 ml/m^3. In the next-higher concentration group, this symptom occurred on 8 days, and also closing of the eyes and salivation (both on 2 days) and rubbing of the snout and jaw on the bottom of the cage (on 3 days). The additional symptoms lacrimation (on 7 days) and hunched position (on 9 days) were observed only at the highest concentration of 2000 ml/m^3. In the main study, groups of 25 female rats (Crl: COBS CD BR) were exposed on days 6 to 15 of gestation to tetrahydrothiophene (purity 99%). The animals were exposed to tetrahydrothiophene concentrations of 0, 250, 750 or 2000 ml/m^3 (analysed values of 234, 782 and 1910 ml/m^3, corresponding to 844, 2822 and 6888 mg/m^3) in whole animal exposure chambers for six hours a day according to the OECD Guideline for Good Laboratory Practice and OECD Test Guideline 414. During the exposure period, the symptoms observed in the animals exposed to 234 ml/m^3 were licking of the inside of the mouth and lacrimation. The animals exposed to 782 ml/m^3 also rubbed their snout and paws on the bottom of the cage and partially or completely closed their eyes. After exposure to concentrations of 1910 ml/m^3, salivation and, towards the end of exposure, extreme agitation were observed in the animals in addition to the above symptoms. Unlike in the range-finding study, the frequency of the symptoms in days was not given (see also the section “Subacute, subchronic and chronic toxicity”). A slight reduction in body weight gains was observed in the first few days in the animals of the low exposure group. This reduction was more pronounced in the animals of the next-higher exposure group during the first four days; water consumption increased progressively. Exposure to 1910 ml/m^3 led in these animals also to an initial reduction in food consumption. Tetrahydrothiophene was not found to have any effect on the number of *corpora lutea*, the number and gender distribution of the living foetuses, the number and gender distribution of the embryofoetal losses, or on individual foetal weights and anomalies. The NOAEC for maternal toxicity is therefore 234 ml/m^3, the NOAEC for toxic effects on development 1910 ml/m^3 (Pennwalt 1988a).

Genotoxicity

In vitro

The genotoxic effects of tetrahydrothiophene (purity not stated) were investigated in the *Salmonella* strains TA98, TA100, TA1537 and TA1538 both with and without metabolic activation. Tetrahydrothiophene concentrations of 0 to 5000 µg/plate were used both in the standard plate test and in the pre-incubation test; the solvent used was dimethyl-sulfoxide (DMSO). In the investigation with the standard plate test, a bacteriotoxic effect was determined only with strain TA98 at 5000 µg/plate and metabolic activation, while in the preincubation test such an effect was determined after concentrations above 2500 µg/plate. Tetrahydrothiophene was not found to be mutagenic under these test conditions (BG Chemie 1986).

Another test with tetrahydrothiophene (purity at least 99%, concentrations of 50 to 5000 µg/plate), carried out according to OECD Test Guideline 471 with *Salmonella typhimurium* TA98, TA100, TA1535 and TA1537, yielded negative results (Pennwalt 1987a).

An SCE (sister chromatid exchange) test with CHO cells (a cell line derived from Chinese hamster ovary) with concentrations of 15.63, 31.25, 62.5 and 125 µg/ml in the absence of metabolic activation and 15.63, 62.5 and 125 µg/ml in the presence of metabolic activation yielded negative results (Pennwalt 1987e).

In a UDS test (test for unscheduled DNA synthesis) with HeLa-S3 cells, the effects of tetrahydrothiophene (purity at least 99%) in concentrations of 2.5 to 5120 µg/ml were investigated both in the presence and absence of a metabolic system. The test yielded no evidence of the induction of UDS (Pennwalt 1987c).

The potential of tetrahydrothiophene to cause chromosomal aberrations was investigated according to OECD Test Guideline 473 in human lymphocytes with and without the addition of metabolic activation (S9 mix from the livers of male Sprague-Dawley rats treated with Aroclor 1254). The highest concentration with toxic effects was

1000 µg/ml without metabolic activation, and 2000 µg/ml with metabolic activation. An increase in the number of aberrant cells (gaps excluded) was observed in the presence of metabolic activation. According to the authors, the reason for this increase, which was within the range of the historical control data, is that no aberrant mitosis (gaps excluded) was seen in the cells of the solvent control (BG Chemie 1989).

In another investigation with human lymphocytes carried out according to OECD Test Guideline 473, tetrahydrothiophene (purity at least 99%) was used in concentrations of 0.24, 0.49, 0.98, 1.95, 3.91, 7.81, 15.63, 31.25, 62.5 or 125 µg/ml with dimethylsulfoxide as solvent. Concentrations above 125 µg/ml had led to droplet formation in the culture medium. The concentration of 125 µg/ml was not found to have toxic effects in the absence of a metabolic activation system, but reduced the mitosis index to 46% of the value determined in the solvent control in the presence of metabolic activation. The induction of chromosomal aberrations was not found (Pennwalt 1987b).

In an HPRT gene mutation test with CHO cells carried out according to OECD Test Guideline 476 subsequent to a range-finding test with tetrahydrothiophene concentrations of 25 to 200 µg/ml, the test substance (tetrahydrothiophene, purity at least 99%) was used in concentrations of 100, 125, 150, 175 and 200 µg/ml dissolved in DMSO. Small beads of the substance collected on the surface of the medium of solutions with tetrahydrothiophene concentrations above 200 µg/ml. Cytotoxic effects of tetrahydrothiophene or a significant increase in the frequency of mutants compared with that in the control cells were not observed (Pennwalt 1987d).

Manifesto (MAK value, classification)

No studies of humans have been published which would enable the derivation of a MAK value. The symptoms observed in investigations of eye irritation and acute inhalation toxicity show tetrahydrothiophene to be an irritant. At the lowest concentration of 51 ml/m^3 (no information was given in the studies about the number of animals), lacrimation and salivation, the first signs of irritation observed, occurred once in 13 weeks in Sprague-Dawley rats. This does not seem to be significant because of the low frequency. An additional sign of irritation, closing of the eyes, was observed at the next-higher concentration of 236 ml/m^3. In another strain of rat (COBS CD BR), no clinical symptoms were seen at a concentration of 125 ml/m^3; lacrimation occurred at 234 ml/m^3. In the 13-week study with Sprague-Dawley rats, histopathological examination revealed light bleeding in the nasal cavity in one male animal after exposure to a concentration of 1442 ml/m^3.

The data demonstrate that Sprague-Dawley rats are more sensitive to exposure to tetrahydrothiophene than are COBS CD BR rats. A concentration of 125 ml/m^3 can be considered to be the NOAEC, since no clear irritation was seen in Sprague-Dawley rats after concentrations of 51 ml/m^3; the NOAEC for COBS rats was 125 ml/m^3. The MAK value has therefore been established at 50 ml/m^3. As irritation is the critical effect, the substance is classified in Peak Limitation Category I. Neither substance-specific data nor findings in humans that substantiate at which concentration sensory irritation may be expected are available to establish an excursion factor. Therefore, the basic excursion factor of 1 has been established. Studies which suggest that the substance should be designated with an "Sh" or "Sa" (for substances that cause sensitization of the skin and airways) are not available. *In vitro* studies did not reveal a genotoxic potential. There are no studies of the carcinogenicity of tetrahydrothiophene. Classification in a category for carcinogenic substances or germ cell mutagens is therefore not necessary. A study of the developmental toxicity of tetrahydrothiophene in rats with concentrations up to 1910 ml/m^3 yielded negative results. As there are no studies with a second species, tetrahydrothiophene is classified in Pregnancy Risk Group D, tending towards C.

References

Bayer AG (1981) *Tetrahydrothiophen* (Tetrahydrothiophene) (German). Gewerbetoxikologische Untersuchungen. Bayer AG, Institut für Toxikologie, Report No. 9670, 07.01.1981, unpublished

BG Chemie (1986) *Report on the study of tetrahydrothiophene (ZNT test substance No.: 86/140) in the Ames test (standard plate test and preincubation test with Salmonella typhimurium).* BASF AG, BG Chemie, Heidelberg, unpublished

BG Chemie (1988) *Prüfung der subakuten inhalativen Toxizität von Tetrahydrothiophen (THT) an Ratten* (Investigation of the subacute inhalation toxicity of tetrahydrothiophene (THT) in rats) (German). Projekt 217280/88, FhG (Fraunhofer-Institut für Umweltchemie und Ökotoxikologie), BG Chemie, Heidelberg, unpublished report

BG Chemie (1989) *In vitro cytogenetic investigations of tetrahydrothiophene in human lymphocytes.* Project No. 30MO140/864180, BASF AG, BG Chemie, Heidelberg, unpublished report

CREP (Centre de Recherche du Groupe Labaz) (1974) *Toxicité subchronique (1 mois) du tetrahydrothiophene (T. H. T.) par voie pulmonaire chez le rat* (Subchronic inhalation toxicity (1 month) of tetrahydrothiophene (THT) in the rat) (French), CREP Grenoble, France, unpublished report

Henschler D (Ed.) (1979) "Tetrahydrothiophen" in *Toxikologisch-arbeitsmedizinische Begründungen von MAK-Werten*, 7th issue, VCH-Verlagsgesellschaft, Weinheim

HSDB (Hazardous Substances Data Bank) (2004) Data bank factsheet, http://www.nlm.nih.gov/pubs/factsheets/hsdbfs.html

Mozier NM, Hoffman JL (1990) Biosynthesis and urinary excretion of methyl sulfonium derivates of the sulfur mustard analog, 2-chloroethyl ethyl sulfide, and other thioethers. *FASEB J 4:* 3329–3333

Pennwalt (1975) *Tetrahydrothiophene, toxicology report.* Pharmacology Research Inc, 21.03.1975, Pennwalt Corporation, King of Prussia, PA, USA, unpublished

Pennwalt (1985a) *Acute dermal toxicity study in rabbits. Test material: Tetrahydrothiophene.* Biodynamics, Project No. 5418–84, Pennwalt Corporation, King of Prussia, PA, USA, unpublished

Pennwalt (1985b) *Acute oral toxicity study in rats. Test material: Tetrahydrothiophene.* Biodynamics, Project No. 5417–84, Pennwalt Corporation, King of Prussia, PA, USA, unpublished

Pennwalt (1985c) *Eye irritation study in rabbits. Test material: Tetrahydrothiophene.* Biodynamics, Project No. 5420–84, Pennwalt Corporation, King of Prussia, PA, USA, unpublished

Pennwalt (1985d) *Primary dermal irritation study in rabbits. Test material: Tetrahydrothiophene.* Biodynamics, Project No. 5419–84, Pennwalt Corporation, King of Prussia, PA, USA, unpublished

Pennwalt (1986) *An acute inhalation toxicity study of tetrahydrothiophene in the rat.* Biodynamics, Project No. 84–7765, final report, Pennwalt Corporation, King of Prussia, PA, USA, unpublished

Pennwalt (1987a) *Ames metabolic activation test to assess the potential mutagenic effect of tetrahydrothiophene.* Huntingdon Research, Report No. PWT 55/87178, 15th April 1987, Pennwalt Corporation, King of Prussia, PA, USA, unpublished

Pennwalt (1987b) *Terahydrothiophene (THT). Metaphase chromosome analysis of human lymphocytes cultured in vivo.* Huntingdon Research, Report No. PWT 58/87411, 16th July 1987, Pennwalt Corporation, King of Prussia, PA, USA, unpublished

Pennwalt (1987c) *Autoradiographic assessment of unscheduled DNA repair synthesis in mammalian cells after exposure to tetrahydrothiophene.* Huntingdon Research, Report No. PWT 57/87481, 15th July 1987, Pennwalt Corporation, King of Prussia, PA, USA, unpublished

Pennwalt (1987d) *An assessment of the mutagenic potential of tetrahydrothiophene in a mammalian cell mutation assay using the Chinese hamster ovary/HPRT locus assay.* Huntingdon Research, Report No. PWT 60/87393, 9th November 1987, King of Prussia, PA, USA, unpublished

Pennwalt (1987e) *Frequency of sister chromatid exchange in Chinese hamster ovary cells cultured in vitro after treatment with tetrahydrothiophene (THT).* Huntingdon Research, Report No. PWT 59/87695, 26th May 1987, Pennwalt Corporation, King of Prussia, PA, USA, unpublished

Pennwalt (1988a) *Effect of tetrahydrothiophene on pregnancy of the rat.* Huntingdon Research, Report No. PWT 52-R/8849, 27th May 1988, Pennwalt Corporation, King of Prussia, PA, USA, unpublished

Pennwalt (1988b) *Tetrahydrothiophene. 90-day inhalation study in rats.* Huntington Research, Report PWT 50/871158, Pennwalt Corporation, King of Prussia, PA, USA, unpublished

completed 03.02.2005

Tetrahydrothiophene

Supplement 2007 (prenatal toxicity)

MAK value (2005)	**50 ml/m³ (ppm) ≙ 180 mg/m³**
Peak limitation (2005)	**Category I, excursion factor 1**
Absorption through the skin	**–**
Sensitization	**–**
Carcinogenicity	**–**
Prenatal toxicity (2006)	**Pregnancy Risk Group C**
Germ cell mutagenicity	**–**
BAT value	**–**
Synonyms	tetramethylenesulfide thiacyclopentane thiolane thiophane
Chemical name	tetrahydrothiophene
CAS number	110-01-0

Developmental toxicity

The following study of developmental toxicity is described in the 2006 MAK documentation.

Groups of 25 female CD rats were exposed to tetrahydrothiophene concentrations of 0, 250, 750 or 2000 ml/m³ for 6 hours a day on gestation days 6 to 15. The analysed concentrations were 0, 234, 782 and 1910 ml/m³. During the exposure period, the symptoms observed in the animals exposed to 234 ml/m³ were licking of the inside of the mouth and lacrimation. The animals exposed to 782 ml/m³ also rubbed their snout and paws on the bottom of the cage and partially or completely closed their eyes. After exposure to concentrations of 1910 ml/m³, salivation and, towards the end of exposure, extreme agitation were observed in the animals in addition to the above symptoms. A slight reduction in body weight gains was observed in the first few days in the animals of the low exposure group. This reduction was more pronounced in the animals of the

The MAK-Collection Part I: MAK Value Documentations, Vol. 26. DFG, Deutsche Forschungsgemeinschaft

ISBN: 978-3-527-32306-7

medium exposure group during the first four days; water consumption increased progressively. Exposure to 1910 ml/m^3 led in these animals also to an initial reduction in food consumption. Tetrahydrothiophene was not found to have any effect on the number of *corpora lutea*, the number of living foetuses, embryofoetal losses, individual foetal weights or anomalies. The NOAEC for maternal toxicity is therefore 234 ml/m^3, the NOAEC for toxic effects on development 1910 ml/m^3 (Pennwalt 1988). As the difference between this concentration and the MAK value of 50 ml/m^3 is sufficiently large, re-evaluation of the data has allowed the classification of tetrahydrothiophene in Pregnancy Risk Group C.

References

Pennwalt (1988) *Effect of tetrahydrothiophene on pregnancy of the rat.* Huntingdon Research, Report No. PWT 52-R/8849, 27th May 1988, Pennwalt Corporation, King of Prussia, PA, USA, unpublished

completed 02.12.2005

Authors of Documents in Volume 26

Frau Dr. I.-D. Adler, Max-Hueber-Straße 6, D–85737 Ismaning
Dr. R. Bartsch, Senatskommission der DFG zur Prüfung gesundheitsschädlicher Arbeitsstoffe, Technische Universität München, D–85350 Freising-Weihenstephan
Frau Dr. U. Baumgärtner, Etterschlager Straße 7, D–82237 Wörthsee
Professor Dr. Dr. H.-P. Gelbke, BASF SE, Abt. GUP, Gebäude Z 570, D–67056 Ludwigshafen
Privatdozent Dr. Dr. U. Knecht, Institut und Poliklinik für Arbeits- und Sozialmedizin der Universität Gießen, Aulweg 129/III, D–35392 Gießen
Frau Dr. P. Kreis, Senatskommission der DFG zur Prüfung gesundheitsschädlicher Arbeitstoffe, Technische Universität München, D–85350 Freising-Weihenstephan
Frau Dr. B. Laube, Eurotoxis GmbH, Hartestraße 8, D–30539 Hannover
Professor Dr. A.W. Rettenmeier, Universitätsklinikum Essen, Institut für Hygiene und Arbeitsmedizin, Hufelandstraße 55, D–45122 Essen
Dr. R. Rossbacher, BASF SE, GUP/PC - Z470, D–67056 Ludwigshafen
Frau Dr. U. Reuter, Senatskommission der DFG zur Prüfung gesundheitsschädlicher Arbeitsstoffe, Technische Universität München, D–85350 Freising-Weihenstephan
Frau Dr. G. Schriever-Schwemmer, Senatskommission der DFG zur Prüfung gesundheitsschädlicher Arbeitsstoffe, Technische Universität München, D–85350 Freising-Weihenstephan
Dr. R. Schwabe, Senatskommission der DFG zur Prüfung gesundheitsschädlicher Arbeitsstoffe, Technische Universität München, D–85350 Freising-Weihenstephan
Frau Dr. G. Stropp, Bayer Healthcare AG, BSP GDD-GED-IC Tox, D–42096 Wuppertal
Dr. C. van Thriel, Leibniz-Institut für Arbeitsforschung an der TU Dortmund, Ardeystraße 67, D–44139 Dortmund
Frau Dr. K. Wiench, BASF SE, DUP/PB - Z470, D–67056 Ludwigshafen
Frau Dr. K. Ziegler-Skylakakis, Senatskommission der DFG zur Prüfung gesundheitsschädlicher Arbeitsstoffe, Technische Universität München, D–85350 Freising-Weihenstephan

The MAK-Collection Part I: MAK Value Documentations, Vol. 26. DFG, Deutsche Forschungsgemeinschaft

ISBN: 978-3-527-32306-7

Index for Volumes 1–26

The MAK-Collection Part I: MAK Value Documentations, Vol. 26. DFG, Deutsche Forschungsgemeinschaft

ISBN: 978-3-527-32306-7